The Weight of the Past

CONTEMPORARY ANTHROPOLOGY OF RELIGION,

A series published with the Society for the Anthropology of Religion

Robert Hefner, Series Editor
Boston University

Published by Palgrave Macmillan

Body / Meaning / Healing
By Thomas J. Csordas

The Weight of the Past: Living with History in Mahajanga, Madagascar
By Michael Lambek

The Weight of the Past

Living with History in Mahajanga, Madagascar

Michael Lambek

Excerpt from Giorgio Agamben, translated by Georgia Albert, The Man Without
Content. This translation copyright © 1999 by the Board of Trustees of the Leland
Stanford Junior University. With the permission of Stanford University Press,
www.sup.org.

THE WEIGHT OF THE PAST

First published 2002 by
PALGRAVE MACMILLAN™
175 Fifth Avenue, New York, N.Y. 10010 and
Houndmills, Basingstoke, Hampshire, England RG21 6XS.
Companies and representatives throughout the world.

PALGRAVE MACMILLAN is the global academic imprint of the Palgrave Macmillan
division of St. Martin's Press, LLC and of Palgrave Macmillan Ltd. Macmillan® is a
registered trademark in the United States, United Kingdom and other countries.
Palgrave is a registered trademark in the European Union and other countries.

ISBN 1-4039-6067-4 hardback
ISBN 1-4039-6068-2 paperback

Library of Congress Cataloging-in-Publication Data can be found at the Library
of Congress.

A catalogue record for this book is available from the British Library.

Design by Letra Libre, Inc.

First edition: December 2002
10 9 8 7 6 5 4 3 2 1
Printed in the United States of America.

le dur désir de durer, Paul Eluard

Contents

List of Figures
and Illustrations

Acknowledgments

The research I conducted in Mahajanga in annual visits between 1993 and 1996 and in 1998 and 2001, as well as a period of writing in autumn 2001, has been supported by grants from the Social Sciences and Humanities Research Council of Canada.

I was received with enthusiasm and tremendous hospitality in Mahajanga and its surroundings. Amana Ibrahim and Zaina (Maman' Hasan) made a home for me (and subsequently for Sarah Gould) and have proved unfailing sources of wisdom and good sense. Aghat Amana served as an able assistant and Roslan, Jamila, Pirzad, Mamý, Amedy, Nahoda, Dôná, Jean, Moustali, and Rôsina provided much advice, hospitality, and companionship.

Among the many people I would like to thank for their interest and courtesy are Ampanjaka Randrianirina Désiré Noël (Ampanjaka Désy), the late Ampanjaka Moandjy Hamady (Maminiarivo), Ampanjaka Amina, the late Ampanjaka Tsialaña (Ndramfoiniarivo), Ampanjaka Philibert Tsiaraso, Ampanjaka Zera, the late Ampanjaka Zava, Ampanjaka Isa (in Ambilobe), the late Ampanjaka Voady (Ndranasatrokoarivo), the late Mady Simba, Tsitandra (Baban' Tombo), Ampanjaka Veloma Soloteña, Natombo Alphonse, Fahatelo Asindrazana Edouard (Doara), Fahatelo Dzaovita, Tefindraza, Bemanangy Honorine Zafitody, Bemanangy Safia Gol, Moalimo Karija Jérôme (in Bemilolo), Rafidimanana Botrolahy (in Miadana), Mariam Ali (Maman' Rokia), Mama ny Hasily, Hasily, Maman' Jamfar, Doudou, the late Amady bin Ali, Josoa, Besomotro, Pierrette Rasoamampiarina Boulay, Zoky Moana, Maman' Doara, William Kadafy, Vazaha Kely, Pierrette (ny Ndramaro), Mama ny Sababaka, Maman' Dorothée, the late Mme Mariam Mkolo Be, the late Maolida Kely, Manantany Tsiriky, Manantany Mahatonga, the late Mariam from Nosy Lava, Fara, Folo (in Ambatoharañana), Baban' and Maman' Fazola, Sebastienne, Marceline, Frank, and Dady ny Gaston. Many other people have been extremely helpful and put up with my intrusions. My sincere thanks also to those consultants—from diverse positions within the ancestral system—who have preferred to remain anonymous. Finally, although I never met her, I feel very grateful to the late Dady Ankosy.

My deepest debts will be evident in the text. They are to Mme Doso (Dady Bonbon) and to Kassim Tolondraza (Dadilahy Kassim) and his wife Tsetsa (Maman' Sera). I have also learned a great deal from Maolida Saidy Mohamed (Dadilahy Lidy). All of these individuals have balanced mentoring and companionship with the Sakalava virtues of discretion and dignity.

I should thank the ancestors as well, especially Ndramandikavavy, Ndramandisoarivo, Ndranihaniña, and Ndramisara for enabling my stay in their domain and my study of their shrine; Ndramboeniarivo, Mbabilahy, Kôt' Mena, Ndramandaming, and Sengy for their particular support and attention; and also, Ankanjovôla, Ndramarofaly, Ndramandenta, Lakolako, Ndramianta, Ndramanisko, Mampiaming, Tsiromasoandro, Kôt' Fanjava, Ndrankehindraza, of course, and many others. *Mankashitraka anareo!*

Researchers in Madagascar are fortunate to be able to draw on a network of wonderful colleagues. For their many kindnesses over the years, intellectual generosity, and at times direct facilitation, I thank Rita Astuti, Sophie Blanchy, Nerina Botokeky, Gwyn Campbell, Bob Dewar, Ingela Edqvist, Gillian Feeley-Harnik, Paul Hanson, Susan Kus, Pier Larson, Karen Middleton, Hilde Neilssen, Kim Raharijaona, Elie Rajaonarison, Jean-Aimé Rakotoarisoa, Michel Razafiarivony, Lesley Sharp, Shirin Sotoudeh, and Pierre Vérin. I must single out Maurice Bloch, a wonderful and inspiring friend; and Henry Wright, a mentor now for thirty years. I am also immensely grateful to Andrew Walsh for his intellectual companionship and careful reading of several chapters; and to Sarah Gould for her acute and enthusiastic feedback from the field.

Emmanuel Tehindrazanarivelo has been an inspiring teacher and friend. He wishes to record that he has had no part in divulging or discussing matters the Sakalava Community holds sacred and considers confidential. I confirm this to be the case. However, I feel compelled to add that I would never have begun or succeeded at this project without him. He directed me to Mahajanga, facilitated my rapid entrée, and intrigued me with his passionate and subtle evocations of the richness and complexity of the Sakalava world. He has not read this manuscript; all errors of fact and lapses of craft and judgment are entirely my own.

Beyond the world of Malagasyists, Aram Yengoyan has remained mentor, friend, stimulus, and sounding board. Many other friends and colleagues, including Filip De Boeck, Jane Huber, Wendy James, Bruce Kapferer, Jack Kugelmass, Jan Ovesen, Michael Peletz, Jonathan Spencer, Paul Stoller, Dick and Pnina Werbner, and the late Skip Rappaport offered support, advice, or hospitality at critical junctures in this project. Bruce Trigger drew my attention to Inca oracles. The book really got underway during a Centennial Visiting Professorship in Anthropology at the London School of Economics. I am deeply grateful to all the members of that wonderful department, and especially to Rita Astuti, Maurice Bloch, and Fenella Cannell for extending the invitation; to Jonathan Parry and Margaret Bothwell for ensuring such a pleasant and productive stay; and to the members of the Madagascar seminar in 1997.

The members of my own department have put up with various seminar presentations (and idiosyncracies) and provided much helpful feedback. Principal Paul Thompson of the University of Toronto at Scarborough and successive chairs of the Division of Social Sciences—David Cook, Ted Relph, Ron Manzer, and Sue Horton—as well as Hy Van Luong, Larry·Sawchuk and other colleagues have supported me in numerous ways. Annette Chan and Audrey Glasbergen know how much I depend on them. Paul Antze, Sandra Bamford, Janice Boddy, Charles Hirschkind, Anne Meneley, and Andrew Walsh provided an inspiring writing group. Bob Hefner read the entire manuscript with great speed and acuity. I am indebted to Bob, Janice, and Kristi Long for expediting the publication and for last minute help with the title, to Diane Gradowski for the figures and to Sarah Gould for the index.

Chapter 3 is a somewhat revised version of "The Sakalava Poiesis of History: Realizing the Past through Spirit Possession in Madagascar," an article that appeared in *American Ethnologist* volume 25, number 2, May 1998 (copyright American Anthropological Association). The acknowledgments made there continue to stand. A portion of chapter 2 was delivered to a workshop at Oxford in 2001 organized by Ruth Barnes, Zul Hirji, and David Parkin. A paper drawing on a portion of chapter 5 was presented to a conference at Avignon, also in 2001, and will appear in *Slavery, Unfree Labour and Revolt in Asia and the Indian Ocean Region,* edited by Gwyn Campbell. A small portion of chapter 10 was used in Lambek and Solway (2001), while another portion appears in a paper delivered to a workshop in Cambridge in 1999, to be published in *Processes of Naming and the Significance of Names,* edited by Barbara Bodenhorn and Gabriele vom Bruck. Ethnographic material toward the end of chapter 11 has been used in two papers delivered, respectively, at the LSE (published as Lambek 1998a) and at Toronto, Virginia, London, and Bergen (Lambek in press). I am much indebted to the various convenors, audiences, editors, and referees.

Jacqueline Solway has been, as always, my most significant interlocutor. I am grateful to her and to Nadia and Simon both for their patience and for their impatience. They sustain me. Simon even offered a blurb for the cover: Sakalava history is no longer a mystery!

During the winter and spring of 2002, the people of Madagascar suffered the effects of political dissension and turmoil, while courageously refusing to let the situation slide into violence. As I send this book to press, I hope it is not too optimistic to predict that things have begun to look up. I dedicate this book to the citizens of Mahajanga with great respect for their courage, integrity, and good humor and with the wish, the ancestors willing, for peace and prosperity.

Stylistic Conventions

Malagasy "o" is pronounced like English "u." English "o" is noted as ô. Velar "n" is noted as ñ. Malagasy "j" is pronounced "dz." "I" shifts to "y" at the end of a word.

Malagasy nouns do not generally distinguish number or gender.

Exchange Rates

fmg = francs malgaches
1994—c. fmg3,500 per U.S. dollar
2001—c. fmg6,250 per U.S. dollar
2001—local rice fmg2,000/kilo; Pakistani rice fmg1,750/kilo

Glossary of Frequently Cited Words

ampanjaka	ruler, royalty, royals, members of the royal clan
ampanjaka be	reigning monarch
ampangataka	intercessor, official who addresses the ancestors on behalf of supplicants
Analalava	seat of an important Bemihisatra polity (Ampanjaka Soazara)
anjara	established responsibility or right; lot or destiny
Antandrano	"Water-Dwellers;" a cohort of Zafinifotsy spirits who died by drowning
Antankaraña	Zafinifotsy kingdom north of Sakalava territory
bemanangy	Great Woman, treasurer and animatrice for a particular section of royalty
Bemazava	royal faction contesting Bemihisatra rights to the relics
Bemihisatra	royal faction that currently controls the relics
Betsioko	cemetery in Ampanjaka Désy's domain where Bemihisatra ancestors, from Mbabilahy to the present, are buried
Bezavodoany	cemetery in Ampanjaka Amina's domain where the most senior royal ancestors are buried; shortened to Bezavo
Boina	the historial northern Sakalava kingdom, including Mahajanga
doany	shrine, sacred place associated with royalty
efadahy	the four senior ancestors whose relics rest at the shrine of Ndramisara in Mahajanga, an abbreviation of "four men"
fady, faly	taboo, forbidden
fahatelo	manager of Ndramisara's shrine; official below the ampanjaka and manantany, therefore literally, "third"
fanompoa	service on behalf of living and ancestral royalty, including labor and prestations
fanompoa be	the Great Service; annual celebration at the shrine of Ndramisara in Mahajanga
fômba	custom, the right way to do things

hasing	sacred potency
Jingô	a clan of Ancestor People
karazaña	kind, often used in the sense of "ethnic group" or "clan"
koezy	greeting of respect to royalty (from Swahili)
mañaja	to honor (someone)
manantany	senior official, commoner delegate of royalty
manompo	to perform royal service
Marambitsy	Ampanjaka Amina's domain, southwest of Mahajanga
mashiaka	violent, punitive, caustic, fierce
mianjaka	to be actively possessed (said of a host); to rise in a host (said of a spirit)
miavong	proud, reserved, standoffish
moas	diviner with special powers (pronounced "mwas")
mosarafa	return gift from royalty to those who serve (from Arabic)
ndrañahary	god, ancestral spirits
Nosy Be	seat of an important Bemihisatra polity north of Mahajanga
pousse-pousse	rickshaw
rañitry	guardian, especially at royal tombs
raza, razaña	ancestor
razan'olo	"Ancestor People;" members of certain clans with specific responsibilities to serve royalty
saha	spirit medium
tantara	historical recitation
tany fotsy	white clay, kaolin
tômpin	owner, master
tromba	spirit who possesses and speaks through a human medium; usually the spirit is an ancestral member of the royal clan
Tsiarana	clan descended from Ndramandikavavy
tsimandrymandry	night-long festivities with spirits, called "all-nighters" in the text, literally "not much going to bed"
valamena	sacred enclosure, sacred fence, literally "red pen"
varavara, varavaraña	door, gate
varavara mena lio	sacred, exclusive entrance, literally "red blood gate"
viarara	sacred knife or saber belonging to Ndramandikavavy
vôlafotsy	silver, literally, "white wealth/money"
vôlamena	gold, literally "red wealth/money"
vonahitry	honor, respect
zafy	grandchild, descendant
Zafinifotsy	"White/Silver Clan;" first conquerors of Boina
Zafinimena	"Red/Gold Clan;" fraternal lineage to Zafinifotsy, subsequent conquerors of Boina; the present-day ruling clan
zomba	temple, palace (from Bantu)
zomba vinda	house-like structure inside the temple in which the relics are kept

Key Personae

Ndramisara	"Lord Diviner," architect of the Sakalava polity
Ndramandisoarivo	conquerer of Boina; shortened to Ndramandiso
Ndramboeniarivo	Ndramandiso's son and successor; shortened to Ndramboeny
Ndranihaniña	fourth relic, considered ancestor to the Zafinifotsy clan
Ndramandikavavy	wife of Ndramandiso, mother of Ndramboeny
Mbabilahy	youngest son of Ndramboeny and, according to Bemihisatra, his successor; short form of Mbabilahimanjaka, also called Ndrananilitsiarivo
Ankanjovôla	Mbabilahy's wife
Ndramandaming	a spirit of the colonial era (1930s)
Ndramianta	older brother to Ndramandaming
Kôt' Mena	a male "Water-Dweller" spirit
Ndrankehindraza	lively, quasi-Muslim spirit, "brother" to Ampanjaka Amina and associated with both Bezavodoany and Nosy Be
Ampanjaka Amina	Bemihisatra ruler at Mitsinjo with domain over the ancestral cemetery at Bezavodoany
Ampanjaka Désy	Bemihisatra ruler at Ambatoboeny with domain over Ndramisara's shrine (Miarinarivo) in Mahajanga
Ampanjaka Moandjy	Bemazava ruler of Mahajanga (since deceased) with domain over the "Northern Shrine" (Doany Avaratra) in Mahajanga
Ampanjaka Soazara	Bemihisatra ruler at Analalava
Honorine	a Great Woman, representing Désy and the ancestors from Betisoko cemetery in Mahajanga
Kassim	a medium of Ndramboeny
Lidy	a medium of Kôt' Mena and several other spirits
Maman' Sera	Kassim's wife, and medium of several spirits
Mme Doso	a medium of Mbabilahy and several other spirits
Somotro	a medium of Ndramisara

Part I

A Poiesis of History

 **1**

Bearing Sakalava History:
A Glossary of Key Terms

Men make their own history, but they do not make it just as they please; they do not make it under circumstances chosen by themselves, but under circumstances directly encountered, given and transmitted from the past. The tradition of all the dead generations weighs like a nightmare on the brain of the living.

—Karl Marx, *The Eighteenth Brumaire of Louis Bonaparte*

Almost unnoticed, the idea of culture, which once connoted all that freed men from the blind weight of tradition, was now identified with that very burden, and that burden was seen as functional to the continuing daily existence of individuals in any culture.

—G. W. Stocking, "Franz Boas and the Culture Concept in Historical Perspective"

BEARING . . .

At the climax of the "Great Service," the annual ritual of political and cosmic renewal in Mahajanga, a procession emerges from the inner sanctum and circumambulates the temple. Sounds of drums, conch shells, and musket fire fill the air. A female spirit medium, possessed by a former monarch swathed in red, leads the way, followed by men holding up, respectively, a jagged knife, two decorative lances, and eight iron spears, each implement glistening with a fresh coating of honey and oil. Two lines of men holding outstretched hands form a corridor through which the bearers stride. Immediately following the spear bearers are four men, each carrying high on his back the well-swaddled relics of one of four royal ancestors. Encased in their ornate boxes and concealed under a cloth wrapping, these "four men" (*efadahy*) have just received their annual bath. They form small bundles, yet the bearers bend and shake under their seemingly immense weight. The weight of the ancestors indicates the force of their *hasing*—their sacred potency.

A crowd of women and men in elegant, colorful apparel to the rear, the procession winds itself once round the rectangular building, between sacrificed cattle lying with open, bloody necks, and then retires back inside. It lasts five minutes, no more, and it is barely glimpsed by the majority of celebrants, but the moment is one of intense joy, exaltation, and concentrated energy.

This book is about historical labor and historical consciousness among Sakalava inhabitants of northwest Madagascar. It inquires into local ways of making history, of "imagining continuity," and of selectively inscribing and engaging the passage of time and formulating historical experience. It expands what might be contained within the rubric "history," beyond deliberate, reflective, autonomous thought and language. Despite the presence of many local intellectuals, the history in question proceeds less by means of abstract theorizing or objective marshaling of facts than by embodying the forces, characters, and arguments of history, bearing them forth and engaging practically with them. This can be no less scrupulous than the best Western scholarship.

Sakalava historical consciousness is constituted by means of a particular art and idiom and informed by a particular ideology. The art is that of spirit possession; the idiom, that of royal ancestors; and the ideology, derived from a form of power that characterized, or is today understood to have characterized, Sakalava statecraft since before the onset of the kingdom of Boina around 1700. My account of this consciousness is resolutely cultural in nature. That is, I seek to elucidate the forms, practices, and meanings that constitute a historicity specific to Sakalava. I mean specific not to anyone who may carry the label "Sakalava," nor exclusively to such people, but to the people who embrace the forms and practices under discussion, who are subject to them, and who master them. I show how, for them, the past remains prevalent, pervasive, and resonant. More abstractly, I show how history, structure, sanctity, power, art, and ethics are interlinked.

Sakalava do not so much study their history as they bear it. This book is about the burden and the bearing; about the manner in which Sakalava conceive, carry, care for, and pass on their history; about history as work, craft, and practice. The "burden of history," as exemplified by the ancestral relics carried on the backs of their servitors, describes the relationships of Sakalava to their past more generally. Rituals are not mere formal events to be passively attended but times at which ancestors are directly encountered. They engage body and soul, and demand considerable investment of time, labor, and money. The past also infiltrates ordinary life, constraining action and directing personal careers. Ancestral things are weighty—potent, difficult, dangerous, and precious (*raha sarotro, mahery*). They require attention and commitment, and they impose all kinds of demands and prohibitions. Ancestors and their living retainers are ready to pounce on those who inadvertently or deliberately ignore them. Sakalava history has an edge and an immediacy; the burden is a live one.

Realized in practice, the concept of the burden is also embedded in language. The word *ampanjaka* (or *'mpanjaka*), which in everyday parlance means "ruler" and, around Mahajanga, "member of the clan of rulers," is translated by G. Feeley-Harnik (1991a: 591) in her scrupulous way as "ruler, literally 'burdener', from the verb *manjaka*, 'to impose/exert/burden', whence *fanjakana*, referring to all forms of government; people descended from royalty or ennobled may call themselves or be called *ampanjaka;* the sovereign ruler is distinguished as the 'great ruler' (*ampanjaka be*)."[1] Such burdening has a temporal quality. The root *zaka* is translated in the classic Malagasy-French dictionary as "bearable" with the sense of "endurance."[2] Moreover, the authors give the verb *mibaby,* "to bear on the back," the figurative meaning of "supporter, endurer." The double meaning of "endure" is highly relevant in what follows.

In describing Sakalava as bearing their history I mean "bearing" in a number of senses that I briefly unpack but that should be understood as simultaneous connotations of a central polyvalent concept.[3]

"Bearing" as Supporting a Heavy Load, Being under Obligation

The burden of being subjected to rule is a heavy one. This weight is not only literal (physical), as when the servitors stagger under the relics at the Great Service, but metaphorical, ranging through the material, moral, and metaphysical. The burden of the ancestors is at once the exigence of a predatory state and exploitative elite and the intransigence of authoritarian rule; the weightiness of Sakalava history, conquest and defeat, hope and suffering, sacrifice and tragedy; the elaboration, pervasiveness, potency, and demands of order, the gods, the sacred; the existential weight of being and becoming, of separation and attachment, freedom and determinism; the heaviness of distance, perdurance, and depth, of fathomless profundity.

Much activity concerning the ancestors is referred to as *fanompoa,* a word that denotes service on behalf of royalty—labor, tribute, and prestations to both living royalty and royal ancestors, and the ceremonies in which it is offered, displayed, and recognized. The ceremony whose climax I have described and that is known as the *Fanomopoa Be,* the Great Service, highlights material tribute.[4] Leading up to the event are a complex series of local activities whose purpose is to collect money on behalf of the ancestors (chapter 7), and the first two days of the Service are devoted to public prestation (chapter 8). The Great Service of the temple at Mahajanga is the largest and most significant annual ceremony in the region, attracting thousands of people from the town and countryside. Pilgrims arrive, and supplicants send money from distant parts of Madagascar, Mayotte, the Comores, Réunion, and metropolitan France.

These offerings are simultaneously gift, tribute, and the requisites of the cult; acts of political agency and political submission; an honoring of royal ancestors who have become deities (*ndrañahary*) and their supplication for blessing and material assistance. The relics in Mahajanga are known as the *efadahy,* the "four men," all associated with the foundation of the kingdom of Boina that controlled northwest Madagascar from about 1700 and has survived in weaker, fragmented, and encapsulated fashion through to the present. Ndramboeniarivo, generationally the youngest of the four men, represents the apex of northern Sakalava power.

The Great Service is the entire tributary mode of production in celebration of itself—the enactment of a historical form of politics and political economy. However, the obligations to royalty and the past are not evenly distributed or the same in kind. The division of labor is described in subsequent chapters. Here, it suffices to note how spirit possession stands out as a distinctive, immediate form of bearing—one that epitomizes Sakalava subjection. To be host to royal spirits is understood as an active burden, one that serves as a model for all citizens. Thus, although the bearers of the relics at the Great Service are not in trance, one consultant likened the experience to the feeling of being actively possessed and one carrier said he felt on the verge of trance. The verb for being actively possessed or possessing, *mianjaka,* contains the same root, *jaka,* found in the words for ruler and rule with the sense of enduring burden. Yet if spirit mediums are burdened by exigent masters, they also incarnate and are empowered by them. Bearing the past is an obligation that simultaneously provides scope for creativity, expanded agency, and responsible judgment.

The past is understood as a primary good necessary to acknowledge in order to flourish in the present. Royal ancestors are the chief source of blessing and prosperity. As they provided sustenance and fruition in the past—such that people are here today—so they will continue to do so. But they did not and do not give it automatically. They must be remembered, and often they must be given things of substance in order to receive a return. The past also imposes; the ultimate sacrifice, which, as Sakalava know, was the source of the success of the royal dynasty whose members they now address, was of human life itself. Nowadays the exchange is never so extreme, or perhaps so balanced—life for life—but the demands remain quite explicit. Before the Great Service, the living ruler (*ampanjaka*) sends letters to royal subjects requesting money. Ancestral spirits (*tromba*) proffer blessing but regularly importune their hosts and others for money, food, drink, clothing, housing, and attention. People attempt to please the *tromba* by sending them home entertained and satiated. Every time one calls upon a spirit, greets a living ruler, or recites a historical narrative, "custom" (*fômba*)— in the form of an offering of money or a drink of rum—must be satisfied (cf. Kus 1997: 210). Cash is not a "modern" imposition or somehow "outside" the system but the primary means to access ancestral spirits, living royalty, and authoritative knowledge about them.

"Bearing" as Giving Birth, Creating, Bringing Forth, and Carrying Forward; as Raising and Caring for

The connotations of bearing are not entirely negative or passive but charged with ethical agency; obligations to the ancestors are not static but conceived as matters of continual restoration, renewal, and rebirth. The relics are bathed annually, shrines regularly maintained, and spirits provided for. A new ancestral spirit (*tromba*) emerges from the extensive mortuary practices following the demise of each member of the royal clan. Gifts are not only carried in subservience on the head but joyously "driven forward" (*maness*) as one drives cattle, themselves the quintessential gift and source of renewal.

Ancestral activities are regulated by weekly, lunar, and annual cycles, marked by openings and closures with connotations of death and birth. The lunar cycle especially is redolent with reproductive imagery, and the moon lends its significance to the round coins used to "open the mouths" of royals in their transitions to ancestorhood as well as to the monetary offerings made regularly to ancestors (Lambek 2001). A common word for heavy (*mavesatra*) also means "pregnant." Sakalava women are supposed to bear their labor pains in stoic silence (Feeley-Harnik 2000); the image of bloody yet uncomplaining parturition is exemplified in the founding myths of the royal dynasty, in which mothers die on behalf of their sons, and is reproduced in cattle sacrifice. The silence of women during childbearing affirms and enhances their dignity. If the queens are portrayed as bearing Sakalava history—in the shape of their sons—quite literally, their absence from the *efadahy*, the four men, whose relics constitute the shrine at Mahajanga, is marked rather than unmarked; female ancestors are, in effect, all the more potent for their public opacity (chapter 2).

Through their artistic exertion, spirit mediums continually bring the ancestors into being—body them forth and give life, voice, and agency to the past. Bearing here is a form of creativity, and creative performance a vehicle for history; history provides a locus for aesthetic action and display. I refer to this crafting as poiesis and discuss it below and in chapters 3 and 11.

Another polyvalent word that describes Sakalava relations with their ancestors and that adds a sense of mutuality is *caring*. Ancestors care for their subjects and "raise" them (*mitarimy*) as parents do children or as people keep domestic animals, but they demand to be cared for in turn. They need to be acknowledged, remembered, restored, and reproduced. Bathing the relics can be likened to bathing infants, and the relics are carried on the back (*mibaby*) exactly as parents carry small children.[5] Ancestors are coaxed, coddled, and clothed. Houses are built for them, and their remains and possessions are curated. At some moments great affection and intimacy between ancestral spirits and their subjects is expressed. Moreover, ancestors too take their place in a hierarchy of generationally based care; younger, more active, and generationally junior ancestral spirits care for and work on behalf of the older ones, much as the living as a whole are engaged on behalf of the ancestors.

Subjects must care in a more abstract sense as well; they care not only for royal ancestors but about the past. They care that ancestral work is performed, and performed correctly. Taking care means both "looking after" and "paying attention." One must take care to listen to the ancestors—to respect them. Respect is shown by maintaining taboos, and these require continuous care: If one does not eat a certain food on a certain day of the week, one must pay attention both to one's diet and to what day it is. Ancestors expect that care will be taken and punish those who are forgetful, as parents might do with children. Taking care is a form of embodied ethical practice—understood simultaneously for one's own well-being and for the collective good, and simultaneously as means and end.

"Bearing" as Enduring Suffering

Bearing history refers in an additional sense to how Sakalava stand or suffer it. Events granted the value of public history are frequently traumatic ones. Despite the fact that most of them took place many generations ago, their recollection is not one of dispassionate objectivity. The past perdures. Not only is the subject often painful, but frequently the means are as well. The queen who passed away during childbirth reproduces her condition in generations of women. Not merely reporting facts, her mediums live her agony in their own inability to sustain live births as well as in their performance of the queen in labor. Mediums also live in fear of the ancestors' considerable propensity for anger; spirits are fierce (*mashiaka*) and frequently punish their hosts and other people who disregard them. The activities surrounding particular shrines and spirits provide webs of meaning that are not only publicly but personally emotionally charged and linked to deep issues of selfhood, illness, disappointment, hope, conflict, and anxiety. Bearing history is not a neutral activity; mediums draw on identification and empathy but also experience great ambivalence.

Although much of what is commemorated is painful, it is also both highly selective (not all traumatic events are explicitly commemorated)[6] and sublimated (private and collective pain is transformed into, and by means of, publicly available art). Spirits are epitomized by the manner of their dying. The poignancy of suffering and death is evident, but the death of each ancestor is also choreographed—channeled into particular styles of costume, gesture, and performance. No one cries anymore over ancestors who died before the living were born.

As they move in and out of presence, spirits evoke the suffering of absence and loss. Like Western literature and social science, Sakalava representations can be said to emerge from or attempt to fill a void of longing. The world is never whole as it was. The question is, How in a given cultural context is this primary human experience recognized and recomposed? How are the ruptures of history and ontogeny brought to speak to each other? Is the striving for wholeness seen as valiant; the recognition of its unattainability as noble? What is the right mix of tragedy and irony, of effort and wisdom?

The idea of history as suffering—as something to be borne—needs to be distinguished from an ideology prevalent in North America today. Sakalava do not commemorate past trauma in order to paint themselves as victims or acquire rights on that account. They want to be understood as powerful. But the Sakalava notion of power is characterized by the way it turns inward; potency is understood in a way that highlights silence, sacrifice, and constraint, but also connection to the wider order rather than individual rights or autonomy from order.

"Bearing" as Ethical Activity

Informing Sakalava historical practice and emerging from it is thus a whole theory of power: potency and subjection, action and passion, toughness and tenderness. The idiom of bearing indicates that history is, in the first instance, embodied. People endure the fierceness of the ancestors "on their backs" but also flourish for it. Spirit mediums are often people who have found themselves personally vulnerable to illness or misfortune but also most able to care for the misfortunes of others. For all parties, not just mediums, bearing history is partly about addressing structural constraint, discovering and accepting the limits beyond which one cannot go but also learning both how to draw from constraint and when to push against the barriers. Bearing history also means acknowledging the inevitability of death; there are even some spirits who are known to first enter a medium shortly before the medium's own death.

Sakalava practice contains a fundamental idea of the necessity of sacrifice (in a broad sense) as a source of power and a means to differentiate, specify, or center identity. Bloch (1986) has emphasized the creation of the authority of a transcendental order in Malagasy ritual by means of the symbolic capture of vitality. The vitality captured is not only that of others, but one's own. Hence, it is at once submission to ancestral constraint and the source of a superior form of potency. At a collective level this is the magic Sakalava attempt to perform on their own history.

The art of maintaining and transcending vitality is generated in the tension between enduring, flexible performance (both as and on behalf of ancestors) and finished product. This tension is manifest in the politics of architecture, specifically whether to build in perishable or solid materials (*en dure*), but it is also evident with respect to royal mortuary practice and the constraints imposed on the personification, embodiment, and agency of ancestral spirits (chapter 10). The ideal is a balance between freshness and ossification, between history located in the act of working and in the work as finished product, between the circulation of value and its encapsulation, and between popular will and central authority.

In all three prior senses—supporting a load, bringing forth, and suffering—bearing history is both painfully necessary and desirable.[7] In the ways that weightiness, caring, and suffering feed into each other, they create the kind of density of unqualified expectation that might be referred to as hegemony. But to leave it at that would be not to take Sakalava experience and insight seriously and to miss the good that the ancestral system engenders or enables—the agency, dignity, responsible practice, interdependence, creativity, and the means provided subjects to acknowledge their ambivalence toward their own subjection.

Bearing history is viewed as an end as well as a means. That is, it is experienced as a virtuous activity. Historical perdurance is an underlying reality and a significant source of value and meaning. Acknowledging and responding to the continuous connection of past and present by means of the royal dynasty is simultaneously a form of political subjection and deeply moral. Indeed, the force of the past, its sacred potency (*hasing*), which is also the source of present-day flourishing, can only be sustained through continuous practice. As one medium put it, "People suffer when taboos are broken and the *hasing* used up."

Obviously, not everyone attends to the burden of history equally. Some people find themselves coerced to participate, whereas others feel freer to go their own way, remembering only occasionally to pay respects at the shrine or give coins towards the Service. At shrines ancestral customs must be followed more exactingly than elsewhere. Ordinary people get a taste of the sort of burdens the mediums and shrine servants carry when they cross into sacred spaces. If they stay away it is more likely for fear of inadvertently violating a taboo than out of resentment at the subordination entailed. Obligation is concentrated in certain people, times, and places, but this framing serves to mark obligation itself and to clarify its significance for everyone.

The burden of the past is not a dead weight. It provides the principles and basis for order, integrity, and responsible practice. Indeed, I will argue that it sanctifies the present. Rather than simply feeding on the vitality of the present, the past provides possibilities for living authentically and with dignity. Sakalava are not "free," but complete freedom, as envisioned by the existentialists, is apprehended as futility, absurdity, and dread. Action is forged in the balance between the "incredible lightness of being" and the incredible heaviness of history—*incroyable . . . mais vrai!,* as is often stated in north-ern Madagascar.

Finally, among the rewards of bearing history is pleasure. Although always serious, the ancestral cult is by no means always solemn or sober and it is rarely somber. Sakalava commemorative performances—even those evoking such traumatic events as defeat, suicide, or murder—are characterized by periods of drinking, ribald singing, sexy dancing, and other forms of play. Moreover, if history can be variously expressed in modes of romance, comedy, tragedy, and satire, elements of each mingle in the Sakalava genre.[8] There is the means for tremendous intellectual, aesthetic, and sensual enjoyment.

"Baring" as Exposing

A difficulty in describing Sakalava practice concerns bearing in a final, homonymous form. Baring, in the sense of laying bare, is something that Sakalava abhor. Dignity is to be found in discretion.

In chapter 2, I describe the Sakalava world as a maze. Ethnographic fieldwork requires following many paths, doubling back, and figuring out how to get around the blockages, silences, and disclaimers. One can easily get into the spirit of this. But when it comes to writing there is an evident tension between the interest of the ethnographer to disclose a world and the interest of many Sakalava to keep parts of it undisclosed or to mark acts of closure, enclosure, and disclosure in spe-cific ways. The issue is simultaneously ethical and epistemological. Excessive clar-ity speeds up the communication of information but also destroys its meaningfulness and, hence, its value. The value of Sakalava "ancestral knowledge" is maintained in part by means of its opacity, the difficulty of reaching it, and the partial nature of any view obtained. The relics themselves, carried so fleetingly, should be neither seen nor fully described. They lie within a series of concentric enclosures, each of which is increasingly restrictive of access. Their brief excursion remains deep within this maze.

I learned early—and repeatedly—that exposure shrivels and demeans its object, that the medium is the message and the journey its own reward, that intimating or ac-knowledging secrets is often more powerful than revealing them. This inhibits the sort of surgically precise ethnographic description conducted by the master structure-functionalists that I admire (though nothing provides as clear a picture as an autopsy), but it is not so far removed from the theory and practice of Western literature, of "fic-tion," as we say. This book is definitely a work of "nonfiction," but it tries to evoke and imply rather than state certain things—periodically to entice by means of glimpses, fragments, and allusions. This is the way that most Sakalava learn things as well. At the same time, I know that scholarly readers are impatient and publishers have word limits. So in most passages, where it does not excessively blunt ancestral power nor violate Sakalava sensibilities, the exposition is more direct.

Sakalava discretion is itself under threat. This is the subject of much debate and dis-concertion (chapters 10 and 12).

. . . SAKALAVA HISTORY

This book is an ethnography of history, but it is not thereby fully historical anthropology. Despite increased expectations of what an ethnography should cover, no book can do everything. If what Sakalava bear is "history," what this book bears—its subject—is something somewhat different, namely, historicity.

The book neither pretends nor aspires to fully document the history that Sakalava bear. An encyclopedic version is impossible for a number of reasons: not only that I am not Sakalava and that my powers of recording are not up to the task, but also because, of course, Sakalava do not agree among themselves, and because it would reduce the dynamism inherent in a dialectical process. Nor do I choose between a Sakalava version of history and an "objective" account—between attempting to be true to local ways of speaking the past and attempting to be true to the past itself. For one thing, Sakalava themselves do not conceive a difference between them. Given my questions, many people, quite justifiably, interpreted my research as historical, and they have anticipated a scholarly master narrative that would compile and adjudicate among the various oral narratives (*tantara*) and written sources and "tell it like it was." I am sorry to disappoint them. I do attempt historical sketches in some places in the text, notably in chapter 4, but I have not conducted a full and properly historical inquiry. I am not even very concerned with determining the accuracy or inaccuracy of what Sakalava report took place in the past. What interests me is not so much the "invention of tradition" (Hobsbawm and Ranger 1983) but, as Sahlins aptly put it (1999), the "inventiveness of tradition."

What people in Mahajanga expected from me was an event history that would focus on acts of royalty, not an account of colonialism or the structural transformations that it wrought. Historical anthropologists with converse expectations are equally likely to be disappointed in this respect. Nor is this, by and large, an account of a "long conversation" (to reprise Comaroff and Comaroff 1992, 1997) between colonizing Europeans and colonized Malagasy; hence it does not attempt to explain its object primarily in terms of "hybridity," "hegemony," or "resistance." None of these choices are wrongheaded, but they simply are not mine.

Historicity

Most historical ethnography begins with the insight that culture is historical (diachronically constituted). The converse, that history is cultural (or structured), has been recognized at least since Lévi-Strauss (1966). Two complementary lines of argument have followed. One of these, epitomized by the work of Sahlins on the Hawaiian interpretation of Captain Cook (1985, 1995), explores how new events are culturally mediated, understood, and absorbed within a given society as part of its history. The question asked is how history happens.[9] The other line of argument is concerned less with how the past structures the reception of new events in the present than how the relationship of past to present is locally formulated and understood in the present—how the past is articulated with the present to give a particular shape and form to time. Both of these concerns are pertinent to what I call historicity or historical consciousness.[10]

Sakalava history has form and substance. Neither a timeless pre-present nor a shapeless disposal bin into which completed actions are discarded, the Sakalava past has density, differentiation, and depth—temporal depth but depth also in the senses that a painting has and that a novel, poem, or intellectual argument have. The Sakalava historical past is not completed but continuous in the present; in grammatical terms it is imperfect rather than perfect.[11] Personages move forward from the further reaches and voices interrupt one another, but the relative temporal precedence of each is maintained, anchored in its genealogical relations within or with respect to the royal clan.

In brief, if Western historicity (at least in the twentieth century) can be summed up by the much invoked figure of Benjamin's allegorical angel of history, that is hardly the Sakalava vision.[12] Instead of turning back toward the mounting ruins of the past onto which the angel mournfully gazes, Sakalava face the future, bearing the past on their backs, carrying it with them.

As Marx suggests, the burden of the past is a heavy one. But its weight need be experienced as "nightmarish" only from a perspective that imagines history as a struggle toward progress and enlightenment. For Sakalava, the burden is not what holds them back but what enables a safe journey. However, this does not imply that Sakalava idealize the past or look at it with nostalgia. They cannot, and have no need to, since it follows them, intrudes, and demands recognition.

In fact, Sakalava historicity is constituted through a dialectical tension between the ways the past irrupts in and confronts the present (largely by means of the *tromba*) and the way it is conceptualized in and from the present and contained by it. If the former grants history a kind of autonomous rhetorical force, the latter approximates the way "history" is commonly understood in the West and to which, some might argue, the term should be restricted. For the past to impinge on the present it must be understood, conceived, and imagined in a particular way; yet conversely, the way that the past is understood by Sakalava is shaped by the way its personages are carried forward and intervene.

It is this dialectical "historicity" of which this book is an ethnography. I explore the specific discursive forms, registers, and modes of practice by means of which it is produced and engaged. It is already obvious that these refer primarily to royal ancestors. Of course, there are other historical idioms and discourses to be found in contemporary Madagascar. I have not attempted to explore all the ways in which ancestral historicity articulates with consciousness produced by Christianity or Islam, state schooling, or academic or official history.[13]

Like my previous books, this one is informed by a kind of structural hermeneutics with a strong leaning toward practice. It is structural insofar as it attends to the social dimension in the production and distribution of historical consciousness and to the cultural categories, rules, conventions, oppositions, and constraints that underlie practice. It is hermeneutic insofar as it attends to the poetics of Sakalava experience, strains to catch the spoken and unspoken meanings that underlie and emerge from Sakalava tradition, and forages in Western tradition for the least inadequate language in which to phrase its translations. It is based in practice insofar as it follows and returns to ac-tual projects, events and conversations, real persons, and the milieux in which they act, routinely, spontaneously, and judiciously. The book describes the practice of people who are not alienated from the past but busy living with it, moving in time with, drawing from, beholden to, and engaged with history. Politics and ethics are central to practice and to this book.

The ancestral past permeates present-day Mahajanga and its environs. On the sur-face, the city looks all too caught up in the contingencies of the present—a present of muddy unpaved streets and tin-roofed houses in the Sakalava quarters, little disco bars and imposing mosques and churches, thronged markets and empty avenues, busy port and perpetually closed textile factory, superb fresh fish and hungry families, and pa-tronage and unemployment. Not usually directly visible, behind gates and fences, in groves on the outskirts and in the distant countryside, at night, in cupboards, in tombs and under wraps, in embodied practice and in the moral imagination, lies the past—all the more powerful for remaining discretely concealed and protected; set apart, yet immanent.

The book describes those "clearings" when and where the past is brought out into the light, where the past explicitly informs the present, and where the present culti-vates its past. The biggest clearing in which the past is made most visible to the largest public is the annual Great Service. But the past does not work only by means of pub-lic spectacle; it extends in "capillary" form into the intimate formation of individual subjects.[14] ·

Historicity is located in the spaces prised open between history and memory; be-tween fixed, official, and scholarly accounts and popular or individual versions, fan-tasies, and defense mechanisms; between order and jouissance, between Marxist and Freudian fetish; and in the sparks or friction where they rub too closely—spaces and sparks, projects and practices, where the past is conserved and curated, but also crafted, performed, personified; received and addressed; invoked, supplicated, appeased, ac-cepted; authorized and ironized. Historicity is constituted in traversals across these spaces and connections among these acts; between the past as power and ideal, imma-nent and transcendent. The book is about historicity as culture.

In sum, Sakalava historicity is not identical to the history *of* Sakalava, neither to what happened in the past nor to what Sakalava represent as having happened, though it is shaped by both of these.[15] It is constituted by the density or thickness of the Sakalava past, and the place and impact that dense past has in and for the (ethno-graphic) present. It is the "imagined continuity" that gives the "imagined community" (Anderson 1991) its characteristic form, outlook, and sense of purpose. Historicity constitutes contemporary Sakalava "identity," linking existence and essence into a sin-gle, dialectical whole.

If it provides history *for* Sakalava—history as Sakalava craft and exercise it—his-toricity is also constitutive of its subjects, so that those excluded from its reach and practice are not markedly or meaningfully Sakalava, at least not in the same fashion. It is then not the history of those who successfully ignored or evaded Sakalava power—people who emigrated and broke or lost their connections, talked their way out of serving, devoted themselves entirely to Christianity or Islam (though most Sakalava have not been converts in an exclusive sense), secularized, or bourgeoisified. However, many upwardly and geographically mobile people have remained strongly attached to the Sakalava center—one has only to note the money flowing in from Malagasy and Franco-Malagasy in the capital and abroad.

Recently, the complex of practices has become relevant for other Malagasy who see Sakalava ancestors as metonymic of all properly Malagasy history, filling in for missing royalty and ceremony elsewhere. In part, through policies enacted by the central government in its attempt to promote a vision of national unity in regional diversity, Sakalava ancestral practice, especially the Great Service, has begun to become objectified as Malagasy "culture," available for the edification of all citizens as well as foreign tourists. *Tromba* and the Great Service have become the model for a particular kind of historicity that has seized the imagination of many Malagasy, including many for whom it is no longer a direct, lived reality.

Although cultural objectification has had an impact on the conduct of the Great Service, which I will describe, my emphasis is more on how Sakalava subjects for whom it does remain a lived reality chart their course through an uncertain postcolonial present. "Tradition" need not imply conservatism and is certainly not in any simple opposition to "modernity." Indeed, rather than rejecting innovation, Sakalava (no less than other Malagasy) draw on the past in order to authorize change and sanctify the present. Sanctifying postulates (following Rappaport 1999), frame and inform action, establishing the means for evaluating, rather than determining, its content. As such, they are not opposed to change but provide the terms through which it can be rendered meaningful, acceptable, and orderly. The sacred is thus the necessary "background," characterized by its constancy and invariance, against which movements in the foreground can justifiably take place.

To say this is to make an argument as to why historicity is central to Sakalava life and, hence, for an appreciation of that life by others. The history that Sakalava bear is no mere charter or ideological superstructure but the ground of value and virtue. That is, it forms a whole.

THE WHOLE AND ITS REFRACTIONS

The weight of the past encompasses social reproduction, hierarchy, power, labor, exchange, pleasure, death, illness, prayer, sacrifice, and the quest for human flourishing. Work on this book has reinforced over and over the insight of Marcel Mauss (1966) concerning the unity, in certain societies, of what Western thinkers have distinguished as "religion," "economy," and "politics" or "government." This is equally a social world from which "art" has neither hived off nor split itself between highly individual expression and aestheticized spectatorship (Agamben 1999) but runs its dangerous course in the thick of life. The social whole is complex and internally differentiated but not along the lines that have produced the various civil institutions or academic disciplines in the West. If analytic categories like "politics" and "religion" distort the subject matter, where to start? The same Durkheimian intellectual tradition offers a heuristic tool in the division of labor, which I address specifically in chapters 4 through 6.

It is important not to misunderstand. By "whole," I mean what is implied by the notion forged by Mauss out of Durkheim of the "total social fact," not the reified or organic whole of a distinct, bounded culture or society. It is whole because of a dense symbolic and social nexus and a confident capacity to attract and encompass, not because of any functional requirements or a policing of boundaries.[16] Indeed, Sakalava practitioners are characterized by cosmopolitanism more than insularity. Individuals move in and out of attachment to the complex that I describe, and it shifts in and out of focus, competing for their attention with other discursive formations, attractions, interests, demands, and interpellations (cf. Lambek 1993).

I focus on social actors who are among the most embedded in and committed to Sakalava history, most deeply subject to bearing it, most comfortable with its emergent properties and unpredictable potential, and most articulate in explicating their understanding, pleasure, care, and, it is not an exaggeration to say, their love (with all the pain and ambivalence such a condition implies) for it. Among the most original and articulate are certain senior spirit mediums, people who have inspired parts of this work, who reappear throughout it, and who are the subject of direct attention in several chapters.

Such exemplary individuals need to be understood alongside another Maussian concept. In a significant essay Mauss (1985 [1938]) referred to the members of certain societies as *personnages*. By this he meant that they held social roles, as though cast in a theatrical performance. Personages perdure from generation to generation irrespective of the particular biological "actors" (individuals) who "play" them. The sociological use of role is a much weaker version of this concept because the personage refers to an integrated totality, not a specific function. In a sense, the personage is an instance of the "total social fact," and a microcosm or refraction of the social totality. The personage or character is also understood as an intrinsic part of a wider order, unlike the modern "possessive individual" (Macpherson 1962), who is seen as unique, distinct from, and possibly in some opposition to society. The personage is intrinsically relational and enduring, and a figure who is not split into the overly objectified individual, constituted through the effects of governmentality, bio-politics, and memoro-politics (Hacking 1996), and the overly subjectified individual whose qualitative uniqueness and autonomy are celebrated in liberal humanism.[17]

The personage represents a kind of fusion of the person with the office, in which the office itself is attributed personal characteristics. Elements of such a system are visible in the British monarchy or American presidency, where incumbents are in some sense identified with their predecessors. But in the Maussian universe everyone ("all free men" [Mauss 1985: 7, cf. Mauzé 1994]) might live their lives as manifestations of such characters. Sakalava royal ancestors are personages "borne" by mediums and custodians. Of course, the latter, as living, rounded human beings, are not identical to the personages.

Poiesis and Phronesis

Two master concepts or analytic conceits of the book are drawn from Aristotle. They are, respectively, poiesis and phronesis. Poiesis—making, creative production, craft, artistry—is useful both because it grasps the creative quality of so much Sakalava practice and because it avoids the rift between ideal and material that has characterized so much thought since Plato. Nor does it presuppose what Marx noted as the historical split between intellectual and manual labor. Not distinguishing "art" from "work," poiesis provides a framework in which neither concept must be given priority. When I speak in subsequent chapters of a division of labor, such labor should be understood to encompass the continuous crafting and presentation of Sakalava history.

Phronesis refers to the ethical dimension of action (praxis) based on the cultivated predisposition to do the right thing in the circumstances. As practical wisdom or character it does not make the radical distinction, so prevalent in modern (Christian, capitalist) thought, between the instrumental or interested and the high-minded or disinterested but emerges as Aristotle's virtuous golden mean. Just as poiesis evades distinctions between material and ideal and overly literal conceptions of truth and falsity, so phronesis overcomes those between interest and disinterest, action and passion, free agent and determined subject. It avoids the pervasive distinction comprehended by such dichotomies as "culture and [versus] practical reason," or "formalist and substantivist," or yet, freedom and necessity, existence and essence, in ever finer declensions of agency and structure, subject and discourse, and so forth. Phronesis affirms dignity and self-respect as central aspects of human practice.

In my usage poiesis and praxis refer not to mutually exclusive phenomena but to distinctions in emphasis; practice has a productive dimension and production is practical. However, their conflation, Agamben notes, is a product of later Western thought. Praxis can be defined as willed action (1999: 75)—a form of doing in which "the will . . . finds its immediate expression in an act" (ibid.: 68). Whereas for the Ancient Greeks, "the essential character of poiesis was not its aspect as a practical and voluntary process but its being a mode of truth understood as unveiling" (ibid.: 69). Agamben defines it as bringing into being or pro-ducing into presence.

When I refer to "creating history" I am not suggesting that Sakalava history is fictional in the sense of being uninterested in corresponding to the real. As I discuss in chapter 3, the concept of poiesis serves as a reminder that *all* history must be crafted or composed, that events, circumstances, and so forth, are viewed in a certain light and put together in a particular way (White 1973). However, Agamben's derivation suggests that empirical truth may not be the only kind at issue. Poiesis unveils the truth inherent in things; indeed, this is sometimes referred to as poetic truth.[18]

The Aristotelian terms are useful insofar as they highlight how Sakalava constructions of history differ from Western ones and yet stem ultimately from the same human concerns, capacities, and forms of engagement with the world. They are helpful also in that they do not mirror the usual analytic divisions by means of which scholars in the Western tradition tend to map out the world. Both terms avoid the primary dualism of Plato, which takes a number of forms. One version, sometimes phrased as poetry versus philosophy (Havelock 1963), has come down to us as tradition versus modernity, savage versus scientific thought, enchantment versus enlightenment, error versus truth—in sum and fundamentally, irrationality and rationality.[19]

Aristotelian language reinscribes the debate from one of rationality to one of holism and from the realm of pure, idealized thought to thought as it is realized in making and doing, that is, thought immanent in, rather than transcendent of, culture and social life. This obviates simple oppositions between mind and body (imagination and embodiment), sacred and secular, reality and mystification. I emphasize the holistic cast of poiesis and phronesis in which art, work, religion, politics, and ethics are subsumed, neither identical or fully commensurate with one another nor distinguished as discrete provinces of action.

However, the attempt to overcome stubborn oppositions should not entail rejection of dualism per se (Lambek 1998a). My account does not, I hope, evade examination of the force of contradictions, tensions, and ambivalence as they arise in, and are transformative of, structure, practice, and subjectivity.

Historical consciousness entails the continuous, creative bringing into being and crafting of the past in the present and of the present in respect to the past (poiesis), and judicious interventions in the present that are thickly informed by dispositions cultivated in, and with respect to, the past, including understandings of temporal passage and human agency (phronesis). That is, historical consciousness or historicity encompasses the complex relations of past, present, and future suffusing and emerging from production and practice, rather than simply the objectified knowledge *of* the past.

Earlier I suggested that history be viewed as cultural and that this book is about Sakalava history as culture. This is not to oppose a static or synchronic concept of culture to a dynamic one of history nor to reject the insight that culture be grasped as fully historical. Once poiesis and phronesis are understood as basic modalities of culture, culture is grasped as dynamic. Understood as "culturing" (White 1969) and, hence, as neither timeless code nor motionless web, it transcends the material/ideal distinction. The culture of history is the poiesis and praxis, the making and doing of, and with respect to, history. As indicated, because praxis so often has been conceived in Western theory as fundamentally instrumental or autonomous, I specify and emphasize the kind of praxis that Aristotle distinguished as phronesis.

"Character" thus embraces both the Maussian "*personnage*," the inspired representations of possessed mediums, and the product of ethical cultivation as it emerges from an Aristotelian account of practice. I examine the imaginative and ongoing crafting (poiesis) of Sakalava historical characters through the registers of material objects and embodied performances.[20] Sakalava characterizations are distinctive and diverse, and performances range from large and boldly sketched ceremonies to compositions that draw on minute details like the folding of a handkerchief to make their point (chapter 11). The renewal of ancestral personages enables a vast bricolage that grows like coral along the genealogical spine of the royal clan. I emphasize also the virtuous practices carried out in the name, and by means of this history, including the offering and reception of money, respect, and *hasing* (sacred potency); and the care of the relics, restoration of the shrines, and mutual concerns of spirits and mediums. I am especially interested in the situated, historically informed judgment (phronesis) that permeates daily life, political agency, conflict, critique, representation, and intervention. I focus on the artistry and moral practice of the senior spirit mediums and end by arguing that in their conjunction the mediums manifest not only historical consciousness but historical conscience.

Throughout, I demonstrate how historical consciousness is both reproduced and left emergent. These twin potentials are rendered through pervasive divisions of labor. In the curatorship of the relics of royal ancestors, in the custodianship of their tombs, and in the care and embodiment of their persons, the past is conserved, authorized, renewed, and rendered available for further thought, production, and action. Through the bearing of distinct royal ancestors rising in their mediums, the past is thrust into the present. No act or scene is fixed or monological. Historical performance always includes a variety of specific, jostling voices—individual personages from discrete periods in the past—and it speaks always in dialogue with diverse interlocutors from distinct social loci in the present. The polylogue is constituted along temporal, geographic, and social/hierarchical axes.

Two material circumstances that contributed to how I carried out the research and have written the book deserve mention. First, the division of labor and distributed quality of "collective" memory mean that no two interlocutors have seen things in quite the same way. Local politics also contributes to divisiveness. Hence, I could not be true to one voice or one vision without being false to others or to the whole. My method has been to measure passion with dispassion. I talked to people across Mahajanga and in various relations to the ancestral polity, but when I found particularly compelling practitioners of things Sakalava I tried to benefit from sustained presence at their elbows and absorb their frames of thought and points of view. I have also tried to stand back and see things from the perspective of a comparative social science—from the standpoint of a committed anthropologist (which I am) no less than from the standpoint of a committed Sakalava (which I am not). Several voices weave through the text, but they do not determine it and none are likely to agree with all of it.

Second, my own labor on this project has been divided into a series of five research trips over the period from 1993 to 1998 plus a brief sixth visit in 2001 as the first draft of this book neared completion. Each trip was relatively short, four to six weeks in duration, but intensive, and timed to encounter key events in the annual cycle. I was able to plunge right in because of linguistic and cultural competence developed since 1975 in neighboring Mayotte and subsequently among Antankaraña and because of the enthusiastic yet always carefully discreet instruction and introductions I received from a Malagasy scholar in Toronto (Lambek 1997; Tehindrazanarivelo 1997). The research is better described as multischeduled than multisited, providing an acute sense of the nature of short-term change and the partiality of single encounters. Successive interlocutions with Sakalava consultants pushed, pulled, and generally extended the way I saw things. Discussions with colleagues in North America and Europe in the intervening periods have done the same.

The trajectory of the book is as follows. In chapter 2, I locate Mahajanga and the shrine containing the relics of the "Four Men" in the light of Sakalava reflections on power and first appearances. Chapter 3 describes a spirit possession ceremony and makes the critical argument that possession viewed as poiesis offers a modality of history. Both chapters address the Sakalava "chronotope"—the specific conjunctions of time, space, and person that form the Sakalava world.

Part II explores the social field constituted through three intersecting divisions of labor. These divisions provide the opportunity for partial articulations of, respectively, the history of the polity and relics (chapter 4), its sociopolitical constitution (chapter 5), and the centrality of the distinctive personages (chapter 6). Part III then focuses on how the divisions of labor come together in the Great Service. Chapter 7 examines performances in the neighborhoods of Mahajanga in which money is collected and history rehearsed, and chapter 8 describes activities at the central shrine that culminate in the Great Service itself.

Part IV attends to the practice of spirit mediums conserving and renovating the shrines (chapter 9), contending with change and conflict (chapter 10), and performing and practicing history (chapter 11). Chapter 12 concludes the argument with a review of the distinctive polyphony of Sakalava historical consciousness, the relationship of power to potency, and sanctity to social change in the ongoing realization of tradition in Mahajanga.[21]

2

Into the Maze:
Surface and Center,
Place, Person, and Potency

*But I repeat: all understanding falls short. It is as if the poem, the painting, the sonata drew around itself
a last circle, a space for inviolate autonomy. I define the classic as that around which this space is peren-
nially fruitful. It questions us. It demands that we try again. It makes of our misprisions, of our partiali-
ties and disagreements not a relativistic chaos, an "anything goes," but a deepening. Worthwhile
interpretations, criticism to be taken seriously, are those which make their limitations, their defeats visible. In turn, this visibility helps make manifest the inexhaustibility of the object. The Bush burned
brighter because its interpreter was not allowed too near.*

—George Steiner, *Errata: An Examined Life*

*. . . [the] task is to demonstrate how those whom power-holders consider to be marginal are central to
themselves.*

—Maurice Bloch, Foreword to *The Time of the Gypsies*

If the poem, the painting, and the sonata can be "classics" in Steiner's sense, why
not places? The shrine (*doany*) of Ndramisara, home of the relics of the "four
men" (*efadahy*) who stand at the foundation of the kingdom of Boina, is such a
place. In this chapter I visit the city of Mahajanga and the shrine located in a
small village on its outskirts. I begin with some of my own perspectives on these
places—perspectives characteristic of first visits—and conclude with certain Sakalava
reflections on first appearances, surfaces, and depths.

GOOD PROGRESS

My first impression of Mahajanga was of stark poverty. In the springless back seat of
a *deux chevaux* taxi that limped from the airport toward the center of town, we
passed along the edge of Tsaramandroso ("Welcome" or "Progress") market. People

On the edge of Tsaramandroso (Good Progress) market, Mahajanga.

City hall, Mahajanga.

sat patiently behind small rows of recycled plastic water bottles and fragments of rusty implements, amid the jumble of market umbrellas and the press of *pousse-pousses* (rickshaws) whose lean but very muscular drivers waited to pull sacks of rice or weary shoppers home.[1] On that occasion I was merely looking for a place to eat between planes. The flight from Mayotte had arrived that morning, and the "Air Mad'" flight from and for Antananarivo (Tana'; figure 2.1) was late and not expected for several hours (perhaps President Ratsiraka had "borrowed" it). There had been no place to change money, and once the throng in the airport cleared I gave in to the importuning of two young men and agreed to ride into town in their taxi. In 1991, currency regulations were quite strict and as I had no Malagasy francs for the fare (or lunch), the drivers promised to take me to a bank. We drove past the crowded market, along a wide boulevard lined with churches, schools, and government buildings and into the older part of town near the harbor where the Hôtel de France and the city's three banks clustered amid silent, weathered two-story buildings with shuttered overhanging verandas. When all banks proved shut, the taxi drove up a side street, past some older stone buildings with beautifully carved wooden doors, and parked adjacent to a wholesale hardware store. With some hesitation at being apprehended, I handed the driver a French bill and waited as he disappeared down an alley. Shortly after, he returned with more Malagasy francs than I had expected. We hastened to a market where the driver's companion hopped out to purchase bread and oranges.

Returning to the airport, along a stretch of incredibly bad road through dusty scrub, the taxi ran out of fuel. Luckily the driver had enough in a jerry can to get us back to a gas station in town before proceeding. In 1991, I had no plans to conduct research in Mahajanga and was relieved to get a flight onward that evening. My plans changed for reasons that had nothing to do with my experience that day and everything to do with a chance encounter a few months later in Toronto with a man from Mahajanga. We have described the encounter and our work together in some detail elsewhere (Lambek 1997, Tehindrazanarivelo 1997), but the consequence was that when I returned to Mahajanga in 1993 to begin this study I was able to move immediately into the home of Amana Ibrahim in the very quarter of Tsaramandroso Ambany (Lower Good Progress) whose market had impressed me with its poverty and noisy chaos.

Amana Ibrahim is a highly respected local intellectual with particular interests in politics and history. The household, run by his extremely solicitous wife, Zaina, embraced several grown children, a couple of grandchildren, and a continuous stream of more and less permanent visitors. I too have called this calm and pleasant haven home on each of my subsequent visits. The house was a modest but comfortable six-room affair with a screened terrace looking onto the street and a kitchen, shower, and latrine in the internal courtyard.

The parts of town inhabited by Malagasy speakers—laid out in rectilinear grids of mostly unpaved roads, paths, and channels of waste water—seemed a far cry from the quarter with the banks and wholesalers, shaded with high buildings, some dating back well over a century, and inhabited mostly by people of South Asian (Karany) descent.[2] Japanese owned the large offshore fishing enterprise that was a major employer,[3] but

2.1. Map of Madagascar

along with the French residents and American missionaries, they were hardly visible to me. Yet despite the evident spatial separation of Mahajanga's quarters, there were multiple social connections traversing them. Many Sakalava I came to know had some French, Chinese, Indian, Yemeni, Senegalese, Somali, or Greek parentage. The kinship and cultural links between Comorians and Sakalava were especially dense.

The Malagasy districts, which mingled houses of better and lesser quality, included people from diverse parts of the country as well as Comorians. "Sakalava" refers to the long-standing inhabitants of the western strip of Madagascar. However, although Mahajanga is the largest city in northern Sakalava territory, Sakalava inhabitants comprise neither the majority nor the largest group.[4] They are numerically dominated by Merina migrants from the central highlands. Without wishing to reify ethnic difference, Merina (Borzan) are Christian and predominantly Protestant, whereas many Sakalava remain nonChristian and others are Catholic. Merina are more highly stratified, but it is Sakalava who retain reigning monarchs. Appearance and comportment of the two groups are quite distinctive and, as subsequent chapters show, their mutual history has not always been happy. Mahajanga also contains large populations of Tsimihety from the northern interior, Betsirebaka from the east coast, and Antandroy from the south.

Amana Ibrahim is originally from the northern Sakalava town of Analalava. He has a huge network of kin and friends, mostly Sakalava, but some who are Antalaotra and Comorian. Like a significant minority of northern Sakalava, he is an observant Muslim (Silamo). None of the immediate family are spirit mediums or have much to do directly with the practices described in this book. However, as Sakalava and intellectuals, they were very interested in these practices from arm's length and happy to introduce me to kin and neighbors who were intimately involved.

Good Progress neighborhood contained many spirit mediums and some members of the royal clan whose connections with Islam were more tenuous. Immediately adjacent to the house, on all sides, however, were families with roots elsewhere in Madagascar who knew little about Sakalava ancestral practice. Although I never met all of the closest neighbors, soon after I arrived Amana Ibrahim invited me to join him on his long walks across the city and I quickly had friends and contacts in several Malagasy quarters.

In the city most people were poor and struggling. I lived in a milieu in which people were somewhat better off than average. Some had government pensions, and they owned their own houses. Others rented, in 1995 paying a minimum of fmg(francs malgaches)20,000 (about US$5) per month for a two-room house and, if they could afford it, water and electricity on top of that. Young people struggled to find work, and many postponed marriage because they could not support a family. Marriage ceremonies themselves could be managed quite cheaply if one did not require an elaborate Muslim wedding. There was some theft but little violence. According to Moustali, a medical student from Anjouan, Islam was the factor that made Mahajanga "the safest town in Madagascar. Muslims look out for each other; if one family isn't eating, the neighbors share their meal." Indeed, Moustali drew on a distant kin connection with Amana Ibrahim to take his meals in the household one year. Other years the family served a severely mentally ill man unrelated to them by kinship or religion.

Like any city, Mahajanga is a heterogeneous place for which ethnicity provides too simplified a gloss. It can be described as composed of a series of overlapping communities of practice.[5] Not all Sakalava interested themselves deeply in the things that interested the people who interested me. Mine is not a study of the city but a study conducted in the city and among only some of its inhabitants. It is not a study of a neighborhood, an ethnic group, class, or local congregation but of a set of creative acts and moral practices. I am interested in the meaningfulness and consequences of these acts and practices within and for an ongoing cultural tradition, as well as the lives of the actors and practitioners. I locate these people in the tension implied by the juxtaposition of the analytic terms "agent" and "subject" and, hence, am concerned as well with how Sakalava tradition constitutes its human subjects and enables their agency.

Most of the people I worked with in Mahajanga traced their origins elsewhere, to smaller places along the lengthy coastal strip that comprises the northern Sakalava domain. They came to Mahajanga to study, work, or marry and ended up staying. My closest consultants were spirit mediums. I also conversed and conducted informal interviews with members of several segments of the royal clan, officials, servants, residents at several shrines, and various others, including people like Amana Ibrahim, who kept their distance from the shrine and spirits. My interlocutors were quite diverse in their backgrounds and in what they knew; I think the study would look quite different had I focused it around one or more of the reigning monarchs and their immediate entourages. As several people pointed out, the traditional value placed upon undisclosure has produced many gaps in public knowledge and left people dependent on objectified narratives and rumors whose validity they often doubted. I try to offset this by examining the distribution of historical knowledge among different kinds of people.

The Sakalava monarchy traditionally subsisted in the hinterland. I cannot pretend to provide the perspective of the countryside here. I visited briefly a number of villages that form significant nodes in the politico-religious network but not the cemeteries or royal capitals outside the city.[6] The city itself stands in a rather paradoxical relationship to ancestral history. On the one hand, it is a major center and the home of the most important shrine in northwest Madagascar. Its inhabitants, on the other hand, are cosmopolitan, diverse, and most open to change. It is not organized as a traditional polity itself, and the interests of several rulers from the countryside overlap within it. People from the bush (*labrousse,* as the entire hinterland, cities and all, was referred to) were sometimes expected by the townspeople to know more about ancestral practices, but they knew less of the current politics of the center. My older consultants, as part of a generation whose migration to town animated and rejuvenated urban religious practice, somewhat mediated this divide.

The Categories and Their Realization: Chronotope as Maze

Events take place in a world. People inherit the world they live in and act with respect to its givens, but they also act on and in that world, give greater attention to some parts, attempt to ignore others, and frequently disagree with one another's acts and interpretations. People both reproduce certain aspects of their world and create new ones. New acts and new ideas need to be realized—made real—and authorized with respect to older ones, to the known, taken for granted, and powerful aspects of the world as given.

Novelty and challenge come from what is conceived, from certain perspectives, as "outside" as well as "within." Recent theory has emphasized continuous tension between world-making, consolidation, and reproduction ("structure") and the external forces that impinge on and transform structures in the making ("history"). The Sakalava world described in these pages has been marginalized and encapsulated by "history," that is, by forces emanating from larger, more powerful worlds of capitalist global economy and nation-state. Yet Sakalava have never been closed to external forces or ignorant of alternatives. Sakalava world-making took place in conjunction with a regional political economy that relied extensively on foreign trade (Lombard 1988; Lambek 2001). American interests, for example, were probably of greater direct importance in Mahajanga at the beginning of the eighteenth century than at the beginning of the twenty-first.[7] It is necessary not to reduce structure to history (or vice versa) but to attempt to understand local cultural production and reproduction mediated by human thought, imagination, experience, and habituated practice. As the world peripheralizes places like Mahajanga, their inhabitants struggle to recenter themselves.

The contours of the world include the culturally informed fundamental categories of understanding as laid out from Aristotle through Kant and Durkheim: time, space, cause, and person. These form the basis for thought, action, and representation, but they are by no means fully discrete. Time is realized through places and persons; journeys are simultaneously temporal and personal passages.[8] The inextricable connection of time and place in narrative is realized in Bakhtin's concept of chronotope (1981), literally "timespace."

If it is difficult to separate time and space; it is also difficult to conceive them as a seamless whole. Where narrative inevitably prioritizes time—through the sequence of events described and the duration of narrative itself—architecture gives greater emphasis to ordered space existing in and through time. Spatial and temporal constructions provide vehicles to signify each other.[9] One way to begin to understand Sakalava historicity is to examine how Sakalava construct temporally informed and saturated places and thereby spatialized time. The spatial emphasis in this chapter will be complemented by the temporal one in the next.

Landscape and architecture have been shaped by rootings and uprootings. Sakalava retain accounts of entering the region of Boina from Menabe in the south at the beginning of the eighteenth century and moving within and beyond it, to Mayotte, the Comoros, Réunion, France, and back. At the height of Sakalava power, reigning monarchs founded their own places of rule; housing was not made of durable materials, and former capitals were permitted to disappear. Royal tombs and tomb precincts were envisioned as permanent structures, though older tomb precincts were supplemented by ones at new locations. Mobile live rulers and their stationary ancestors were mediated by ancestral relics—detached pieces of the latter that were moved along with the former and, as mobile roots, served as symbolic foundations for the monarchy and the authority of the reigning monarch (Feeley-Harnik 1991a). The mobile living and the fixed tombs were also mediated by ancestral spirits who appear in new bodies and at new times and places. When a medium I knew moved to France and wondered whether there was any need to pack the clothing belonging to her spirits, one of them rose and said, "Do you expect me to travel naked?"

The importance of relics and their link to spirit possession were reported in the 1600s by Portuguese Jesuits, Father d'Azevedo and Father Mariano. In 1741 Dutch sailors described the Sakalava monarch consulting the relics before deciding whether to trade and receiving a response through spirit mediums.[10] During the period of political insecurity following the incursion of the Hova (Merina) under Radama from 1824, Sakalava rulers moved frequently. By the imposition of French rule in 1895 they had split into a number of related lines, each bearing authority over a more limited territory. However, the number of relics did not grow much and, so far as I know, relics are no longer regularly produced. Royal authority has become more constricted, and as succession to rule has become less desirable, possession of relics by ruling members of the branches (except in Mahajanga) has become less critical.

Concomitant with the fissioning of rule has been the anchoring of all branches in the unitary source of the earliest royal ancestors, as materialized in their relics. Having captured the relics from the reigning monarch, the Merina first kept them in a shrine in the fort they constructed on a hilltop above Mahajanga harbor. In 1839, a British merchant, Leigh, was told that the relics were located in the fort to prevent Sakalava, who held them in veneration, from attacking (Allibert 1999: 150); no doubt, possession of the relics was a way of keeping Sakalava subject (Vérin 1986: 376, no. 3). Leigh was more surprised to see the Hova themselves venerating the relics.[11] It is evident that religious ideas and cultural practices transcended political and ethnic difference and enabled the survival of the relics (cf. Rakotomalala et al. 2001).

The relics have moved several times within Mahajanga over the course of the last 175 years through a succession of places whose order was frequently recited, in order to stay on the outskirts in a place where purity could be maintained. Since 1973 they have been housed in a location known as Miarinarivo under the control of the Bemihisatra faction of Sakalava royalty, taken there by force from Mandresiarivo. During the 1990s the latter shrine village, denuded of relics, temple, and sacred enclosure, still maintained its complement of permanent guardians and was the site of veneration for the Bemazava royal faction.

In the production and reproduction of sacred places Sakalava have been attentive to the following, not entirely independent, processes: rooting in the past; centering, enclosing, and encircling; and progressive hardening and preservation. Rooting entails establishing foundations through the burial of humans, blood, and artifacts; the planting of trees and fences; and the repetition of ceremonies. The Great Service (*Fanompoa Be*) is not said to be completed each year but to have grown (*nitombo*).[12] The idea of growth emphasizes that the annual cycle is not a detemporalizing "eternal return." Centering, largely through the construction of a series of successive, ever more inclusive enclosures, produces purer, more sacred, more powerful, and more dangerous spaces as one moves inward. Hardening and preservation concern the bones and relics, as well as the material of their various enclosures.

An apt metaphor for the Sakalava chronotope is a maze. By a maze I do not mean chaos or confusion but, rather, that movements are not direct or consecutive along a linear path and that centers are screened. The paths between places and times do not always follow the straightest routes, and the whole is never visible, either in reality or in representation (as a map).[13] Surfaces and vistas are deceiving; there are unexpected twists and turns, obstacles, and encounters. In a maze the journey is the thing; the center exerts its power as a goal, an idea, and an ideal; but if the goal is reached it is revealed as simply another place and moment in the maze. The center is not a conclusion or an exit; after the center there is simply more maze. The ultimate lesson of the maze may be that there is no center and no exit.[14]

Sakalava history is constituted as a chronotopic maze of enclosures and disclosures. Bodies, voices, persons, objects, and lessons are diffused and congealed, and displayed and concealed behind fences, screens, and gates; around corners and over mountains; within and by means of narrative fragments, snatches of song, flashes of character, and irruptions of the past in the present. The maze amazes. It attracts, lures, provokes, inspires, and entraps. In order to cross each threshold one must remember something. Each time one turns the corner, one forgets something. Rules and restrictions—which foot to put forward, which item of clothing to leave behind—are the means forward and perhaps the string or trail of crumbs to guide one back.

The image of the maze grasps—visualizes—the chronotope in spatial form. Temporality is partly constituted in the tension between the visible and the invisible, presence and absence, voice and silence. Reversing the emphasis, Sakalava temporality helps *place* Sakalava in their contemporary world, producing that home in which Sakalava are "at home in the world" (Jackson 1995) but also the cracks through which the uncanny—the *unheimlich*—seeps. However, my aim is to deliver less a phenomenological account of Sakalava experience than a cultural account of the vehicles, forms, and acts through which such experience has been produced, mediated, articulated, and appropriated.

THE SHRINE AND ITS VILLAGE

Mahajanga is full of life. The ever-crowded shopping streets are crammed with clothing stores, pharmacies, bars, and discos. Dense food markets are supplemented by omnipresent stalls of packaged and fresh snacks. Residential compounds, serving several families or one large one, are full of guests and boarders, pulsing with the comings and goings of neighbors and kin. Schools, churches, mosques, the stadium, and public squares fill for their respective activities, and the celebrants of ritual events squeeze into front rooms and courtyards or under awnings set up in the streets. Work takes place in the few factories, in fishing and maritime transport, in the hospital and government offices, in the markets, shops, and stalls, and in bits and pieces in and out of houses, wherever the informal economy can grasp a toehold.

But for many inhabitants of Mahajanga, the true center lies elsewhere—*off-center*—with respect to the daily secular world of home, work, and market.

A few kilometers northeast of the port sits the shrine village known as Doany Miarinarivo (Level a Thousandfold) or Doany Ndramisara (Shrine of Lord Diviner), containing the relics of the "four men." One approaches the village, climbing gently along the sandy roads of a suburb of Mahajanga. There is a standpipe, and then the path rises farther into a park-like area filled with shady trees. It is noticeably cooler in the shrine village, and if it is not a festival day, quieter and less dusty. There is a sense of tranquility and ease. People sit chatting under the luxuriant mango trees or move slowly about their business. Running is forbidden. The houses are simple rectangles with thatch roofs. Those inhabited on a daily basis have furniture, such as beds, but those belonging to ancestors are virtually empty with only a low altar at the northeast corner and a shelf or two. Some have small plots of manioc, corn, vegetables, and fruit trees.[15]

The village appears itself as a kind of relic, at least as a traditional, rural Sakalava village. It is full of old trees. Houses are constructed of traditional materials. There are no metal roofs, brick walls, or cement floors; no paving, shops, bars, or stalls; and no electricity, latrines, pools of dirty water, or garbage.[16] Radios are permitted but not television. Running water is available at standpipes located on the western fringes of the settlement. "Modernity" per se is not the issue but respect for the purity of the relics. Defecation is forbidden within the village proper, and inhabitants with diarrhea or other severe illnesses are moved elsewhere. As someone said, "There are many taboos in Ndramisara's town." One cannot enter on a Tuesday (*thalata*) from morning until

Looking north toward the south face of the sacred enclosure, Doany Ndramisara, Mahajanga. The "red blood" door is to the east and the roof of the temple visible inside.

the stars come out or on a Thursday before noon. One cannot eat peanuts (*pistaches*), sell rum, or sit with legs outstretched toward the shrine; one cannot venture to the most sacred eastern side of the hill at all.[17]

Sakalava reverence is exhibited not only by tranquility and dignified composure. The shrine community is also the scene of frequent *tsimandrimandry* ("all-nighters"), designed to amuse, entice, and warm up the ancestors on the eve of important festivities and characterized by heavy drinking, music, dance, and sexual play. From the ancestral (non-Muslim, non-Christian) point of view, these are not "dirty" activities and so do not have to be avoided.

Even the carousing spirits are circumspect once they cross the threshold of the temple grounds. The temple sits along the eastern flank of the hill within the large and imposing enclosure known as the *valamena,* literally, the "red pen," composed until 2001 of closely planted tall wooden posts with sharply pointed ends. The sacred enclosure has a large wooden door along the western side toward the southern end and a smaller door on the southern face toward the eastern end.[18] Proscriptions within the fenced enclosure are more elaborate and more strongly enforced than those outside: no tailored clothing or sewn tube-like body wrappers (*salovaña*) for women; no trousers for men; no footwear, hats, undergarments, glasses, handbags, or accessories, and (anthropologists, nota bene) no cameras or notebooks. Men wear simple waist wraps with or without shirts, and women wear longer cloth wraps tied above their breasts. No one menstruating, ill, or unbathed is admitted. No personal greetings among devotees take place within the enclosure.

Visitors enter the enclosure through the large gate on the western face. The southeastern gate, known as the red blood gate (*varavara menalio*), the taboo gate, or Ndramisara's gate, is reserved for distinct categories of people (living royalty, spirits, ancestral servants) and special occasions. This gate is forcefully pronounced *dangereux!* The main door itself is only open on auspicious days of the month (when the moon is visible in the night sky) and of the week, and only during certain daylight hours. People who have come to pay their respects (*mikoezy*) to the ancestors are ushered in by the gatekeeper. Like everyone, they mark their progression to more sacred space-time by carefully placing the right foot over the threshold first.[19]

The ancestors reside in the temple (*zomba*), a rectangular building in the southeastern quadrant of the enclosure and, like the fence, containing a public door along the west face and a sacred one at the southern end. When I began to visit Mahajanga, this building was constructed of raffia wood planks (*mavañaty*) with a high tin roof. The courtyard also contained a smaller storage building, a few small trees, and the horns and craniums of cattle.

Entering the temple from the western door, again with the right foot, the visitor encounters an interior space of similar but far more substantial proportions—larger and airier—than a Sakalava house. But from a height of some seven or eight feet down to the floor it is partitioned into an L-shaped space. The northeast sixth of the temple is curtained off by a long white sheet, the *safiday,* that "wraps the men." Accompanied by the official known variously as the doorkeeper (*mpitamvaravaraña*), intercessor (*ampangataka*), or simply "palm of the hand" (*lambantañana*), visitors move in a crouch as far north and east as possible, while still staying on the near side of the curtain. If a crowd is paying their respects, each supplicant who wishes has a turn at the front. Supplicants tell the intercessor what they are seeking and hand over a bit of money. Seated on the mats like the supplicants, the intercessor intones in a loud voice to draw the attention of the ancestors and addresses them by name. Then he utters the supplicant's name and wish, places the money in a container just under the curtain, and offers the supplicant a fragment of kaolin (*tany fotsy*).

One day I followed Madame Doso to pay our respects. She told the priest my name and what I did. He addressed the four ancestors on my behalf, repeating what she had told him and stating the amount of money I had brought. I put my money in the tray and received a piece of kaolin (white clay). "Kaolin is the ancestors' 'medicine' (*fanafody*)," said Mme Doso. "Take it home and eat a bit before every challenge; your wife too. Eat some before speaking to your boss and he will shake in his boots."

For those not habituated to the shrine, it can be a frightening place. My friend Dôná explained how he had been inside the temple only once in his life. He had gone on behalf of a woman whose child was missing and sought help from the ancestors but was afraid to enter herself. Dôná worried that the "skeletons" behind the sheet would pop out. He gave some money and, sure enough, the child was brought back safely.[20] Others were reluctant to enter out of simple self-consciousness. Anxiety was enhanced by stories that the guardians were waiting to pounce on people who committed errors and fine them. Each time I crossed the threshold, I was certainly aware of numerous eyes observing which foot I put forward. Punctiliousness was thus a feature of worship at the shrine, and it separated the habitués from the novices.

A shrine servant places a mark in wet kaolin on the face of everyone who has paid their respects in the temple. The mark is either an upturned crescent moon and dot on the forehead, for mediums, or a vertical line down the forehead and nose.[21] Seating in the temple also differentiates people. The mediums sit as far to the north and east as they can, and once their spirits rise they arrange themselves in order of precedence. Shrine servants cluster at the less sacred southern end, where on festive occasions the men beat drums hanging from a post or blow a large conch. Servants also step behind the curtain in order to look after the objects there and traverse boundaries more or less at will.

Behind the curtain and directly in front of the supplicants sits a miniature raised wooden house, the *zomba vinda* (palace/temple of destiny), with cross poles carved in the shape of horns. The relics lie in the *zomba vinda,* whereas other sacred objects, such as spears and drums rest beneath it. The relics are installed within silver-plated reliquaries coated with fermented honey wine and castor oil with which they are bathed and polished annually. The reliquaries should not be closely described.[22]

The shrine has three officials. The *fahatelo* manages the affairs of the shrine and oversees work and money. He is the person who officially receives and redistributes gifts sent for the Great Service and who talks with visitors and represents the shrine to the external world. He lives to the south of the temple enclosure and should be matched by a more senior official, the *manantany,* who lives north of the enclosure. For several years the latter position was empty. The intercessor (*ampangataka*) lives to the west of the main gate, controls the thresholds, ushers devotees across them, intones their requests to the ancestors, and accepts their offerings. There are many occasions when the *fahatelo* plays this role as well.[23] As a royal person once explained to me, "Even royalty have to 'request a pathway' (*mangataka lalaña*) of the *fahatelo* to enter the temple, pay their respects, and leave money for the ancestors." The officials periodically determine the distribution of the daily offerings among themselves and the remaining workers at the shrine and also open the *zomba vinda* to check on the relics.

The spatial ordering, whereby east is most sacred and north is the direction of highest rank,[24] determines the dispositions of the houses belonging to the spirits and living rulers as well as the officials. Figure 2.2 shows this arrangement.

Concentric Enclosures

This picture has illustrated some of the ways in which centers are constituted: first, with fences—centers exist by virtue of their concentric enclosures; and second, by increasingly constricted access and attention to gates, doorways, thresholds, and acts of crossing. Doorkeeping is simultaneously temporal as well as spatial since access is

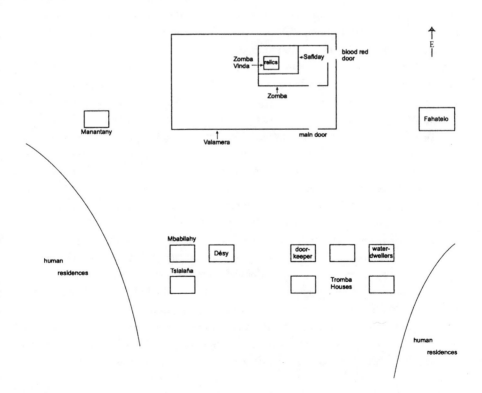

2.2. Plan of the Doany Ndramisara (Doany Miarinarivo), Mahajanga

temporally regulated. The portal is further linked to the mortal through blood sacrifice at the threshold.

The Doany Ndramisara is constructed as a series of containers within containers.[25] The shrine village contains the sacred enclosure (*valamena*), within which is the temple (*zomba*), which contains the curtained enclosure, which in turn contains a smaller "house" (*zomba vinda*), where the reliquaries lie. The reliquaries themselves are containers and enclose layers of their own before the relics themselves are finally reached. The closer one gets to the relics, the more constrained the access, the more secret and dangerous the contents, and the more sacred. Concentricity maps degrees of sanctity, producing in spatial and material form the hierarchy of sanctification which Rappaport proposes is a characteristic of religious discourse in general (1999: 263–76). Conversely, the highly salient barriers substantiate the differentiation among "kinds of people" and their specific obligations and privileges with respect to the sacred.

The nature of concentricity was highlighted when repairs and renovations to the shrine were planned. The shrine managers began by rebuilding the temple itself. This infuriated a former attendant, who said, "Repairs should always be made from the outside in, otherwise you leave the contents unprotected." He concluded forcefully, "One builds a courtyard before a house!"

Successive containment has a temporal dimension in more than the order of construction. In penetrating the spaces consecutively, worshippers move back in time to successively older objects and practices, as evident by the associated taboos. This temporality is also constituted through a kind of "paring away" of objects and habits acquired latterly in Sakalava history. This too corresponds to Rappaport's notion that the hierarchy of sanctity corresponds to a hierarchy of ostensible invariance; the more sacred, the more enduring and the less changeable (ibid.: 328–43). Events at the shrine are marked by periods and moments of greater or lesser penetration toward the sacred center; the deeper the enclosure one penetrates, the shorter the amount of time one stays.

The quality of concentric enclosing is also apparent in the performances at which *tromba* (ancestral spirits) arrive. In the course of a possession event, the more junior spirits arrive earliest and leave last. The most powerful and senior spirits do not appear until several hours into a ceremony, usually just before dawn, and they stay briefly. Spatially too, they sit facing toward the sacred east, arranged in precedence from north to south, their juniors forming a semicircle behind them. They are thus temporally and spatially framed, enclosed, and most difficult to approach.

The hierarchy of sanctification implies a status hierarchy as well. Not only are the spirits ranked by their respective genealogical positions, but shrine personnel and visitors are distinguished according to how far and along which paths of entry they can penetrate. The doorkeepers thus have an important function. Mbabilahy, the ancestral spirit who has a managerial role in the affairs of the Doany Ndramisara, is portrayed as the holder of the key to the *zomba vinda*. This does not imply that there is such a key literally in his possession but that he must be called up in a spirit medium and his agreement reached before a specific festival is begun or a new event or change in direction is taken at the shrine (chapter 10). Mbabilahy has to be consulted before events "because he holds the key." On one occasion when he was called up inside the temple, he said, to show his agreement, "I will open the door." Thus, agency and authorization are expressed in an idiom of enclosure and progressive movement inward.

Spatial enclosures are also used as an idiom for kinship. The smallest kinship unit is often referred to as a house (*trano*). Legitimate offspring are those born "in the house" (*antrano*) as opposed to "outside" (*antany*). Royal offspring who were born "outside" cannot subsequently be buried inside the sacred enclosure (*valamena*) of the cemetery ground. This prospective spatial exclusion in turn articulates their ineligibility to succeed to the position of ruler.

What is sacred is never obvious and can never be approached directly. Sacred centers are actually marked by being "out of center" with respect to abstract coordinates. Shrines follow virtually pan-Malagasy geomantic codes that mark the east as the most sacred direction, followed by the north (Kus and Raharijaona 1990). The Doany Ndramisara is located to the northeast of Mahajanga, and the *valamena,* the sacred enclosure, is on the east side of the shrine village. The temple (*zomba*) is located along the eastern side rather than the geometric center of the enclosure, and the *zomba vinda* sits in its northeast quadrant.

Similarly, the shrine is "higher" than the rest of the landscape. To go to the shrine is always referred to as to "ascend" to it, and to leave it is to "descend." Centers are often located in out-of-the-way places, on the edge of towns, deep in the bush, on remote hilltops. They are hard to get to, and the secluded terrain is enhanced by practical difficulties: taboo days to enter or leave, unavailability of food, and so forth. The Doany Ndramisara is relatively easy to reach and situated in a pleasant grove, but Bezavodoany (Deep Fog Shrine) in the distant countryside is pictured by city dwellers to lie in a terrifying forest where wild animals lurk and deliberately show themselves.

This obstacle-strewn off-centering has temporal as well as spatial form. Pilgrims to a shrine must wait for an auspicious time of the month, day of the week, and hour of the day to approach and leave it. Thus, they must *take time* and immerse themselves in the tempo of the sacred. At the Doany Ndramisara those wishing to enter the *valamena* and *zomba* must first ask permission of the "doorkeeper," who then accompanies them and speaks on their behalf. Additionally, contact with the sacred is always mediated by gifts. One cannot pay respects without leaving money with the "palm of the hand" or call up a spirit without leaving money on the medium's altar, nor can one request a recitation of sacred history (*tantara*) from a knowledgeable person without an offering of cash or liquor, referred to discretely as "custom" (*fomba*). The transfer of money in the right context is an act of honoring the recipient and the occasion and of marking the act as a sacred one. Handing over the money is like stepping into the next enclosure.

The spatial and temporal marginality, yet centrality, of the shrines is matched by the practical inconvenience with respect to housing and codes of conduct, dress, and comportment. Shrines constitute, in effect, an alternate habitus. Removing garments and paying attention to footing and posture bring to the ordinary supplicant some of the experience of what it is like to be a spirit medium. Such paring away or adding on, is reminiscent of Turner's discussion of liminality (1969). Entering sacred space or living with a sacred being requires consistent attention to one's behavior.

Person and Cause

In the Western imagination mazes are envisioned as largely empty places in which lone heroes wander, occasionally to confront a lurking beast. The Sakalava maze, however, is populated, and even partly constituted, by a whole series of characters. Place and time are not constituted or experienced in the abstract or at a remove from the persons who inhabit them. Specific characters bear chronotopic entailments. Moreover, the Sakalava world is a relational one; Sakalava are not lonely.

Times and places are conceived with respect to persons and realized in the ways they are populated and serviced. Conversely, the chronotope has implications for the kinds of persons who inhabit it. A critical distinction is between the living and the dead, and the servants, mediums, and spirits conceived to mediate between them. Routine practices and energetic performances are necessary for such mediation, and they shape, in turn, the persons who carry them out.

The category of person has intimate connection with that of cause. Although the book is not about causality or accountability per se (cf. Lambek 1993), it is implicit in much of what follows, especially as it is linked to particular conceptions of power and political authority. A main force is *hasing,* a word I translate as sacred potency. *Hasing* is manifest temporally and spatially and a distinctive feature of times and places. Ancestors and their relics, and secondarily, living members of the royal clan, are invested with it. But *hasing* is also maintained and produced through obeisance and obedience and is thus a product of the acts of subjects like those who supplicate the relics (Bloch 1989a). Collective and individual flourishing require correct propitiation of the ancestors. Access to *hasing* provides the means for flourishing but, itself, depends on correct behavior. Unmediated and excessive access is dangerous, and attention must be paid to keeping *hasing* at some distance (cf. Gell 1995). *Hasing* is a meta-human force, but its channeling depends on human subjection and agency. The Sakalava world is a deeply political and moral one.

To began exploration of Sakalava conceptions of persons, chronotopically situated, I turn to one of the people who lived at Ndramisara's shrine in the mid-1990s and was, in fact, the chief spirit medium of Ndramisara himself.

THE MAN WITH THE BIG BEARD

Fieldwork takes place in time and over time. In the course of my research in Mahajanga since July 1993, the poverty has not changed; indeed, it has probably increased. But much else has undergone virtually continuous change. I can illustrate this best by going right to the shrine and the medium of Ndramisara, who lived there for some years.

Most of the inhabitants of the shrine village were servants of the royal ancestors, many of them compelled to live there. Together with their spouses and children, they had to respect the taboos of the shrine with regard to bodily comportment, housing, and material consumption. But their dress and appearance were no more "traditional" than those of most people in town, especially when they were not on duty.

In the mid-1990s the exception was a man popularly know as Somotro (Beard) or sometimes as Besomotro (Big Beard). The medium of Ndramisara was a colorful figure. For tourists and other visitors he seemed to signify the quintessence of the place. Not only did Somotro have a long black beard and mustache among people who shave regularly, he wore his hair braided in hanging locks. Unlike other men, he never wore trousers but only a bright red smock and red waist wrap (sarong). When he

Somotro practicing divination in his house at the shrine.

went into town he carried a staff of blonde wood associated with ancestral authority and held by several of the oldest *tromba* but not by their mediums.

The spirit has the royal prefix Ndra- in his name and *misara* means to divine. The shrine is in Ndramisara's name not only because his relics are one of the four sets contained therein but because he founded the shrine and the system of power that coagulates in and emanates from the relics. He is not the most senior royal ancestor but someone who served as their diviner. In part, he is an outsider—the lowly man who, in the inversion characteristic of liminal space, becomes the means of access to power. So Ndramisara, the historical figure and *tromba;* and Somotro served as his medium.

With his staff, garments, and unruly hair and beard, Somotro reminded me of an Old Testament prophet or the Sri Lankan priestesses described in *Medusa's Hair* (Obeyesekere 1981), but he seemed more relaxed than either of these characterizations would suggest and often had a twinkle in his eye. I assumed that he was of great importance at the shrine. It seemed rude to go directly to the most sacred figure, so I postponed conversing with him until I could appreciate what I thought I would be able to learn from him. In the meantime I observed him sitting under the mango trees chatting with visitors, servants, and royalty, assisting in the temple when another medium entered trance, or taking a turn beating the drum. I heard from others that occasionally Ndramisara spoke through him but I never saw him in a state of active possession.

At the occasion of the Great Service he was the center of attention for visitors who must have seen him as personifying its mysteries. Foreign tourists photographed him. A Japanese man, who was apparently studying the event, took a series of portraits. But it was not just foreigners who dwelled on his appearance. A Sakalava woman in town explained how Somotro wanted to wear trousers and cut his hair like everyone else. He beseeched the ancestors, but they refused him. The woman said she was afraid of him and stayed away if she saw him on the street. But she expressed several times how sorry she felt for him.

Men are expected to wear waist wraps at the shrine, and no spirit medium can wear trousers when he is calling up his spirit. Male clients, too, are expected to change into waist wraps when spirits are present or at least to cover their trousers with a waist wrap. However, other mediums of Ndramisara wore ordinary shirts and trousers in town. No one else had such long facial hair, though some other mediums of Ndramisara sported short locks. This style is sometime referred to as *rasta,* but as Somotro himself pointed out, it is not. *Rasta* do not braid their hair, whereas the mediums do. Nor is the style feminine, because women knot up their braids (except at the shrine, where they too are expected to let them hang loose). The style is, in fact, characteristic of Ndramisara himself, who is reputed to have had four sacred locks (*taly masing*), and approximates that of Sakalava men from past centuries. Today, most men, including mediums, keep their hair very short.

Somotro was an idiosyncratic figure. When he first arrived on the scene, he caused quite a stir. His dress and hairstyle consolidated his presence at the shrine and served as a continuous reminder of his status. He was reputed to know medicine and received many private clients, for whom his attire was probably an attraction. His appearance did not, however, provide him with any advantage with respect to the powers at the shrine.

For a long time I assumed the contrary. Somotro was a medium of Ndramisara who is clearly a central figure at the shrine that bears his name. In addition, he was sometimes referred to by others as *Lehiben'saha,* or Chief Medium. Ndramisara, himself, is no ordinary ancestor but is recognized as a *moas,* a diviner with special powers. Unlike other spirits, he neither eats nor drinks. And unlike other spirits, his mediums, almost without exception, are men.

In 1995, I discovered Somotro's practice as a diviner and got to know him a bit better. A friend wanted a consultation for her niece. Somotro was not her first choice, she said dismissively, because one knew in advance that his calculation (*sikidy*) would diagnose *tromba.* We settled on mats in Somotro's small consulting room. Somotro sat in the southeast corner, holding forth with much good humor. Within reach were a red cloth bag containing the seeds he used for divination (*voan' sikidy*), a small diving mat, black from the seeds, a white ceramic dish with a circular mirror lying face up, kaolin, an incense burner, a notebook, a pad of pink paper, and a pile of herbal medicines.

Clients each gave Somotro a small coin and announced their name and request. My friend introduced her niece, Amina, and briefly described her swollen knee. "Is that really your name?" Somotro asked Amina. He stirred the seeds vigorously with his right hand, drew two smaller piles from the larger one, determined whether each was composed of an even or odd number of seeds, and marked the answer with either one or two seeds on the mat. He repeated the procedure until he had four vertical columns of four binary figures each.

While examining the seeds, Somotro reminded me that he wanted an English phrase book because, he said, he received many English-speaking clients. Indeed, in the room were two young men, not quite dressed like Malagasy. One had dreadlocks and responded to my query that he was from Seychelles; the other, more conservatively attired in neatly pressed jeans, did not want to say where he was from. In any case, he used accented English and depended on his friend to translate. Somotro, himself, enjoyed speaking in French, delivering himself phrase by phrase in a deliberately "cute" manner.

Somotro asked Amina where she was living when she fell sick, and when he heard the town asserted that her condition was provoked by a "bandit" (*fahavalo,* used here as a euphemism for sorcery). Amina immediately began talking about her visit there and confirmed what had already been her strong suspicion. Somotro wrote the names of four medicines and handed her the pink slip.

I stayed on after my friends left and listened as Somotro continued his unfinished business with the two young men. He asked the English-speaking stranger to place something gold into the white plate. The man added his rings and, to my amazement, a "gold" credit card from his wallet. As they both placed their hands on the plate, Somotro called upon the gods (*ndrañahary*) to help him. Somotro said the young man might be troubled by an African spirit, and the client acknowledged that he had an African grandparent. Somotro had questions of his own, to one of which the young man replied that the round trip fare to Jiddah was US$1500. He then admitted that he, himself, was from Saudi Arabia.[26] He said he had learned about Somotro from a fellow passenger on a plane from Jiddah to Nairobi.

Reflecting none too modestly on his international clientele, Somotro told us about a Japanese man whom he cured of a chest illness with kaolin who took a sack with him to Japan, having first asked Somotro why Madagascar did not mass-produce this "*antibiotique.*"

Somotro took only a very small coin for divination. However, satisfied clients gave *cadeaux* (gifts) if they wished. It was from the latter that Somotro earned his living. He received fmg25,000 at the end of this consultation and sent someone to a shop off the shrine grounds to fetch a large bottle of Coke; *bien glacé,* he ordered. I noticed empties along the wall. Once it arrived, he poured some off for his wife and child and consumed the rest of the bottle during the remaining consultations.

Clients arrived throughout the afternoon. Some had questions and received pink slips. Others were returning with pink slips and acquired medicines. Most could not resist putting down a coin for a new question. Somotro addressed a range of domestic, work, and health problems. The Malagasy clients showed quite a surprising class and ethnic diversity, but my main point here is to illustrate the kind of hybridity and heteroglossic conversation (Bakhtin 1981) that took place with the medium of the senior royal ancestor in the sacred village that contains his temple.

Somotro liked to talk, and at one point began a long digression about his life, which I supplemented with further questions on subsequent occasions. He talked easily but without great emotional depth. He explained how he tested new situations and only came to believe (*maneky*) once he had proof. He knew he had a spirit since he was a young boy, about six years old. The spirit saved him (*izy m'sauvé zaho*) several times from drowning and from sharks. Once he saw a large monster (*kaka be*) but did not tell his fishing companions.

Somotro grew up on the northeast coast at Vohemar; he was of Tsimihety background and raised a Catholic by a grandmother, though she was buried in Tsimihety fashion. If Somotro were to die in Mahajanga he would be transported home, to his father's tomb.[27] Somotro was the first medium in his family. He did not know why the spirits picked him. However, ever since childhood he was "punished" if he did bad things; he knew his spirits expected him to work only for the good. He had four spirits—a Zafinifotsy Water-Dweller (Tongtôñarivo), two sky spirits (*tromba añabo*), and Ndramisara. Ndramisara was the one present from childhood, though he only spoke later. At the time, Somotro did not care about it (*tsy mañahy*). But the spirit made him sick, and so he left Vohemar at age twenty-one for Diego and subsequently various other places across the North in search of health. He worked as a stevedore in one place and at a coffee and vanilla plantation elsewhere. "But my *patron* [that is, Ndramisara] must not have wanted me to stay in those places," he said. He had been married several times but had no children until he married his current wife in 1989. By 1995, when he was forty-seven, he had both his health and two children. It was evident that he was very close to the children; they crawled all over him as we talked.

I hoped to pursue the personal meanings of his remarks, but Somotro turned the conversation to economics. "Which currencies have the most weight," he wanted to know, "and what causes the changing values of currencies?" When I proved unable to articulate the mysteries of international finance, we returned to his biography.

Ndramisara rose (in Somotro) in Ambanja and instructed him to come to his shrine in Mahajanga. He told him how to dress and fix his hair. Another medium told me that when Somotro first arrived at Mahajanga, Ndramisara rose at the shrine (in Somotro, I think) and announced that he had brought Somotro here and that people should call him up whenever there was an issue.

Somotro began practicing divination in Nosy Be and Ambanja. He explained it is a gift, a "*don*," rather than a learned skill. He never studied with anyone because most people who worked as diviners were ill-intentioned (*tsy tsara fo, kafiry*); he would not trust one to teach him honestly. Instead, he learned by experience. Despite the mix of idiomatic expressions from French and Muslim discourses evident in these remarks, Somotro revealed an identification with Ndramisara, who is generally described as a *moas,* a potent diviner, and whose very name means "Lord Diviner." *Moas* are always said to have a *don* and are frequently described performing magical feats, such as carrying water in buckets with holes, and so forth. All other *moas* of whom I have heard lived in the distant past; Somotro is the only person I know with a self-ascribed gift.

Somotro's life actually fit well with the idea of Ndramisara. Both were liminal persons. Ndramisara was sometimes spoken of as an outsider, possibly from Africa or Arabia (one man suggested Kuwait). It is appropriate that as an outsider it was Somotro who attracted outsiders to the shrine.

Somotro was, however, notable for his down-to-earth quality. He discussed spirits with me but was more preoccupied with finding thatch for a guest house where he was planning to host visiting clients. He also wanted to construct a more solid house in town for his family. He would continue to live at the shrine but visit them and occasionally sleep there. He wanted his children to have a better life. He said, "It's not the whole family who is possessed, but only me, so why should they be constrained by it?" He put the children in school from the age of three and encouraged them to copy cursive letters on a slate. Somotro, himself, left school at an early age because he was sick. "What kind of sickness?" I asked. "Just sick, my whole body (*vataña jaby*)." He went to the hospital but was not cured. A diviner diagnosed *tromba.* He was really very sick and so he accepted it (*naneky*).

Somotro was proud of his international reputation but also aware, as he told me, that some people were afraid of him. Yet he said that Ndramisara was not particularly fierce (*mashiaka;* most mediums describe their spirits as fierce). "As long as you listen to him, things are okay; the sky *tromba* are more violent," he said. In fact, Ndramisara did not rise often. When I asked Somotro to tell me more about Ndramisara, he protested that he had only recently arrived in Mahajanga and did not know the historical accounts (*tantara*).

Somotro's Departure

Somotro was possessed by one of the most important ancestral figures and was living at the shrine dedicated to that ancestor. But for Somotro, his possession appeared to have more personal than public relevance. He described his presence at the shrine as the result of a quest and the command of his spirit. Moreover, although to many others he was a kind of tragic, pathetic, or frightening figure, Somotro seemed to take his own condition quite matter-of-factly. For insiders at the shrine he was not so central. My friend had been somewhat critical of his divination, and in any case his was a private practice, conducted *in* the shrine village but not *of* it. From the perspective of the shrine authorities, his presence there as a medium of Ndramisara was fortuitous (as is the presence of any medium) but not necessary. Indeed, they were not always happy with what Ndramisara had to say when he rose in Somotro on shrine business.

When I returned to Mahajanga in June 1998, I was surprised to find that Somotro had left the shrine. People said that he had returned north whence he had come. As one person put it, "His 'contract' was over." She thought that it had lasted 10 years.

Salim, a senior spirit medium, related that long before he ever arrived in Mahajanga, Somotro had wanted a child. According to Salim, "His *tromba* said he would conceive only if he displaced himself. 'If you want a child, you must go where I send you.' He had given birth here in Mahajanga [and the child had passed the age of high infant mortality] and so his *tromba* said, 'You have what you wanted, now go home.'"

Salim admitted that he had always disapproved of the fact that Somotro had continued to wear red clothes and to dress his hair when he was out of trance. "*Tromba* are weighty (*maventy poids*), more than ordinary living humans (*olon'belo*)," said Salim. "By continuing to dress like them when you are not in a state of active possession, you compare yourself to them." In a sense, then, Somotro had been acting as though he were as important and powerful as the *tromba* themselves—as though he *were* Ndramisara and not simply possessed by him.[28] Salim declared it a lie that the *tromba* had demanded that Somotro dress his hair (*mitaly*) and abstain from trousers. "He did it of his own choosing (*jery nazy*), in order to draw attention to himself (*pour réclame*) and people were wrong to have felt sorry for him." The proof for Salim was that once Somotro had decided to leave Mahajanga he shifted rapidly to street clothes. Indeed, he was seen leaving town dressed in a jeans suit and had cut his hair and shaved off his beard. Salim explained that Somotro had not received a regular salary and so had to use "strategy" (*hevitry*) to attract clients and bring in an income. Salim felt it unlikely that Somotro would appear at the next Great Service; he would be embarrassed to return once he no longer dressed in his old manner.

This is a revision of the public interpretation of Somotro's story. What appeared to have been evidence of Somotro's sanctity was now understood to indicate precisely the reverse. If Salim interpreted Somotro's departure as the result of having succeeded in the quest that brought him to the shrine, other rumors floated in Mahajanga. I heard from both mediums and spirits that Ndramisara had been unhappy at the changes taking place at the shrine, and there had been differences of opinion over a number of issues (chapter 10). One of the Water-Dwelling *tromba* said that Somotro left because he had heard, second hand, that the managers of the shrine had criticized him. They did not like what Ndramisara had to say when he rose in Somotro and were pleased to see him go. As a medium who, herself, kept at a distance from the shrine said, "The managers can always find a medium who says what they want to hear rather than what is true (*maring*)."

She reminisced, "People were astonished when Somotro first arrived. They knew Ndramisara didn't like trousers, but other mediums after a time were granted permission to wear them. And no other mediums ever had a beard or hair like that. People stayed away from him at first but gradually got used to him. Recently Somotro had been making a lot of money and the manager, and even the king, were envious and also perhaps threatened that his popularity and authority were growing too great. Somotro couldn't stand the intrigue," she concluded, "and knew the shrine managers didn't like him. So he instructed people to call up Ndramisara in him and request permission to leave. But," she added, "it was time for Somotro to go; he was bringing too many strangers into the shrine. And he made lots of money. Rumor has it he even returned to Ambanja in a privately hired taxi!" She predicted that he would get bored there.

Bored or into trouble. Another medium began by asking whether I had heard the news? "Shortly after Somotro arrived in Ambanja, lightning hit a house and killed a child. Somotro was accused of being a witch (*mpamosavy*) and was jailed." The medium, herself, attributed this development "not to any witchcraft on Somotro's part but rather to the anger (*heloko*) of his *tromba*. The *tromba* had not wanted Somotro to leave Mahajanga, but having received the child he had come in search for, he had disregarded Ndramisara's opinion and left anyway."

An entirely different angle was offered by someone who had once worked at the shrine alongside Somotro. He whispered "Somotro became rich by providing 'medicine' for some drug dealers from Jamaica to 'close the eyes of the *douanes*.' They bought marijuana (*jamala*) cheaply here for resale in Jamaica. People only learned of his illegal activities after he left." Since there is plenty of marijuana in Jamaica, this story appears implausible to me but it draws on both the (false) impression of *rasta* locks and on Somotro's evident transnational ties.

This consultant asserted, "There had been no bad feeling with the shrine authorities. Having had the child he wanted, Somotro had simply requested permission from Ndramisara to leave and had received it. He had just finished building a large solid house in Ambanja and had invited several people from the shrine to a moving-in celebration. He had also promised to send the shrine a cow when he was well settled."

Before he left, Ndramisara rose in Somotro and assured people that he would be replaced, and indeed he was. I was directed to an unusually obese man who sat with his entourage in the house that had once been inhabited by the doorkeeper. The new medium handed me a neat business card:

Professeur RALIBERA Emile
Voyant et Guérisseur Traditionnel Malgache
Mpitsabo amin' ny fombandrazana Malagasy
Lot 71 Béryl Rouge
Tanambao v Toamasina

Coordinateur et animateur National
des Cultes royaux et ancestraux
(Doany NDREMISARA—EFADAHY Manakasina
MAHAJANGA)

I was astonished by the objectification of ancestral practice that the card represented and by the distance the medium had come. Toamasina is on the east coast, and the Betsimisaraka background of the medium is a far cry from Sakalava tradition. I thought he had a nerve announcing himself as the national coordinator of the Great Service, but he appeared to be well liked by everyone at the shrine and was highly recommended as a source of knowledge about divination and history. Like Somotro, he was a medium of both Ndramisara and some of the sky spirits. But his manner of presenting his knowledge was quite different, highly didactic, earnest, and involving the use of books. He announced his intentions to improve everyone's understanding of celestial and ancestral matters. What he shared with Somotro, and even exceeded, was the emphasis on outward show.[29]

First Appearances

For about a decade Somotro was the most striking figure at the shrine, and he appeared somehow to embody the values and secrets that lay at its core. Yet by 1998, he had gone. If Professor Ralibera stays, he promises to be equally flamboyant. Curiously, among all the mediums, those bearing the founding figure of the Sakalava politico-religious system appear to have the most overtly modern inclinations. Somotro had his transnational clientele; Ralibera, his national ambitions. Ralibera did not braid his hair but he had other ways of asserting his appearance.

I have begun my account of the shrine with Somotro since he was the first person to strike outsiders or to strike them most forcefully. His story illustrates the way spirits can be linked to personal affliction and appear to take over the lives of their mediums as well as the adventitious quality of the appearance of spirits and mediums. Somotro's practice also illustrates the way the sacred center is simultaneously a transnational crossroads and the struggle people in Mahajanga have to connect their center to ever more comprehensive flows without rendering it fully permeable and hence ungrounding or decentering it. Somotro thus brings a personal dimension to the broader question of this chapter with regard to the way Sakalava chronotopes are constituted within the flux of history.

In addition, Somotro provides an allegory for the way foreigners approach Africa, mistaking and exaggerating the significance of their own impact yet being equally fascinated by the ostensibly exotic.[30] Somotro was compelling because in his person and practice he seemed to combine both extremes. Yet from a Sakalava perspective, no single living person is key and the most spectacular character may actually be peripheral. As the discussion of concentric enclosures suggested, outward forms are designed to conceal inner contents. Thus what is on the surface may be less important than first appears. Somotro was neither as "traditional" nor as authoritative as outsiders like myself first imagined. Yet he was not fully malleable either, and there were people at the shrine who were certainly happy to see him go.

Somotro's story (and, I am guessing, Ralibera's) illustrates both the deceptiveness of the brilliant surface of things and the transience and ambiguity of events, interpetations, and memory—useful reminders at the beginning of a study about historical consciousness. Somotro illustrates how what is most visible is not necessarily true or most important. Indeed, his very visibility actually worked against his authority. To this moral attaches a whole theory of power.

POTENCY AND SELF-CONSCIOUSNESS

On my own departures from Mahajanga, Amana Ibrahim and his family deluged me with gifts: coffee beans and peanuts, tablecloths, spicy lemon pickle, and woven hats. One year they gave me a rectangular cotton cloth of the kind very common along the Mozambique Channel with an image and a saying printed on it. The image was an idyllic scene in red and white of canoes sailing along a coastline replete with palm trees and sandy beach. The saying, which Amana Ibrahim read to me with evident pleasure, was a local proverb: "*Ny feno tsy mikobaña* (full containers don't rattle)."

The meaning of the proverb is that those who need to brag about their abilities are actually indicating a lack. Authority that draws attention to itself is self-defeating. True power and authority are quietly self-evident. They are simply recognized by others; they do not advertise. People who are the "real thing" do not show off; those with true power do not call attention to themselves. So, above all, sit the relics, silently and in concealment, drawing supplicants to themselves effortlessly.

For Sakalava, to draw attention to oneself is to lose authority; self-advertisement is a sign of weakness or insufficiency. A truly authoritative person makes no ostensive effort. This view of power, which has its counterpart in Java (Anderson 1990), is akin to a Western theory of virtue. The person who is simply virtuous does not display the fact and does good for the sake of goodness, not for the praise or prestige that accrues to the virtuous. In this sense, power may be said to be one of the Sakalava virtues.

There is a quality of potency rather than action here. Members of royalty demonstrate who they are by what they do not do. Ideally, monarchs do not displace themselves for their subjects; rather, their subjects should come to them. This is what happens at the shrine where the relics remain immobile and are regularly visited. On the other hand, *tromba* often need to be called up. This is potentially embarrassing and it undermines their authority. The reason that money is deposited on the altar when calling up a spirit is to compensate the spirit for being asked to move on the supplicant's behalf, as well as to restore the power and face that the spirit loses by acquiescing. Respected humans, such as spirit mediums, are also given gifts of money if they are asked to travel to an event.

Ampanjaka Soazara distinguished herself by being the only monarch in the early 1990s to disregard President Ratsiraka's request to come meet him in Antananarivo. Other monarchs may have been rewarded with expensive gifts, like cars; however, as in a fairy story, it was Soazara who proved herself a "real" *ampanjaka*. It was not that she resisted the Malagasy government or refused what might have been interpreted as a bribe but that she was uncompromising when it came to her royal status. By displacing themselves, the other monarchs indicated their lower status relative to that of the President. Similarly, rulers do not visit each other, since to do so is to admit lesser power.

Drawing on work in Imerina, David Graeber distinguishes between "two kinds of social power: the power to act directly on others, and the power to define oneself in such a way as to convince others how they should act toward one" (1996: 12). The former is a kind of stored potential that may be invisible to others, and the latter the power to move others to action by display. Sakalava ideology gives much greater weight to the former, epitomized by the obscurity of the relics. However, this is clearly based on a prior definition that has convinced subjects to act in certain way. Today, both living royalty and ancestors must draw some attention to themselves if they are not to escape oblivion. Since royal power can no longer draw directly on wealth, autonomy, or violence, it must be rendered by some symbolic means. Power cannot be entirely unselfconscious. The strategic problem for those who wish to be (considered) powerful is precisely how to convince others of their authority without drawing attention to the fact that they are acting in order to convince and draw attention.

One way to announce authority indirectly is relative to the greater exhibition of lesser authority. Within the space of *tromba* performances, a rhetorical function of junior spirits is to set off a contrast with their seniors. Junior spirits are noted for being much more accessible, sociable, and simply noisier than senior ones. They rise more frequently and stay longer. They are also more mobile; they parade around and interact with spectators whereas senior spirits are less voluble and much more stationary—waiting for supplicants to approach them. Junior spirits often display a humorous braggadocio, whereas senior ones are relatively silent and stern. Junior spirits mirror their audience and demand recognition and response whereas senior ones are indifferent. Powerful *tromba* may refuse to rise in their mediums when called or appear only briefly, at the furthest hour of the night. The seniority distinction is paralleled by a gender one; junior male *tromba* rise to play (*misôma*) whereas females do not. The reason given is not that the latter are secluded but that they are more intransigent (*mashiaka*). In effect, female ancestors are more potent than male ones.

At the shrine the vocal/aural modality of "rattling" is less immediately obvious than visual and proprioceptive modalities. The shrine is a maze that plays on the tension between enclosing and disclosing—between the visible and the concealed. From the road one can see a ways in toward the trees but not through them. The sacred palisade (*valamena*) can be partially seen through, depending on how one stands, though it is forbidden to go up close and peer overtly between the posts. The further to the inside one penetrates, the more opaque, but also the more suggestive, is the next screen.

Graeber notes a distinction between action and reflection: The powerful actor is largely invisible to himself and to others, whereas the reflected person is an object of display. This is partly a matter of degree of self-consciousness. To be self-conscious in the presence of another is to acknowledge the other's greater power (1996: 6–9). In the face of the potency of the immobile, invisible relics, visitors to the shrine are made highly self-conscious through their own mobility, attention to the various taboos, and the fear that everyone is watching and waiting for them to violate a taboo. Giving up ordinary patterns of dress, greeting, and so on, induces the sense of being controlled rather than being in control. The opposition between the source and the potential violators of the taboos is highly marked. Many people offer self-consciousness—their fear of making a mistake or of being caught making a mistake—as the reason they are afraid to visit the shrine. This self-consciousness is simultaneously a consciousness of the difference between themselves and the powerful and, hence, in itself, a powerful tool in the construction of power.

The hiddenness of the relics (Graeber 1996: 16), their "not-being-seenness" or quality of undisclosure, is manifest and highlighted even on the one occasion that they are brought out. At the climax of the Great Service and immediately after their bath, they are carried on the backs of their four bearers. But they are hardly carried very far or for very long—merely around the back of the temple and rapidly in again. It is evident to the spectators that they cannot see very much. Most celebrants are on the far side of the courtyard fence; even those up close see only vague lumps buried beneath woolen coverings on the bearers' backs. The containment and concealment effected by the maze produces and ensures the stillness characteristic of true power.

Partial Disclosure

Somotro's story illuminates a Sakalava theory of power, but it also serves as a kind of allegory for anthropological practice. First impressions may be reflections of what has been designed to impress. And, indeed, wise consultants watch to see whether ethnographers have the wisdom to see beyond them. But the lesson goes beyond the object of display, to the anthropologist's own temporality. Somotro's departure is matched by the anthropologist's, yet events keep unfolding. The anthropologist thinks that his or her time in the field is critical (as it is bound to be *for him or her*), but it is only a brief moment in the history of his or her subjects. It is all too easy to mistake temporal surface for longitudinal depth.

Somotro left between my visits in 1996 and 1998. In 1998, there were diverse and conflicting accounts of what had transpired less than two years earlier. If individual accounts of the immediate past are already at evident arm's length from actual events and necessarily reduce their complexity, how much more selective must be accounts of the more distant past. Historical narratives and performances are "real" social facts and play a substantial part in Sakalava life, but they are not transparent records of the past. Sakalava "history" as discussed in this book is primarily history as Sakalava enacted and recounted it in the mid-1990s and not necessarily as it transpired or as it was enacted and recounted over previous centuries.

I have argued that Sakalava historicity is a function of space, time, and person that can be imagined as a maze. Two properties of the maze are that the journey through it is never direct and that at any point along the journey only a limited perspective is available. These qualities characterized interactions with consultants, whose conversation was often both indirect and limited. Their motivations included both consideration of the rights and risks of disclosure and the attempt to teach me a lesson about the nature of mazes. But they also had to do with the division of labor and distributive quality of Sakalava historical memory, which is the subject of subsequent chapters.

3

The Sakalava Poiesis of History: Realizing the Past through Spirit Possession

Time never begins or ends; life always does.

—Northrop Frye, *The Double Vision: Language and Meaning in Religion*

But where one considers different historical characters from the standpoint of a total development, one could encourage each character to comment upon the others without thereby sacrificing a perspective upon the lot. This could be got particularly, I think, if historical characters themselves (i.e., periods or cultures treated as "individual persons") were considered never to begin or end, but rather to change in intensity or poignancy. History, in this sense, would be a dialectic of characters in which, for instance, we would never expect to see "feudalism" over-thrown by "capitalism" and "capitalism" succeeded by some manner of national or international or non-national or neo-national or post-national socialism—but rather should note elements of all such positions (or "voices") existing always, but attaining greater clarity of expression or imperiousness of proportion of one period than another.

—Kenneth Burke, "Four Master Tropes"

When people in Mahajanga speak of *tantara,* formal tradition or history, it is of accounts concerning Sakalava royal ancestors and especially the establishment of the kingdom of Boina that they refer. Throughout the 1990s and into the twenty-first century, spirit possession has provided a dominant and popular register for presenting these ancestors; *tromba* are deceased members of the royal descent group, stretching back in an unbroken line from those who passed away a year ago to those who lived in the seventeenth century and earlier. The mediums, who come from all walks of life and all manner of kin groups, form part of an otherwise clan-based division of labor through which living and dead royalty have been served.[1] Ancestral spirits, their live descendants, mediums, tomb guardians, shrine managers, suppliants, and the general public interact in complex ways.

In this chapter I tackle head-on the way that spirit possession can be understood as a form of history. Using a broadly Aristotelian framework I propose poetic form as a means for distinguishing historicities. I analyze Sakalava performances of possession by royal ancestors as the creative production of a kind of history, distinguish it from a dominant occidental model of history, and elaborate the chronotope on which it is based and the heteroglossia and historical consciousness it enables. I argue that Sakalava spirit possession has a strongly realist bent and briefly consider the question of historical truth.

The comparative study of historical consciousness is beset by the way in which such consciousness has been seen as the defining feature of Western civilization and, hence, by oversimplified binary oppositions. Western history has imagined its distinctiveness and rationalized its disciplinarity by means of such misleading dichotomies as real versus mythical and objective versus ideological. Yet since any historical work is composed, its form is amenable to poetic criticism, and such criticism, from Aristotle on, has seen beyond these tired oppositions. Thus, to mention only two powerful figures, Auerbach (1968) has illuminated how "realism" is relative to shifting narrative conventions, whereas Bakhtin's concept of the chronotope (1981) clarifies the articulation of different temporalities. These provide tools for specifying the Sakalava case while linking it to other forms of history. I show, too, how suffering, a critical feature of Aristotle's poetics, is no less central to Sakalava history, not only as an object of representation but as a means of establishing that history as both real and true.

SITUATING SAKALAVA HISTORY

Mahajanga, July 9, 1994

Ndramarofaly, Lord of Many Taboos [Violations], hunches on a low wooden stool near the altar table on the sacred northeastern side of Mme Doso's front room. He clutches his spear and taps it impatiently. The *tromba* (manifestation of a deceased member of the Sakalava royal clan) wears a kind of red loin cloth that is drawn up to cover the breasts but leave exposed the ample shoulders and thighs of his female medium. Several other spirits are arranged in a semi-circle with Ndramarofaly around the altar. One of these is Ndramandaming, comfortably seated in an arm chair and, in further contrast to Ndramarofaly, dressed in several layers of clothing, including a man's European dress shirt with gold cuff-links and a jaunty felt hat. Ndramandaming, presently occupying the body of a thin, elderly female medium, pours beer from a bottle into a glass, while Ndramaro, Lord Many, as his name is affectionately shortened, quaffs rum from the bottle. Abruptly shaking his spear threateningly in my direction, Ndramarofaly asks the company what I'm up to. Mme Doso, the medium who is managing the event, attempts to calm him. Ndramandaming whispers to me that Ndramaro is afraid of me. "Ndramaro is *sauvage*," Ndramandaming explains, "and has never before seen a 'European' (*vazaha*)."

At the time, I took this to be a spontaneous gesture of kindness on the part of Ndramandaming's medium, designed to put me at ease and explain what was going on. On later reflection I realized that the remarks of both *tromba* fit into a larger picture that had nothing to do with me personally. Each spirit—the fierce and fearful, and the worldly—played the part expected of him. In fact, I, too, had become an actor or figure, in the collective poiesis of history. Present in Mme Doso's room were spirits of different historical periods. All members of the same royal descent group, in which Ndramarofaly is a direct ancestor of Ndramandaming (FaFaMoMoFa; figure 3.1), they express the distinct historical experiences that different generations of Sakalava have undergone. To their largely non-royal mediums and audience the spirits are beings to whom one offers respect and from whom one requests blessing. In addition to their manifestly political and religious functions, the spirits serve as icons of history. In this context, Ndramaro expressed Sakalava hegemony before the arrival of the Europeans, whereas in his comment on Ndramaro's wildness, Ndramandaming spoke as someone who had served as "governor" of the province under the French colonial regime and was a Christian. His perception of Ndramaro is not to be taken as the point of view of the present any more (or less) than was Ndramaro's perception of me.[2]

The spirits juxtapose distinct historical epochs. The juxtaposition is a part of their very constitution—by means of contrasting signifiers of comportment, clothing, furniture, drink, dialect, and so forth. But more than this, the space of performance enables the simultaneous display of successive temporalities. Sakalava history is thus additive in that, in principle, later generations do not displace earlier ones but perdure alongside them.

Moreover—and this is a central point—this conjunction of temporalities, including the present, allows each period to serve as a locus of commentary on the others. Thus, the provincial governor is able to speak condescendingly about his wild (but immensely powerful) ancestor. Multiple voices and alternate points of view are expressed and made available for consideration, without being subordinated or silenced by others.[3] This is a condensation of historical time within the space of the present but emphatically not thereby a flattening or confusion of historical voice.

What this means is that historical consciousness is not reducible to a single attitude but arises through the interplay of multiple voices. It is neither monological nor static but open, "best understood as the active process—sometimes implicit, sometimes explicit—in which human actors deploy historically salient cultural categories to construct their self-awareness" (Comaroff and Comaroff 1992: 176).

Scenes of the kind that took place at Mme Doso's occur throughout the city of Mahajanga and across much of northwestern Madagascar.[4] History is not an explicit subject of contemplation at these gatherings but is, rather, their means or modality. In this case, the spirits had arrived in order to contribute to the amassing of money to be sent to the Doany Ndramisara on the occasion of the Great Service. In its general outline such activity fits very closely to Bloch's famous analysis of the production of sacred authority, *hasina,* among the Merina (1989a) in which honor and money are exchanged for blessing.[5] The spirits present at Mme Doso's were at once descendants of the kings whose relics are honored and equally the ancestors of the living members of the descent group who receive the material benefits of that honoring. Thus, they

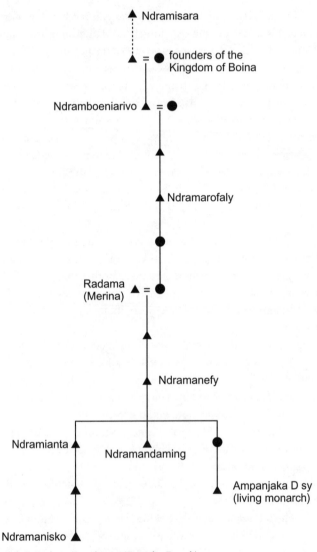

3.1. *Simplified Royal Genealogy (Bemihisatra/Betsioko Branch)*

mediate between the further past and the present and, within the present, among the living, between members of the royal clan (*ampanjaka*) and the people (*boeni, vahoaka*). Ndramaro, the most senior spirit present in the room, ended his intervention by blessing me, as I crouched before him, and placing white clay, the material sign of having made the transaction of honor for blessing, on my palms.

Royal ancestors are thus far from an abstraction or undifferentiated body. Like the living, they are distinguished according to seniority. One problem Sakalava face is that while senior generations of ancestors have greater authority than more junior ones, they have a lesser understanding of the present and so their decisions are not always welcome. At such times the spirits receive gentle tutelage in how things have changed. Negotiation with the spirits enables Sakalava to work through the ways in which various generations view present issues and thus to produce historically informed responses. A suitable response is pragmatic yet acknowledges the concerns of earlier generations. Today in Mahajanga the public issues in question are largely restricted to activities at the shrines, notably renovations to the tombs and buildings and to matters of succession (chapters 9 and 10). Institutionally, the scope of the sacred has been increasingly circumscribed, yet its past breadth remains resonant.

History, then, is intrinsically and explicitly connected to the political and the sacred. Far from representing a place to which completed battles can be consigned (White 1987: 79), Sakalava history has an insistent relevance in the present. It provides the substance of sanctity and authority and serves as a vehicle for the reproduction and legitimation of hierarchy. These functions constrain the heteroglossia of political expression. Spirits speak from, indeed *as,* their respective times, yet virtually all of the spirits are members of the noble stratum and descent group or defined with respect to it. Whatever the discordance in historical experience, almost all express something about the power of royalty—a power realized at each performance. Thus, there is an assertion of totality, in which the diverse voices are embedded in lineal and political continuity.

One could equally turn things around and say that in all this explicitly politico-religious activity, history itself is always an implicit, emergent subject. To invoke or evoke a particular *tromba* is always to invoke the past and to evoke a sense of both its connection with and disconnection from the present. There is, as Ricoeur puts it, a "secondary referentiality" or "figurative reference" to the "structure of temporality" that provides events with an aura of "historicality."[6] Such historicality, as manifested in the spirits, mediums, and architecture of the shrines, is the very stuff of being Sakalava (not an essence but a mode of existence). Sakalava history neither denies the evidence of discontinuity in favor of a reversible time, as Lévi-Strauss (1966) argued for "cold" societies, nor takes the manifestation of irreversible discontinuity to represent underlying reality, as may be the case for "hot" societies (cf. Bloch 1996). Like any history worthy of the name, it speaks to the relationship of continuity to discontinuity; it provides "imagined continuities."[7] But it does so from a rather different prefiguration of the ground than that underlying Western historiography.

The heteroglossia contributes to the often carnivalesque quality of historical performance. A few days after the incident described above, Mme Doso instructed me to bring to the next gathering of Ndramaro and his descendants some *mason' piso* (literally, "Cat's Eye"). "What is *mason' piso?*" I inquired. Nothing less than an expensive greenish bottle of that favorite drink of French colonials, *pastis.* It is only Ndramaro who does not know that *pastis* is the real name and who, seeing his descendants imbibing a drink that did not exist in his time (or perhaps I should say, in the time his presence has come to epitomize), scornfully calls it "Cat's Eye." *Mason' piso* thereby comes to condense the discordance between the precolonial and colonial worlds and the distinction in historical consciousness that colonialism established but that, equally, Sakalava historical poiesis manages to comprehend and thereby, perhaps, transcend.

Conversely, people outside the spirit milieu—and in Mahajanga they are plenty—may recognize *pastis* but not *mason' piso.* Implicitly, then, the appellation also signifies the distinction between those who cultivate this particular art of history and are able to enjoy a specialized wordplay (and drink), and those who try to stay at arm's length from the activities of the *tromba* and mediums. As noted in chapter 2, Mahajanga has a heterogeneous population crosscut by divisions of class, ethnicity, political connection, religion, and knowledge and interest in spirit affairs. The severe poverty does not override the cosmopolitanism of the city's inhabitants, including many of those deeply involved with Sakalava history. Citizens of Mahajanga draw on cultural resources from the Sakalava politico-religious domain, from pre-Sakalava and other Malagasy traditions; from Islam, Catholicism, and Protestantism; from Comorians, South Asians, and others who live among them; from the French-influenced school curriculum and national culture; from the global culture available on television and from the experiences of family and community members who have traveled abroad. All of these are significant; none unifies, encapsulates, or excludes the others. The knowledge of and tolerance and empathy for diverse points of view is evident in the multiple voices of the spirits. In Bakhtinian terms (1981), one could say that there is a tension evident between the centrifugal pull of the multiple voices and the centripetal pull exercised by the powers that attempt to harness the spirits to their own ends.

HISTORY AS POIESIS

In Western usage history has a deliciously ambiguous relationship to the past. In historical inquiry we go back in time, yet history itself marches ever forward, leaving the past behind, in ruins. The western sense of historical time establishes definitively the "pastness" of the past. History thus has a peculiar relationship to memory, one that may even be perceived as antithetical. Memory consists in clinging to remains thrust aside by history's relentless advance. To adhere to past habit is to be "out of date." In its most sophisticated conceptualization, as transference, (implicit, unconscious) memory is the inappropriate retention and activation of past dispositions in present circumstances. This tension between time and memory may not always be successfully mediated by the profession of history. The role of the historian is a Sisyphean task, ever straining to see what continues to recede, to preserve what is in the process of dissolution, and to reintegrate what has been definitively prised apart. But the historian has success of a kind, writes a book, and, in the very act of rescuing it, assists in the process of the reduction and objectification of the past, in disciplining it and pronouncing its epitaph. Historical writing keeps the past in focus by fixing the distance of the spectatorial gaze. Scholars may argue, but it is always one objectification or another that comes to stand in place of memory. Historical scholarship, through its own distanced gaze, in effect rationalizes the distantiation and alienation, the acute sense of discontinuity, that it recognizes as produced by the "march" of history.

It is this kind of opposition between history—the dispassionate representation of the past—and memory—the subjective continuity with it—to which non-Western historicities like the Sakalava invite alternatives. To grasp this, clearly we cannot restrict our attention to the facts and narratives that Sakalava enunciate about the past but must include all of the registers through which historicity is realized. I call this broader creative production poiesis. In its sense of "making" or "composing," poiesis comprises what in much of social thought have been separated and opposed not only as the material and the ideal, production and creation, but as ritual and narrative, the making and the made.[8]

In Sakalava understanding, the past is far from escaping the present; the past calls upon the present with a literally imperious insistence, yet moves continuously in and out of focus. Historical distance is measured, yet the distances are traversable in both directions (not only the present back toward the past, but the past forward to the present) and the gaze is not unidirectional. For example, the colonial governor responds to the past but addresses his remarks toward me in the present. The relationship with history is largely one of direct engagement, even embodiment, rather than of distanced objectification. Yet, at the same time, a discursive space opens up; moreover, the reproduction of the Sakalava past is highly formalized and is no mere subjective preoccupation with an infinity of particulars. The content concerns discrete personages and events arranged according to a chronicle.

As a record of past events, this *is* history; what we are not used to is the form. Hence my emphasis is less on *what* Sakalava historicize than on *how* they historicize—on how actors and acts are distinguished and emplotted; how the very relationship of the past to the present is constituted, legitimated, and realized. My perspective here is not that of an oral historian, concerned with evaluating what Sakalava say in order to construct a more authoritative version, although I do not discount such an approach.[9] In sum, my interest lies less with the content per se than, to borrow Hayden White's helpful phrase, with "the content of the form" (1987).

I use the word "poiesis" explicitly and in preference to a currently more fashionable term, namely, "mimesis." Mimesis is the term by which Plato opposes identification with the content of what one is performing or imitating to reason, that is, to rational, reflective thought. In Havelock's rendering of the opposition (1963), mimesis entails an immersion in the concrete and in the flux of endless events, whereas reason distantiates, enabling abstraction and freedom from time conditioning in order to reflect upon it. To be sure, Plato assigns all poetry and narrative to mimesis, but this depiction of artistic performance and reception as a kind of unreflective "sleepwalking" is far too extreme. That is, although the Platonic concept of mimesis is helpful in pointing to the centrality of identification, to circumscribe the distinctiveness of the sort of practice I have begun to describe by such a label is to give far too much away. We do not have to make a radical choice between abstract reason or distanced representation and subjective identification.

That mimesis is not sufficient to describe possession is made apparent by the articulation of the spirits with one another; their synchronization is obviously a form of objectification. Moreover, mimesis per se cannot account for the silences. Ndramaro's fear of me and the ensuing direct contrast with Ndramandaming neatly evades direct reference to the entire mercantile period (during which Ndramaro lived), a time in which Sakalava power was maintained by the slave trade with Yankees and Europeans.

Aristotle offers a ready alternative to the pervasive and seductive Platonic dualism. He presents a trichotomy, the terms of which are not defined by direct opposition or in the mutual exclusion of each other and therefore do not lend themselves to evolutionary typologies demarcating whole epochs or modes of thought.[10] History may be conceptualized with reference to all three of Aristotle's kinds of "intellection," that is, contemplating (*theoria*), practical deliberation or doing (*praxis*), and productive creation or making (*poiesis*) (1976: VI, ii). And although poiesis certainly does not preclude a measure of identification by artists with their subjects, such identification does not fully explain the artistry or its reception, which is always mediated by particular cultural forms. Indeed, Aristotle begins his *Poetics* (1947) by distinguishing various modes of imitation according to their means, objects, and manner.[11]

Aristotle's terms are abstractions, and it is clear that in practice any one of them entails the others. Thus, whereas liberal academic history (or anthropology) might characterize itself in terms of *theoria,* and Marxism in terms of *praxis,* these are choices of emphasis. Sakalava differs from European history most strikingly in its creative means, and this is what I emphasize here. Yet the concepts are mutually complementary and overlapping; poiesis must always be understood in the context of the other two modes. I have alluded to the practical in my own acceptance of blessing and subsequent purchase of *pastis* and in the legitimating functions served by history. Whereas spirit possession does not have the aim of studying past events as such, the contemplative is evident in Sakalava attempts to evaluate or reflect on performances and narratives or piece them together.

To begin with Plato's opposition means to remain within the confine of questions of objectivity and of truth narrowly conceived. It limits our focus to questions such as, Is the Sakalava version of their history correct? Are Sakalava, as historians, sufficiently detached from their object? (Or, in the current critique and inversion of the model, Are we, as anthropologists, sufficiently connected?) These questions are restrictive in large part because they stay suspiciously close to the legitimations of Western historiography. Inevitably, other histories are judged inferior, and inevitably, their underpinnings are ignored.

Yet to part with Plato is not to surrender to relativism but, rather, to suggest the lines along which different histories, both within any given cultural context and between them, can be compared. It is, in part, White's argument that a poetical base is immanent in *any* historical account of the real that has emboldened me to speak about Sakalava imaginative performance as explicitly historical.[12] White argues that history cannot be written without what he calls "tropological prefiguration of the historical field" (1973: xii) and without particular strategies of formal argument, emplotment, and ideological implication. He demonstrates the distinctive styles of particular nineteenth century European historians and philosophers of history according to specific combinations of these modes. These, in turn, are built from limited sets of universal tropological structures and rhetorical strategies that provide means of comparison.[13]

The critical point is that recognizing the poetical dimension of history does not require conceptualizing it in terms of the mythical, the irrational, the nonempirical, or the unreal. It serves not so much to oppose historical constructions like those of the Sakalava to Western ones as to provide a bridge of comparison. The emphasis is simply on the creative side of any production of history, without additional baggage. My argument is clearly not that the dominant Western and Sakalava modes of representing the past are identical; but the flamboyance of Sakalava practice with its puns and liquor, its royal corpses rising to life and mimetically enacting their deaths, does not discount it from being a discourse of the real.

I hope that rendering the centrality of poiesis enables me to evade the twin risks of ethnocentrism—the exoticization of Sakalava achievement, on the one hand, and its reduction to Western terms, on the other. Turning this around, I want my account to preserve and illuminate all of the complexity, sophistication, authority, and significance of Sakalava history, as well as its difference from the kind of work we are accustomed to describing as history.

Objective and Embodied Historical Knowledge

I wish neither to reduce Western or Sakalava history-making to a single mode or dimension nor to oppose them to each other.[14] Nor is to speak of poiesis to deny the contemplative or practical sides of Sakalava history. There are many occasions when people objectively seek out correct genealogical relations, sift between distinct versions of a story, reflect on the sources of such divergence, or simply try consciously to compose a coherent, consistent narrative. Mediums and their performances are subject to critical evaluation. Moreover, mediums themselves are often contemplative and critical when out of trance. They often told me they did not know which version of a story about the past was correct and tried to reason it through or seek the answer elsewhere. It is of considerable interest, however, that the knowledge of history, much like its embodiment, is subject to a division of labor regarding obligations of commemoration and rights to speak. Thus, many spirits and mediums can report on specific segments of history that are relevant to them, but very few have a view of the whole. Nevertheless, when compiled, the pieces fit together remarkably well. Moreover, certain spirits are known to be better historians than others; thus, I was referred to one spirit, Ndramanisko, who had been a student of history during his lifetime. Spirits and mediums are also constrained by what they have the right to speak about; more marginal people may have looser tongues but lesser authority.

Moreover, if becoming possessed is a kind of mimetic surrender to history, when the spirits rise they are understood as rational historical agents. Although rarely asked directly about the past (such questioning would be perceived as rude inquisition about their personal lives, which were not always exemplary), spirits may be called upon to discern things in the present or give direction to current enterprises.

Conversely, despite the significance of contemplative historical thought among Sakalava, a picture of history as abstract knowledge is insufficient. It cannot grasp the medium (replaced in each generation) who lives permanently at an isolated shrine in the bush, bearing the *tromba* of the wife of one of the founding kings—a woman murdered by her husband over 250 years ago.[15] The medium signifies her burden by strenuously avoiding the town, the main shrine, and all of the sites associated with the husband. When these barriers are traversed, her physical distress is pronounced. The pathos of this rarely seen and rarely discussed, but always existing, figure in the Sakalava drama simply could not be so powerfully evoked through words alone. Perhaps because I have not met the medium, her image crystallizes for me the passion of an embodied poiesis of history.

Chronotope and Character: The Content of the Form

A distinction between the mythical and the real (or the ritual and the commonsensical) can be faulted for its dualism—for missing the diversity of forms operative in any cultural arena and for homogenizing the mythical as internally consistent and all of a piece. It also ignores the fact that the commonsense "real" world, itself, is culturally constituted and, to a degree, contingent and variable, capable of foregrounding only certain features of the environment and certain connections among them at the expense of other features and connections. This Boasian point is reinforced by Auerbach's remarkable analysis of the development of Western literature (1968), which demonstrates that realist accounts, themselves, come in many forms. Any representation must be selective. But the principles of selection—how widely or how narrowly the lens is focused—need not be ranked according to an external logic. We can therefore compare the "realism" of a writer like Stendahl with that of a writer like Virginia Woolf. This is obvious as well when we shift from literature to historical writing, some of which focuses on moments of time, some on the *longue durée,* some on political events, some on family life, and so on. One of these forms is not necessarily more "realist" than another, although they all follow certain canons of historiography.

What each kind of writing has to offer is a particular set of stylistic conventions or constraints by means of which a time and space for the action is laid out, with a concomitant formulation of a specific "ratio" between actor, act, and scene (Burke 1945a). Thus, the depiction of the social milieu characteristic of the nineteenth-century novel or historical account retracts in later fiction like that of Woolf, or in psychohistory, as the focus narrows further upon the actor, and time and scene are virtually internalized.[16]

It may be useful to speak of these varieties of narrative and performance, following Bakhtin (1981), as providing alternative chronotopes, particular configurations of time and space, or "differing ratios of time-space projection" (Holquist 1981: xxxiii) that organize and emerge from particular cultural productions. Bakhtin applied the concept to literature but argued that such an "intrinsic connectedness of temporal and spatial relationships" could be found in other areas of culture as well; "the image of man is always intrinsically chronotopic" (Bakhtin 1981: 84–85). Thus, instead of making unilateral and binary distinctions between the "mythical" and the "real," we can delineate the play of chronotopes at hand.[17]

The Chronotope of Sakalava Possession

The chronotope connects the former lives of deceased royalty, their genealogical links to one another across time, and their actions in the present, shaping these around the places of their living and dying, as well as their tombs, and the mediums, contexts and occasions in which they are called to perform in the present and around the spaces that are inaccessible to them. To begin with, the space of action is not restricted to the original lifespan of the ancestor. In that spirits perform in the present, their chronotope includes contemporary action. Moreover, in stories about the past it is not always immediately apparent whether the live person or the *tromba* was the agent, and the distinction is not always salient.

In my experience, most Sakalava historical accounts are not now explicitly articulated in the form of a lengthy chronicle, as a series of events in sequence.[18] Nor are most people in the spirit milieu concerned with establishing an abstract overview of the spirits as a group. Genealogies are remembered in segments. Each spirit must be able to name its parents and offspring; and in this way the whole genealogy is reproduced, with occasional gaps, but always piece by piece. Nevertheless, the pieces do presuppose a chronological grid (cf. Feeley-Harnik 1978: 411). Some nobles and intellectuals keep written genealogies, and it is commonly asserted that each shrine contains a book with a written record; but I have met only one person, a devotee of *tromba* activities who had traveled extensively in the region, who could recite complete sections of the royal genealogy directly from memory (and he was famous for being able to do so). The versions and pieces largely confirmed one another (with certain exceptions derived from either waning engagement or conflict of interest over certain portions). The narratives concerning the lives of royalty exist also in fragments.

If the underlying chronology is implicit when it comes to recitation, the situation is different in embodied performance. The conjunction of spirits from different times is never confused; their relative chronological positions are always carefully maintained, and each generation acknowledges its ancestors and recognizes its descendants.[19] The genealogical distinctions are evident in the ways that spirits manifest themselves and human supplicants approach them. Spirits of earlier generations are treated with greater respect and fear. During a night's performance the most senior spirits arrive, if at all, only just before dawn; their presence is more fleeting, and their pronouncements bear more weight. Their dialect is antiquated, making their speech harder to understand. For the mediums, the experience of active possession by genealogically senior spirits is more intense and exhausting.

Genealogical relations thus form the basis for performative coherence—the relationship of the parts to each another (Becker 1979). Most critically, the temporal and genealogical dimensions are cast in a spatial idiom. The seating arrangement of the spirits provides an accurate map of genealogical seniority in which the northeast has precedence (cf. Kus and Raharijaona 1990). Offspring sit to the west of or below their parents and sometimes briefly on their knees. In addition, junior spirits are generally more mobile than senior ones, moving around within the performance space and sometimes even outside of it rather than sitting still and waiting for others to come and greet them. This spatialization of time and hierarchy (or temporalization of space) is evident equally in the microcosm of Mme Doso's front room and in the macrocosm of the western Malagasy landscape, where it is configured by means of a network of shrines and royal burial grounds, and the movement of supplicants, tribute, and ritual artifacts between them.

The relationship of time to space is organic rather than technical here in the sense that the times and places of individual events are not simply interchangeable with one another (Bakhtin 1981: 99).[20] Rather than subsisting purely sequentially (by calling up the spirits successively), the times of the past are evoked in parallel, redistributed on a spatial grid—across the landscape or within a room. The spatial distribution and temporal sequencing of present-day ritual events are iconic of past events and relationships. When spirits from distinct epochs arrive together, persons and events formerly separated in time are juxtaposed in the same spaces. Conversely, spatial exclusions, where two spirits cannot appear together or on each other's territory, signal either past conflicts or delicate balances of power. Consequently, spatial relations today make contiguous what was once temporally distinct, and separate what was once contemporaneous.

To say that the chronotope is organic is also to say that it is meaningful, and specifically that it contains complex political messages. Gatherings of the spirits speak to the power, endurance, and cohesiveness of the royal line, whereas separations indicate tensions. Thus, Ndramianta, a spirit generally found among his close kin from the colonial era in gatherings in town, is conspicuously absent from larger events at which Ndramisara, the founding figure of the entire monarchy, is likely to appear, and he is excluded from the main shrine, which is Ndramisara's. Ndramianta may never come face to face with Ndramisara because, when he was alive, he refused to acknowledge Ndramisara (who lived some 200 years earlier), referring to him scornfully as "Makoa." According to one version of the past, Ndramisara was a powerful diviner, who came from East Africa; but the epithet "Makoa" later came to connote a slave (since many slaves were taken from the Makua of present-day northern Mozambique). Ndramisara (qua *tromba*) responded by exiling the arrogant Ndramianta. Today, their avoidance is not only a reminder of the ambiguity surrounding Ndramisara's origins (and hence the possibly extra-monarchical source of royal power) but brings forward for appraisal past attitudes to slavery and the racial expressions of the colonial period elite. Ndramianta is portrayed as a Muslim colonial official, wearing both a European jacket and a fez.

It is relevant that for every public performance the number and identities of the spirits to be invited is carefully planned beforehand. The roles are "cast;" specific mediums are invited to perform as vehicles of particular spirits. In most cases, each medium is assigned only one spirit, even if that medium also receives other spirits. This ensures that all of the spirits anticipated can be present simultaneously. Each medium arrives prepared with the necessary clothing, which is stacked neatly on a table near the altar until it is needed. Those in the know can thereby anticipate who will be appearing.[21]

Sakalava do not deny irreversible temporal succession—indeed, they emphasize its poignancy—but they set succession in a plane of simultaneity. Most smaller performances include a group of relatively closely related spirits rather than those from a broad range along the genealogical span. Each spirit arrives as the royal personage was at the end of life, yet each interacts in well-defined kinship relations with others. Thus, Ndramboeniarivo, as the quintessential son, always bows to his mother and may sit briefly in her lap, even though she would no longer have been alive when he was the chronological age at which he is portrayed. She spoils him and treats him with much greater affection than she does the father. This same son relates as father to his own children and as grandfather to the rest. In this way a series of "family romances" depicts relations of love, jealousy, and conflict as eternal, even though not everyone knows the stories of the original actions that underlie them.

These relations of contiguity among spirits continue to be relevant offstage and are partially reproduced among their mediums. The medium of Ndramboeny refers to and addresses mediums of Ndramboeny's mother as "Mother," and so on. This is complicated by the fact that each medium has more than one spirit. The spirits who can appear together in the same medium are consistent with what is evoked in the performances. A medium who has Ndramboeny's mother often has his father as well. But mother and son never possess the same medium; that would be incestuous. Conversely, just as Ndramboeny's wife never appears at the same performance event as Ndramboeny, so she would never possess the same medium. The mediums themselves are thus understood as loci of the spirits in much the same way as houses or tombs. Such considerations constrain the marital and residential patterns of acknowledged mediums. The spirits permeate their mediums' lives, and the mediums represent them even when they are not specifically enacting them. Thus, to the degree that the mediums' social and psychological realities are shaped by the myth models, the distinction between the mythical and the real becomes spurious.[22]

The end of this general, though never complete, displacement of the syntagmatic and the diachronic onto the paradigmatic and synchronic fields is that narratives and history are broken down into individual personages—personages who evoke a particular moment in a longer sequence of events but who can represent that moment at any time. Ultimately, and with the exception of those parties who must avoid each other, all past monarchs could in principle be present together in the same room, in the present.

It will be evident, then, that the central "literary function" of Sakalava possession is character. The chronotope brings together a cast of characters who establish their relationships to one another by means of greetings and the like but who generally subsist side by side in a kind of "parallel play." Often, they come together on the basis of having been buried in the same location, with the space of death taking precedence over genealogical proximity. Each has a past life that provides a source of identity in the present, but only specific elements of this past are relevant to present identity. These have to do with parentage, procreation, manner of dying, and place of burial but may also include aspects of personal achievement and comportment. For example, the highly educated Ndramanisko was recommended to me as a good source of historical narrative (*tantara*); another likes to sing; a third is remembered for a love affair. In many instances, characters evoke stories, generally traumatic or poignant ones, of which they were once part. They may speak to specific political achievements, alignments, or compromises from the past, but not everyone will recognize the allusion. The narratives are condensed in the spirits' appearance and character traits. Sometimes the narrative function is reduced to a name or nickname.[23]

These Sakalava historical figures are not "private persons," yet they are individuals. Their story "does not, strictly speaking, unfold in *biographical time*. It depicts only the *exceptional*, utterly *unusual* moments of a man's life, moments that are very short compared to the whole length of a human life. But these moments *shape the definitive image of the man, his essence, as well as the nature of his entire subsequent life*" (Bakhtin 1981: 116, his emphasis).[24] Time leaves a trace but only as a result of exceptional events, determined by chance, in which the character's guilt, error, or commitment can play an initiatory role.

Characters are portrayed as unchanging in their own time. In their contemporary appearance their biography changes little; the events of their lives become fixed in a kind of permanent display, except insofar as they address the concerns of those who consult them in the present and insofar as the biographical time of their mediums is intimately entwined with them. Yet the *tromba* also evoke a sense of the human cycle; each of them represents a moment in the genealogical succession of the noble line, and they are frequently present with parents or offspring. The individual completion yet cyclicity of the characters is reinforced by their punctual rather than durational appearance (Geertz 1973: 393). Entries and exits are explicitly marked (just as the entries and exits of the living into and out of shrine enclosures are marked) and largely anticipated. There are times of the lunar month at which both spirits and shrines are inaccessible. Additionally, each spirit is unavailable on the day of the week on which the person it manifests died. This affects the scheduling of large events; some spirits will necessarily be absent, whichever day is selected. Similarly, the lives of the mediums are regulated in that they observe stronger taboos on the death days. Spirit mediums pay close attention to the calendar and always know what day it is.

It is because action is not the main object of representation in the spirit performances that plot is not the critical function. Hence, despite its evident theatrical qualities (cf. Leiris 1980), possession differs from Western models of drama derived from Aristotle. In possession, the action is secondary to the characters, helping to constitute them rather than the reverse; nor is it character *development* that is at issue; nor even is the discrimination of good from bad characters critical to their depiction—most are complex and amoral with the potential for positive and negative action. The spirits, especially the most senior, are presented as *mashiaka*—punitive, demanding, and fierce—eliciting fear but also respect. Ndramarofaly is remembered as among the most violent of all the monarchs; while alive he was reputed to slit open the bellies of pregnant women to see how babies were made, and by some accounts he was banished or even put to death by his kin in order to stop him. Yet he is among the most popular spirits in present-day Mahajanga, where his violence is contained and frequently dis-

placed by solicitude for his numerous descendants and for clients undergoing therapeutic initiation into mediumship. Similarly, that Ndramianta insulted Ndramisara is
presented without a strong moral gloss.

What, then, accounts for possession's appeal or interest? Where does the tension lie?
In possession it is the spectacle, or stage appearance, which was the least important element of drama for Aristotle, that takes precedence over plot. Performances are vivid
and concrete. The arrival of spirits is marked by mental dissociation and physical disruption; it contorts and batters the bodies of mediums, interrupting the ordinary flow
of time; but it is always particular to the spirit. Each arrival is elicited or marked by a
particular praise song (*kolondoy*). The presence of the historical character requires a
change of clothing and demeanor and is accompanied by music, tobacco, and drink.
There is sharing, joking, and sometimes dancing; the application of clammy white clay
soon marked with the runnels and smell of perspiration, overlaid with large doses of
perfume; fear, amusement, affection, and exhaustion. Relations between individual
characters and those who witness their arrival are marked by a combination of intimacy and distance. People show respect and reverence for the *tromba* but also laugh
with, and sometimes at, the antics of the more junior ones, sit close to them comfortably, and hold and are held by them. The spirits' lives literally interpenetrate with those
of the mediums and others who attend ceremonies. Pity and fear, the emotions required by Aristotle for a good plot, are certainly aroused in the spectacle.[25]

The most salient events of the characters' lives to be inscribed in performance are
their deaths, although this is by means of a series of negations. The deaths of royalty
can only be referred to euphemistically; royalty are said to "collapse," and a specialized
vocabulary and a set of attendant practices are available for dealing with the corpse
and the pollution of death. Yet in possession the manner of their dying is embodied;
in addition, the day of the week on which it occurred and the foods they ate during
the last day are remembered by taboos impinging on their mediums (Feeley-Harnik
1978). It appears that at each occasion of active possession the characters go backward
through their deaths in order to re-establish themselves as social actors. Mediums prepare to enter trance by applying white clay to the parts of the body whose traumata
led to death. Often they must retreat under a shroud during the transition. When the
spirits emerge they dress themselves in their personal clothing and then begin to enjoy
life with a vengeance, counterposing exuberant vitality and full presence to their previous absence. They partake of their favored substances: chewing tobacco, rum, cigarettes, or perfume. Only when bodily established and fortified do they acknowledge
those around them. Greeting behavior is critical as it re-establishes the spirit's social
position and provides the next piece of the emerging tableau. Yet identities are no
longer the same as they were in life. When spirits announce their names, the places
they identify themselves as "owning" or coming from are in fact where they died; not,
as in life, places of birth, residence, or even an ancestral burial ground. The opposition
between death and life is at once implicit, understated, and yet exaggerated and highly
salient.[26]

Once a character has fully appeared, its function is largely to represent itself—to acknowledge and be acknowledged as such. *Tromba* are characterized by the beginnings

and endings of their lives. Their births determine their places in the social hierarchy and their positions in the genealogy. But it is their deaths that primarily determine their "moral purpose," described by Aristotle (1947) as what characters seek or avoid. In performance time, the *tromba*'s personal histories begin always with the foundational moment of their deaths. This means that the features by which good Aristotelian plots are organized—where characters learn from earlier episodes of their lives, where there is moral development, or suspense, where justice prevails—are either givens, part of the background, or irrelevant.

Or not quite—Aristotle identifies three critical elements of plot: peripeteia (the change from one state to its opposite); discovery (the change from ignorance to knowledge); and suffering. If the first two are not found in a plot unfolding over the lives of the individual characters, they are central to the contrasts between one character, taken as a completed whole, and another (as we saw in the opening episode) and to the relationships between spirits and mediums and the other people living in the present. A good spectacle is one in which numerous spirits rise and subsequent interaction is lively. Peripeteia and discovery are also central to the practice and interests of individual supplicants to the *tromba*. The third element of plot identified by Aristotle, namely, suffering, is fundamental in that spirits are portrayed by means of their deaths and manifestly affect the bodies of their mediums. I will have more to say about this shortly.

In sum, poiesis proves useful for delineating the distinctive ways that historical experience is shaped. The centrality of character is of particular note, as well as the

Raising Ndransinint' to a standing position in order to fix the clothing of the newly arrived spirit, at a tromba gathering in an urban courtyard in Mahajanga.

Assisting a Betsioko spirit to enter trance under a fluttering royal shroud. The woman in the foreground is clapping the beat to a song of welcome.

manner in which characters are constructed primarily by means of synecdoche (use of the part to represent some quality presumed to inhere in the whole) as the refinement of the once living person. Indeed, the "heart" of characters is frequently to be discovered in the form of their deaths, and posthumous praise names further distill the quality. The personage is, in turn, synecdochal for his or her epoch, epitomizing its central qualities. It is striking how similar all of this is to what White distinguishes as the Formist type of historical argument:[27]

> The Formist considers an explanation to be complete when a given set of objects has been properly identified, its class, generic, and specific attributes assigned, and labels attesting to its particularity attached to it. The objects alluded to may be either individualities or collectivities, particulars or universals, concrete entities or abstractions. . . . When the historian has established the uniqueness of the particular objects in the field or the variety of the types of phenomena which the field manifests, he has provided a Formist explanation of the field as such. The Formist mode of explanation is to be found . . . in any historiography in which the depiction of the variety, color, and vividness of the historical field is taken as the central aim. . . . To be sure, a Formist historian may be inclined to make generalizations about the nature of the historical process as a whole, as in Carlyle's characterization of it as "the essence of innumerable biographies." But in Formist conceptions of historical explanation, the uniqueness of the different agents, agencies, and acts which make up the "events" to be explained is central to one's inquiries, not the "ground" or "scene" against which these entities arise (White 1973: 14).

THE PAST IN THE PRESENT

The suspense or urgency of possession cannot lie simply in the representation of a known and potent past but, rather, in its uncertain implications for an unfolding present. Whereas spirits are in a synecdochic and, hence, integrative relationship to the past (White 1973: 34), they bear a kind of ironic relationship to the present. If irony is the trope that negates on the figurative level what is affirmed at the literal one, spirits are intrinsically ironic, at least vis-à-vis the manifest bodies of their contemporary human hosts. Moreover, there is always a tension between the spirits being bound to the limits imposed by the past in which they are embedded and the possibilities for new action offered them by the present. Ndramandaming's address to me is an example of such ironic intervention.

Acts in the present do not appear sufficient to undermine or transform the identity or narrative function brought forward from the past.[28] And yet, although spirits remain in character, their performances differ in one fundamental respect from the sort of historical narratives described by White or Ricoeur in which action and plot are quickly fixed. In the performative mode, characters continue to act. History is not simply past but continuous in the present; not merely subject to additional or alternate readings but relatively open; well known but, at the same time, "in suspense." That Ndramandaming addressed me, or indeed any of the supplicants, was a new event.

A small incident further illustrates the possibilities, as well as returning us to a specific historical theme. The scene was a gathering at the house of one the senior spirits at the shrine, on the night of June 30, 1996, to acknowledge the spirit's receipt of new clothing after an insult received from the shrine managers (described in chapter 10). Senior spirits are the last to rise. Because of their genealogical precedence or advanced age at time of death, some are seated on chairs. But there were only two chairs present, and these were already occupied by Ndramandaming, the provincial governor, and his father, Ndramanefy. Ndramandaming is portrayed as immobile; his medium enters trance seated in a chair and leaves it from the same position. Ndramanefy gave up his own chair for the senior spirits, but he refused to allow the company to take Ndramandaming's. When people lowered his son to the ground, Ndramanefy walked out in protest. Once outside, his medium regained consciousness and awaited the end of the ceremony sitting placidly under a tree.

This event was subsequently much talked about by other mediums. They felt that Ndramanefy had acted inappropriately; in his overconcern for his paralyzed son he had not shown proper respect for his ancestors. Clearly, this was not part of any fixed scenario. To understand the import it is necessary to realize that Ndramanefy is quite Merina in his dress and is a Christian, a descendant of the union between a Merina king and Sakalava queen (though his medium at the time was a Muslim woman). His son, Ndramandaming, as I have indicated, is depicted as a provincial governor during the French occupation. The issue, then, is who carries more authority and should be granted greater respect; the earlier rulers, who manifest Sakalava autonomy and power, or the later rulers who were incorporated into the Merina and colonial states and who, here, also represent the westernized sector.[29] This remains a live concern in present-day Sakalava politics and even on the national scene where there is debate about decentralization. Moreover, the disregard of modernizing junior generations for the authority of their ancestors is an issue to which all Malagasy can relate. Ndramanefy's action thus condensed a great deal and, while speaking as the past, nevertheless evoked issues live in the present. The import is complex, and once again, it bridges time while maintaining a sense of temporal depth.

If Ndramanefy's act was inappropriate, he must have known it would also be futile. In their domain the older spirits have to be shown respect since they are the ultimate source of authority accruing to subsequent generations. Here, then, it was a gesture designed to say something, and it carried specific ideological implications. The Malagasy/Western (*gasy/vazaha*) contrast that has been noted as highly salient by students of the Merina (Bloch 1971, Edkvist 1997) is complicated in the Northwest by the fact that the traumatic experience of the nineteenth century that marked the collapse of autonomy was not French colonialism so much as invasion by the Merina, whose King Radama attempted, as he famously said, "to make the sea the limits of my ricefield." Thus, Antankaraña history emphasizes alliance with France in opposition to the Merina (Lambek and Walsh 1999; Walsh 2000). The conflict is epitomized throughout the Northwest by the ubiquitous Antandrano (Water-Dwelling) spirits, a group of *tromba* who are depicted as having drowned escaping Merina soldiers. Yet the line of Ndramandaming is the product of a union between Sakalava royalty and Radama himself; Ndramanefy is portrayed as virtually Merina. Living members of this branch of royalty (Bemihisatra, buried at Betsioko) underplay the conflict; their argument is less one of Sakalava distinctiveness from Merina than of the pan-Malagasy quality of royalty. But in fact, Ndramanefy's act of supposed resistance to the senior Sakalava spirits served slyly to emphasize the latter's continued superiority to the Merina interlopers.

Thus, in my reading, it was an ironic performance, a pointed message to the living ruler about the importance of maintaining Sakalava tradition and an expression of resistance not to the ancestors but to the living. The performance referred implicitly to the fact that the entire event had taken place because the shrine managers had ignored instructions of the senior spirits. The episode also illustrates the possible refinements of character deployment; Ndramanefy acted toward ends that were not strictly his.

Despite such disruptions, Sakalava spirit performances approximate what White calls the comic mode of emplotment in which "hope is held out for temporary triumph of man over his world by the prospect of occasional *reconciliations* of the forces at play" (1973: 9).[30] These are symbolized in festive occasions, which for Sakalava are literal enactments of festivity such as the main pilgrimage or family gatherings of spirits. At such events the drowned spirits, themselves, are among the gayest participants. Yet always in counterpoint to the comedy or harmony of ritual are the tragedies of the individual narratives, summarized in the modes of dying of the protagonists. Comic resolution is implicitly conservative, "inasmuch as one can legitimately conclude from a history thus construed that one inhabits the best of possible historical worlds, or at least the best that one can 'realistically' hope for, given the nature of the historical process" (White 1973: 28).[31]

The attribution of conservatism, here in the face of actual change, is reinforced when we consider how commoners must acquiesce to noble history. The perdurance of royalty is counterposed to the mortality of commoners, who must serve the *tromba* or become ill. What is apparent is less a "parallelism" of two "worlds," as noted for other African systems of spirit possession (Boddy 1989; Stoller 1989), than their impingement and mutual dependence. Royal perdurance used to be predicated quite literally upon the mortality of subjects; in the past, the "falling" of the monarch was eased by a human "mat" or sacrifice. Royals perdure, yet they perdure in the act of their continuous separation from death. Non-royals die, yet they serve the royals in the course of their lives and, in extremis, by means of death. In the distinctions between kinds of people, Sakalava experience and play out throughout their lives those oppositions between permanence and impermanence, ancestors and living that other Malagasy enact at funerals or reburials.

Many societies commemorate the past, but often this is the reenactment of a fixed moment—a prototype that is, itself, dehistoricized. What makes their history so interesting is that Sakalava do not simply commemorate a past event or even a past time, but the thickness, cumulation, and passing of time; what they repeat is not just a founding moment but a series of time slices and exemplary figures which, juxtaposed, evoke the depth of history. History is understood not in the sense of a static, timeless "past," nor of a simple repetitive endless cycling, nor as a collection of discrete, decontextualized images, but as the series of steps from then to now, always in complex relation to the uniqueness of persons and the irreversibility of death.

There is less an effective displacement of one time by another than their cumulation and mutual resonance. Unlike the dominant mode in the West, where the past recedes and the historian goes back, the Sakalava past is carried forward. The heroic figures are not the historians but the *tromba*. The diverse, consecutive periods of the past are brought together with each other—and with the present—so that each epoch speaks with reference to the others. This cumulation is ultimately less a confrontation than a comic reconciliation; representatives of the different epochs do not directly challenge each other so much as they mutually address the living. It is their copresence in the present that invites interest and provides a lively sense of historical depth and transformation. If the past lends its authority to the present, the mutual presence of multiple generations of historical personages suggests ultimately an acquiescence to history. Might we say also that it is a kind of "working through?"[32]

Additionally, the inclusiveness of Sakalava representations may be compared to the novel. Despite its hierarchy, possession does not presume a supreme authority. Each voice is self-assertive, undercutting, and ironizing the others. This colloquy seems akin to Bakhtin's depiction of "the novel's joyous awareness of the inadequacies of its own language . . . aware of the impossibility of full meaning, presence, it is free to exploit such a lack for its own hybridizing purposes" (Holquist 1981: xxxiii).

Epilogue: Truth and Suffering

Poiesis and history may seem to represent antithetical claims, and it may seem that in emphasizing form I have evaded questions of reference. After all, modernist poetry derives its meaning intrinsically, whereas history is supposed to derive its meaning extrinsically. Yet the point has been that poiesis must be discernible in every genre of cultural production.[33]

One model would distinguish scientific truth—discovered by a process of knowing—from poetic truth—created by a process of making. These are known, respectively, as the correspondence and the coherence theories of truth. As Grant puts it, while the former defers to the fact and attempts to render it with fidelity, "In the coherence theory . . . the epistemological process is accelerated or elided by intuitive perception. Truth is not earned by the labor of documentation and analysis but coined, a ready synthesis, and made current—as is any currency—by confidence, 'the confidence of truth.' Evidence is replaced by self-evidence." He continues, "In the first case the truth is true *to* something, in the second it is true as a line or edge is said to be true when it is straight, flawless—*containing* the truth, not simply representing or alluding to it" (1970: 9).

Sakalava would appreciate the metaphor. Indeed, whole coins signify and enable royal authority and speech. To become *tromba,* royalty must be buried with coins in their mouths, and before a *tromba* can first speak truly it must rinse its mouth and spit out a coin. Silver coins are immersed in the mixture of white clay and water that *tromba* use to provide blessings. Coins also signify the moon—the lunar cycle being an idiom of life and death like the spirits and regulating access to them (Lambek 2001). But of course there is no reason that Malagasy practices be bound by the limitations of Western genres and dichotomies. Sakalava possession maintains a productive tension between the literal and the allegorical in a way that more highly rationalized genres do not.[34] In fact, there is evidence that for Sakalava, representations of the past need to meet both correspondence and coherence criteria of truth value (cf. White 1987: 40).

In 1994, I had the idea that each spirit or its senior mediums would be the authoritative repositories of their stories. I had heard from someone else excerpts from the story of the king whose *tromba* possessed Kassim, and I hoped to gain from this senior medium a more complete account. Who would know better, I reasoned, than the medium? Moreover, I felt that Kassim's account would complement those of the mediums of other spirits who took their own points of view on the past.

My assumption proved to be only partly correct. I spent much time with Kassim as well as a medium of one of the main female *tromba,* hoping to speak to their respective spirits and gain permission to learn more about their lives. Kassim proved to be a wonderful consultant on related matters—his voice reverberates throughout this book—and demonstrated his support of my work by putting me in contact with other experts. However, my attempts to elicit stories about their spirits from the mediums or to meet these particular spirits in person were always put off; either the medium was too busy, or it was a taboo day for the spirit, or some necessary ingredient was missing.

I began to assume that the mediums would not speak because they simply did not know the stories or were afraid of having the limits of their knowledge exposed. My hostess denied this, asserting that the mediums were afraid that they would suffer if they spoke. So when her husband asked how I was getting on, I explained that I had finally realized that the mediums were precisely the people who could not suffer their stories and that I was embarrassed for having put them in an awkward position. "Nonsense," he replied, "they just don't know the facts."

Each of these explanations is correct in its own way, and each is interesting for what it says about the clash between Western and Sakalava assumptions about historical knowledge. First, it is evident that there are limits to the mediums' knowledge. They do not know all of the details of their spirits' lives, nor could *tromba* speak as much about themselves as could an ordinary living person. But it is only in a Western conception of history as the amassing of infinite and precise detail that this becomes relevant. In the Sakalava view it is sufficient that the spirits come from the past and that they evoke and particularize it. The spirits bear certain images of the past, but most people are not explicitly interested in the past "for its own sake," or at least were not until the rise of historicist tendencies and spectatorial distance produced by national schooling (and possibly anthropologists).[35]

The other explanation brings to the fore what underlies both, namely, that the relationship to history is not one of distanced objectivity. For spirits whose lives were associated with excessive acts of violence, their mediums are precisely the people who cannot tell their stories, because to tell them would be somehow to suffer them. The sign of the truth of the story is not its retelling but its silence or the punishment that accompanies narration. Senior *tromba* and their stories are weighty and difficult (*raha sarotro*) and known to be so even in the absence of the details that might demonstrate the reasons for it. The difficulty of speaking about the past is a salient index of its significance.

It is noteworthy that these mediums maintain silence precisely about their own spirits. Their knowledge is intimate, potent, and embodied and therefore not ready, necessary, or appropriate for dissemination. It is literally too painful. If the mediums do not reproduce the past primarily in discursive form, then what is their function; of what are they the vehicle? That they suffer when they speak the intimate events of their spirits' lives exemplifies and reinforces the fact that the "history" relevant to them *is the trauma,* not the stories about it. They do not "represent" this suffering; they suffer it. Suffering is a primary means of embodiment, as well, perhaps, as the true subject of history. In any case, it is the *presence* of the spirits and the presence of the past, rather than objective knowledge *about* the spirits or about the past that is critical. This presence, even in suffering, even in silence, is pregnant with meaning.

The medium's situation highlights the burden of history as well as the organic division in historical labor. Kassim lives a life of which the embodiment of a specific king forms an intrinsic rather than extrinsic part—one that Kassim can never afford to forget because the clay of his (re)creation is nothing less than his own flesh. In addition to having been an honored and powerful king, son of the founder of the kingdom and ancestor of all successive rulers (aspects of his identity that are widely known by the general public and contribute to the tremendous respect he enjoys), Kassim's *tromba* is also remembered by those in the spirit milieu for having brutally killed his wife in a jealous rage. He is the husband of the queen whose medium lives in virtual isolation in the bush and mimetically reproduces the effects and consequences of her murder.

Evocation of a myth, to be sure; but what could be more real—or more realist a form of portrayal—than suffering?

Part II

Structural Remains:
Contemporary Divisions
of Historical Labor

Mechanical Division:
Structure and History
in the Northwest

En tous les lieux que nous avons visités dans les pays sakalaves, nous avons trouvé les événements et les noms qu'elle rappelle encore vivants dans la mémoire des populations, et excitant en elles un enthousiasme et une vénération qui ne peuvent être que le résultat de la foi; nous avons entendu les Sakalaves invoquer ces noms dans tous les actes importants de leur vie sociale, rappeler avec orgueil ces événements, à ceux qui semblaient ignorer leur grandeur passée, et, en présence de ce témoignage de tout un peuple, il nous était difficile de rester complètement incrédule.

—M. Guillain, *Documents sur l'histoire, la géographie et le commerce de la partie occidentale de Madagascar*

La structure Bemihisatra est historique, l'histoire des Bemihisatra est structurale.

—Jean-François Baré, *Sable Rouge: Une monarchie du nord-ouest malgache dans l'histoire*

How is the burden of Sakalava history shared? The division of labor can be understood along three intersecting dimensions. This chapter and the next explore structural elements that, following Durkheim, I refer to, respectively, as mechanical and organic. A third division, which I call particularist, is organized around service to ancestors constituted as specific historical personages and is discussed in chapter 6.

The Durkheimian characterization of divisions of labor is applied rather loosely and serves also as the vehicle for discussing a number of other issues. This chapter offers a short account of Sakalava political history and provisionally considers the relation of structure to history. Social inequality and political subjection are explored briefly in chapter 5. However, these expositions are primarily structural in nature;

exploration of actual practice, conflict, and the manner by which authority is established, reproduced, and struggled over is postponed for later chapters.

A mechanical division of labor is segmentary. The Sakalava polity is reproduced in
various localities and at various levels of inclusion. These segments are not identical to
one another but are based on the same model and recombine elements in an overlapping series of repertoires. Thus, although the shrine at Mahajanga is central, there are
shrines throughout the countryside, some with relics of their own; likewise, there are
many local monarchs. As Durkheim would suggest, it is both the parallels between the
local versions and the annual ritual or coming together at the main shrine that express
and reinforce their common identity. The mechanical division, as I refer to it here, is
primarily a matter of political loyalty based on distinct localities.

The organic division of labor is functional and hierarchical, comprising the descent-
based distribution of tasks associated with service to the royal ancestors. These include
responsibilities of being their live descendants, producing new ancestors, guarding their
material remains and places of residence, managing the shrines, and so on.

The particularist division of labor refers to the fact that certain key tasks are also
partitioned by personage, so that people are assigned to the labor of historically specific ancestors. Thus, one is not a tomb custodian or spirit medium, in general, but custodian or medium to particular ancestors located within the generational matrix.

These divisions have implications not only for what one does, but for what one knows
or what one can lay claim to knowing and the kind of authority with which one can
perform, speak, or remain silent. Memory and knowledge are thus highly distributive.

The divisions of labor should not be taken too literally; they provide an analytic
framework by which to organize the description and to suggest the multidimensional
complexity of the Sakalava ancestral order.[1] They do not imply functional integration;
whatever might once have been the case, such integration could not describe the situation of a modern, highly differentiated society. Indeed, they can also be read as alternate modes of belonging. The mechanical division describes belonging through
places and the particular leaders and ceremonies associated with them. Places are
linked through a segmentary hierarchy of greater inclusiveness. Organic solidarity entails belonging through labor and the "kinds" of people who do it. Recruitment to
kind is primarily by means of descent. The third mode of belonging is mediated by
loyalty and subjection to particular personages—the individual royal ancestors—themselves genealogically ranked. Together, these modes produce a "heterogeneity of hierarchical belonging" (Taylor 1998: 40). This may be contrasted with unmediated forms
of belonging in which uniform individuals are seen as direct citizens of the state or
consumers of the market (both also present in Mahajanga). As Taylor points out, the
hierarchical and the modern are intrinsically linked to very different conceptions of
time and history.

ANCESTORS IN THE PRESENT

The divisions of historical labor are, of course, abstractions from a totality. What unifies them is concern with a single set of ancestors who articulate "the force of a 'time

of origins', a higher time, filled with agents of a superior kind, which we should ceaselessly try to reapproach" (Taylor 1998: 41). In fact, there is a virtual conflation of ancestors and history in the public domain (exclusive of "official" Malagasy national history). Sakalava ancestors are distinctive (and differ from those found in other areas of Madagascar) in several respects:

1. The ancestors in question are virtually all members of the royal clan (*ampanjaka*). Conversely, all members of this clan become ancestors. Unlike in many parts of Africa, successful procreation is not a prerequisite for ancestorhood.

2. Royal ancestors have public relevance: They serve as ancestors for everyone. They represent the collectivity but can also be addressed by individuals of all social statuses. Living *ampanjaka* do not have exclusive access to them. Thus, while Feeley-Harnik (1978) correctly emphasizes the privilege that *ampanjaka* have—namely, that only their ancestors count, that only royalty have or make history—the responsibility for ancestors and the interest in them is much more widespread.

3. They appear as *tromba* (spirits). Sakalava spirit possession has close parallels to ancestral and mortuary practices described elsewhere in Madagascar, but possession brings its own entailments. Possession works through the subjectivity of spirit mediums and transforms their bodily and personal practice, enables dramatic performances, and provides a unique system of communication (Lambek 1981).

4. They are not an amorphous mass but differentiated as individual personages. Narratives, material structures, and *tromba* performances realize and elaborate character, albeit within a comprehensive code by which it can be expressed. Ordinary people, as well as mediums, have particularized relations to specific ancestors from the common stock.

5. The chronological sequence and genealogical relations among these differentiated ancestors are critical and continue to order their relations to one another as well as the way they are perceived and addressed by the living. Ancestors cannot simply replace one another; their individuality is determined through their relations with others. These relationships, constituted as kinship, affinity, love, and rivalry, have coherent temporal and spatial dimensions. This is why one cannot write about Sakalava without accompanying genealogical charts and maps.

6. Sakalava genealogical hierarchy corresponds more or less to Rappaport's hierarchy of sacred postulates (1999), such that the older the generation of the ancestor, the more sacred and the more capable he or she is of sanctifying others. Indeed, Sakalava ancestral history *is* sacred, and the sacred is intrinsically historical.

Feeley-Harnik has argued persuasively that Sakalava work on behalf of living royalty was severely curtailed under French colonialism and that in response to this loss of temporal power, labor went underground, so to speak, redirected toward the realm of the dead—the royal ancestors (1984). She suggests further that the ancestral relics that were retained by living monarchs as sign and sanctification of their authority to rule (and that were inherently transportable and mobile) gave way in importance to royal corpses, permanently anchored in tombs. This part of her argument does not completely fit the situation in Mahajanga. It is true that labor is focused on ancestral rather than living royals, but here the primary focus is on the shrine containing the royal relics, the *efadahy manakasing,* the "four sanctity-producing men."[2] Upkeep of the cemeteries is also a priority but not at the expense of the relics. I will demonstrate below how relics, cemeteries, and spirits are interrelated.

In Mahajanga, two branches of royalty, the Bemihisatra and Bemazava, have been quarreling over control over the relics (which represent common ancestors, predating the factional split) for many decades and since well before Malagasy independence. At present, Bemihisatra have sole control over the relics. This means that the Bemihisatra monarch, Ampanjaka Désy (Désiré Nöel), of Ambatoboeny, and his associates have been the chief beneficiaries of the symbolic capital and material income generated at the annual Great Service (*Fanompoa Be*) held to bathe the relics. Bemazava, led in the 1990s by Ampanjaka Moandjy (since passed on) of Mahajanga, meanwhile became increasingly impoverished and disempowered. Thus, although the overt focus may have shifted from living to dead royalty, control over the relics still has tremendous importance for living royals and their subjects.

The sanctity, power, income, prestige, and authority of living royals derives mainly from their relationship to the ancestors. However, this relationship is neither exclusive to living royalty nor unproblematic for them. The ideology in which the relationship is phrased is that living monarchs must work on behalf of their ancestors along with everyone else. It took some time for me to realize that deceased royalty, that is, the *tromba,* also continue to work on behalf of ancestors from the generations that preceded them. In Mahajanga, this means that all of the spirits (through their mediums) in the many generations (some eight to ten) below those represented by the relics take part in the Service (*Fanompoa*), alongside human supplicants as contributors to their mutual ancestors rather than as recipients. That is, the primary distinction at the shrine is neither between royals and commoners nor between the living and the ancestors, or humans and *tromba,* but between the most senior ancestors (the "four men") and all successive generations. The position of the reigning monarch (*ampanjaka be*) is ambiguous here; he initiates and produces the *Fanompoa* with the rest of the descendants, yet he also stands in for the most senior ancestors at the moment of receipt.[3]

Each generation honors preceding ones. Generational precedence, rather than death, is the critical attribute. In fact, this reinforces Feeley-Harnik's picture. It is not merely the recipients of labor who have gone underground, but many of the laborers themselves. That is, much of the collection of money for the Service and much of the management of the restoration of the various tombs lies not in the hands of the living, but of the ancestors (and, hence, of the spirit mediums).

In sum, if the focus of Sakalava service has shifted from the living to the dead, it is not simply that labor is performed by the living, or tribute collected from the living, on behalf of the dead. In addition, (1) living rulers are beneficiaries of the service car-

ried out on behalf of the relics; and (2) the living work together with the more recently deceased on behalf of the more distantly deceased.

Despite the collective labor and the principle of genealogical precedence, the current system is not overly rationalized. I made the mistake of trying to clarify a precise
decision-making or bureaucratic structure; but titles do not always imply specific roles
or duties, nor do they always correspond with officials in a precise hierarchy. The discrepancy between the facts and the anthropologist's intentions inevitably produces a
kind of tension, since constructing a model is itself a form of rationalization. If the
models that follow have a fault, it is that they congeal what is fluid. What the models
try to grasp is a coherent system able to reproduce and transform itself in the absence
of a central brain or management.

For most Sakalava, who are neither professional historians nor professional anthropologists, nor yet constitutional lawyers, the sheer *force* of the past is more significant
than any abstract model or comprehensive picture. Structure is relevant only insofar as
it maintains, constrains, and preserves ancestral power, providing a conduit enabling it
to continue to flow safely into the present. Sakalava are more interested in grasping
and fending off power than in reproducing structure for its own sake. The articulation
of the parts becomes visible only when they are activated, especially when things and
persons move between places. This is evident at the Great Service, when each contingent does its part, at royal funerals, and in the comings and goings of *tromba*.

The remains of the past, still capable of irrupting into and affecting the present,
must be cared for. Ancestral practice lies in these acts of curation, cultivation, protection, and irruption, carried out in an ethic of reciprocal responsibility (Walsh in press).
The complex, potent, richly textured, and continuously surprising lived reality that
these activities constitute captures Sakalava imagination, as it captured mine. Nevertheless, I will begin with the outcroppings of structure that constitute the social environment in which Sakalava pursue these preoccupations.

WEATHERED REMAINS: MECHANICAL DIVISION IN THE REGIONAL ORDER

By mechanical division I mean the existence, side by side across the landscape, at differing scale and levels of visibility, of so many reproductions of Sakalava order. In its
heyday, the Sakalava kingdom experienced continuous growth and change, so there is
no clear baseline against which present formations can be measured. The structure visible in the present is the product of the formation, expansion, and subsequent erosion
of a particular politico-religious organization and model of rule. As such, it can only
be understood by turning (briefly) to that history.[4]

In the sixteenth and seventeenth centuries the region was characterized by small
kingdoms trading with Muslim ports that formed part of the Swahili network and
oikumene. Mariano (1613–1614) described traders who spoke Swahili and looked like
black Africans. By the early 1600s Portuguese were trading piasters, metal jewelry,
cloth, beads, pewter bars, and glassware for slaves, livestock, rice, millet, arrowroot, bananas, yams, sandalwood, turtles, ebony, fabrics, raffia, honey, and wax (Vérin 1986:

113, 191). Further south, they began to offer firearms, and this fueled a cycle of warfare and Sakalava expansion. During the seventeenth century the Zafinifotsy (White, or Silver Clan) apparently emerged from the southern Sakalava kingdom of Menabe to impose its rule over the North. They continued to govern Antankaraña and, as Zafindrabay, parts of present-day Tsimihety country, but Boina was conquered again shortly thereafter by the fraternal Zafinimena (Red, or Gold Clan).

The Zafinimena consolidated power by rapidly instituting their own relationship with American and European ships, circumventing the Muslims and gaining direct access to firepower, which they used to round up slaves in exchange. Lombard cites a remarkable passage in which Tsimanato (as Ndramandisoarivo was known during his lifetime), newly arrived in Boina, lit a large fire on a headland to attract two passing American slavers (owned by Frederic Phillips of New York) and convinced them to trade with him rather than his older brother in Menabe. The ensuing passage deserves quotation:

> The captains having asked him whether he had any slaves for sale, he replied that he didn't have many, but that if they would lend him some of their men to accompany him to battle he would furnish them with as many as their ships could hold. . . . This proposition was accepted and . . . Tsimanato left for battle accompanied by twenty whites; with their aid he seized a large town where he took many prisoners, from among whom the captains were authorized to take those they wanted, without payment, but under condition that their men would accompany him on a second expedition, which was then done. He returned with many thousands of slaves and large herds of cattle. The captains of the two ships chose around six thousand slaves who, along with an abundance of fresh foodstuffs for the crossing, cost them no more than two or three barrels of powder and a few rifles.
>
> The king proposed that they leave him twenty of their men, promising to furnish for nothing a full new cargo of slaves on their next voyage. These men asking for nothing more than to stay on land, the ships left them there and departed. When they returned the following year, they received their full cargo, conforming to their agreement and they took with them those of their countrymen who wished to return home, leaving those who preferred to stay and who helped Tsimanato force the Antalaotses [Muslim traders] and Vazimba [autochthons] to submit completely to his authority to the extent that he became the king of the entire country of Methelage (from the bay of Boina to the bay of Mahajamba). (Lombard 1988: 40–41, citing A. and G. Grandidier, Coll. ouvr. anc. concern. Madag., tome III, pp. 616–18; my translation).

So much for the origins of the northern Sakalava polity. However, one of the significant features of Sakalava expansion was that in general they attempted to incorporate rather than massacre or enslave local populations (Prud'homme 1900: 11). Hence, their rise did not produce the sort of continuous instability created by the slave trade on parts of the African coast, though it may well have done so in the hinterlands. The trade in slaves was also balanced by many other exports that Madagascar had to offer, notably, food supplies; and eventually, Sakalava and Merina became net importers of slaves from Mozambique.[5]

Once firmly established, the Sakalava began to exercise their trade through the Muslim ports and convinced a group of Antalaotra (Malagasy-speaking Muslims) to establish a new town at Mouzangaye (Mahajanga), which by 1792 contained "more than 6,000 Arabs and Indians with their families" (Du Maine 1810: 27). Gradually supplanted by South Asians in the overseas trade, the "Arabs" (largely Antalaotra and Comorians) served as middlemen along the coast and into the interior. The Sakalava also changed many of the place names and imposed their chronotope on the landscape of Boina.

At its most powerful, during the eighteenth century, the Sakalava monarchy of Boina comprised most of northwestern Madagascar. There was one supreme ruler, yet the monarchy appears to have subsumed local polities into which the kingdom was prone to fissioning.[6] Rulers sent out their siblings to expand their authority but worried about reining them back in again. Fragmentation became a reality when the Merina penetrated the royal capital in 1824, the reigning monarch fled, and the sacred relics were removed. The fragments became relatively fixed under French colonial rule (1895–1960) and have remained so within the independent Malagasy state. During the colonial and postcolonial eras the kingdoms were further encapsulated and eroded by alternate forms of citizenship, religion, education, and competing means to access livelihood, wealth, and power.

At present, there are many Sakalava principalities across the Northwest, of greater or lesser size, power, local autonomy, and reputation; a number are well known in Madagascar and in the ethnographic literature, but others exist in the interstices and on the fringes of the larger ones (figure 4.1). Despite the absence of a comprehensive principle determining the order among the principalities, some branches are senior to others and living rulers can also be compared with respect to their generational status. They are unified insofar as the population recognizes the place of each ruler on the broader royal genealogy and the principles of hierarchy and historicity that make such ancestry relevant. They venerate the same senior ancestors, supplicate the same relics, and recognize the same *tromba* who appear in mediums across the North. But the principalities are autonomous insofar as each is able to reproduce itself. The best term for the Sakalava political system is thus the "segmentary state" in which "the basic relationship between the center and the peripheral units remained that of ritual suzerainty rather than political sovereignty" (Southall 1999: 36).[7]

These polities are each the domain of a specific lineage tracing descent back to the main royal line and whose reigning monarch is acknowledged by others. The domains are territorial in that the land outside them is the province of another ruler. It is a principle of Sakalava politics that there cannot be two reigning monarchs in one domain, and rulers generally keep clear of each other's territory. As Anderson (1991) and others have noted more generally for premodern polities, Sakalava rule (*fanjakana*) is conceived as radiating from sacred centers rather than uniform within fixed borders (though as most polities face the sea, centers and boundaries tend to be marked by rivers and estuaries).

Territorial limits are not determinate. Sakalava rulers do not have full authority over everyone within their territory but merely over those inhabitants who accept

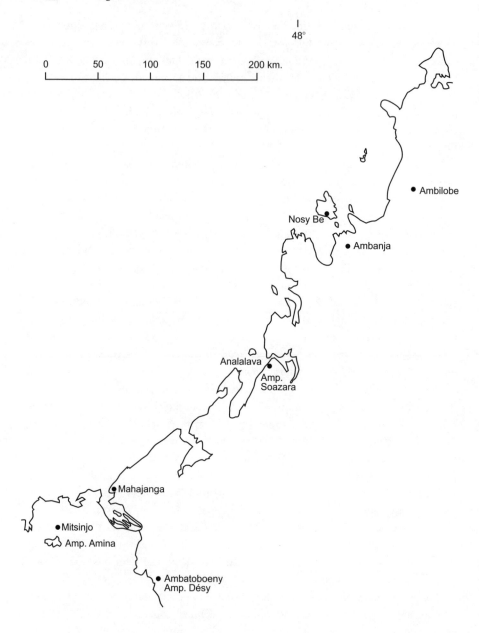

4.1. Major Royal Capitals and Rulers in the Northwest

their rule.[8] Not all inhabitants submit to the same degree; some may feel greater loyalty to a different polity, whereas others simply disregard the principle of Sakalava rule.
Some say that they have rejected the monarchy in favor of Christianity or Islam, or
that they belong to a descent group that has long held the privilege of not having to

support the monarch. Many juggle conflicting or incommensurable identities and obligations (chapter 5). Most inhabitants of a given territory who do not acknowledge the ruler are identified with other "ethnic groups;" conversely, some of these "others" do acknowledge Sakalava rule. Those of South Asian or European background recognize the ancestral polity when it suits their interests. Most inhabitants acknowledge certain Sakalava rights, recognize the power of royal ancestors, observe local taboos, and contribute to Sakalava rituals. In general, acceptance of the ancestral order has become increasingly a matter of personal commitment, much as adherence to institutionalized religion in any plural setting. However, as the following chapter will show, commitment is not always experienced as "free choice."

Specific to Mahajanga is the presence of the most powerful and best-known ancestral relics in the North. These were once in the possession of the supreme rulers of Boina and were a necessary sign of royal legitimacy. In the 1820s the relics were separated from those in the most direct line of rule. After resting under Merina control they were left in the hands of the two branches of royalty who remained in the Mahajanga region and were able to assert claims over them. Rulers of all the northern principalities recognize the sanctity, significance, and priority of these relics and participate directly or indirectly in their veneration. Officials at the shrine work hard to assure that this representation continues annually at the Great Service.

The symbolic capital amassed in, and constituted by, the past has arguably become the most vital resource of Sakalava rule. This is found in the bodies, tombs, and relics of ancestral monarchs and in the enthusiasm with which they are maintained by subjects over the long term and in the course of annual service (*fanompoa*). Additional ancestral wealth is stored in gold and silver artifacts, jewelry, and especially coins. Those monarchs who are wealthiest in modern currency or investments and those with modern education, professional qualifications, or employment are not necessarily the most potent in ancestral terms. Increasingly though, Sakalava have recognized the importance of these qualifications in selecting their rulers. Conversely, the symbolic capital located in shrines and ceremonies has been recognized by the Malagasy state as "culture" and become subject to heritage management and internal and international tourism (Walsh 1998; Lambek and Walsh 1999; Rakotomalala et al. 2001).

Whatever the differences among their respective rulers, the principalities are linked by sentiments of mechanical solidarity: pride in common origins, historical destiny, and ancestral power. Rulers assist one another in conservation projects, and they belong to an association that tries to promote their collective interests within the nation-state. They engage in a complex politics that attempts to negotiate among the interests and demands of the past and the present, their subjects, the state, business, and transnational organizations. They attempt to develop extensive political networks, and they sometimes come into conflict with one another; but for the most part, each is engaged in cultivating his or her own piece of Sakalava inheritance.

The biggest rivalry is perhaps found in Mahajanga itself, where Bemazava and Bemihisatra factions have competed bitterly for control of the sacred relics. Bemihisatra success colors much of the activity in Mahajanga and, inevitably, the way I write about it.

Kinds of Places

Territory is not conceived as neutral space to be divided up; it must be cultivated and turned into Sakalava space by means of names, shrines, cemeteries, and dwellings. An autonomous polity contains sacred places and the right sorts of people to manage them. Each has a capital in which the reigning monarch lives, a royal cemetery, and ancestral shrines (*doany*).[9] The Ampanjaka of Port Bergé, who maintains a capital but buries in the cemetery of her more powerful cousins at Betsioko, does not reign over an autonomous kingdom. Cemeteries and shrines are sacred in Durkheim's sense of being set apart: fenced, gated, and subject to taboos. They are sources of blessing, but inadvertent contact is dangerous and polluting, both to the places and to the tres-passers.

The residence of the active monarch shifts as one ruler succeeds another. Ampan-jaka Désy, the Bemihisatra ruler who has domain over the Mahajanga shrine, resides in the town of Ambatoboeny, some 130 kilometers away (figure 4.1). The late Am-panjaka Moanjy, then leader of the rival Bemazava faction, maintained a modest resi-dence in Mahajanga, from the porch of which his wife sold tea and fried cakes to shoppers in the Tsaramandroso market.

Although rulers have ultimate responsibility for the upkeep of royal cemeteries, they must live at some distance from them. Members of the royal clan stand in oppo-sition to death; they are polluted by it and must not associate with it. Royal cemeter-ies have resident *ampanjaka* from a distinctly junior line (see below). Cemeteries are, in principle, permanent; they have dedicated custodians who continue to live there from generation to generation, ideally even after the cemeteries are no longer accept-ing bodies.[10] Custodians inhabit houses belonging to the respective monarchs buried in the precinct.

A given ruler may have more than one cemetery under his or her domain. Ampanjaka Désy has authority over the important Bemihisatra burial ground at Bet-sioko (figure 4.2). Ampanjaka Moanjy shared authority over a pair of old cemeteries at Mahabo (Manarantsandry), as well as a somewhat newer Muslim one at Boeny. The most sacred burial ground in the region, containing the genealogically most senior an-cestors, is Bezavodoany. This lies within the domain of Ampanjaka Amina, who also maintains two newer royal Muslim cemeteries.[11]

Shrines (*doany*) range from the large communities in Mahajanga to tiny fenced structures with no dedicated resident custodians located in villages and under trees. They are places where *tromba* gather, supplications are made, and offerings collected for sending on to more central shrines. They are always associated with royal service.[12]

Like cemeteries, shrines are locations at which the past is perpetually present and, hence, at which its rules must be respected. Shrines are associated with specific royal ancestors and may contain their relics or artifacts.[13] In this way the countryside is res-onant with history and provides its own genealogical map.

People's loyalty is first and foremost to the ancestral places and the objects in them as representative of the royal ancestors. Living rulers gain loyalty for the same reason, that is, as manifestations of ancestral order. But living rulers are impermanent and as

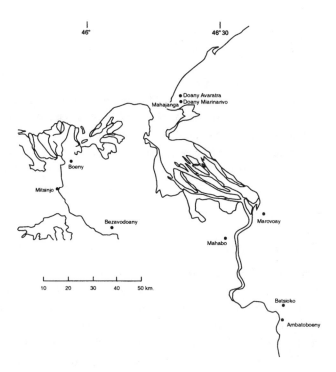

4.2. Major Royal Cemeteries and Capitals in Boina

yet unfixed in that order, whereas the material remains and the *tromba* perdure. There is a shifting balance of power between living rulers and ancestral custodians. Today, an architectural politics of conservation prevails (chapter 10). Cemeteries and shrines depend on regular service (*fanompoa*) in labor, cash, and kind. Upkeep of residences of living rulers, at least within Mahajanga, is less formalized. Living rulers are sometimes subsidized by politicians offering patronage in return for votes.

Because Mahajanga is the site of a shrine whose relics transcend the local principalities and because it is a large city, it contains representatives of various other towns and principalities. These include male officials (*manantany*), Great Women (*bemanangy*) (chapter 7), and spirits from various cemeteries.

I will make two final points about place. In the past, Sakalava polities were economically successful only to the degree that they had access to and control over one or more of the Muslim ports of trade. The sovereign of Boina appointed the Muslim governor of Mahajanga and remained superordinate in their alliance. Second, Sakalava perceive an ontological connection between land and person, self, and community.[14] This is probably felt more strongly in the countryside, where most people were born or raised and where they will be buried, than in the city in which most of them spend their youth and adulthood.

THE "FOUR MEN" IN THE HISTORICAL ORDERING OF BOINA

It should be clear by now that one cannot elucidate Sakalava "structure" without reference to history. This is not only because the structure has "weathered" the forces of history but because for Sakalava, history itself *is* the very structure. That is, just as the logic of structure can only be grasped from what remains when we take into account the effects of history, so is the substance and renewal of structure derived from ancestral history. Conversely, the history that is reproduced is that which is readily structured. Such "history as structure" is evident with respect to the "four men" (*efadahy*) whose relics comprise the main shrine at Mahajanga.[15] In this section I examine each of these personages in turn, though I reserve the older pair for the end. Brief contemporary Sakalava portraits are supplemented by European sources.

Father and Son

One relic belongs to Ndramandisoarivo. He left the southern kingdom of Menabe at the accession of his older brother around 1685 to establish his capital at Tongay, west of the Betsiboka estuary, around 1700, and consolidated his rule over Boina. He began the tomb precinct of Bezavodoany, where he lies with his wife, son, and several descendants. When he arrives as a *tromba,* Ndramandiso (as I shorten his name) is a bent old man who is raised to an elevated seated position. Although recognized as the conqueror of Boina and extremely powerful, Ndramandiso is portrayed as relatively complacent and mild-mannered. He never stays long and has rather little to say. He announces, "I am the master of Boeny (*Za' tômpin Boen-ee*)," but he is easily intimidated by his wife.

All Sakalava royalty of the Zafinimena line in northwestern Madagascar claim descent from Ndramandiso. However, fully legitimate royalty must be descended from *both* Ndramandiso and his wife, Ndramandikavavy, which is through their only joint offspring, Ndramboeniarivo (figure 4.3). They are the only ones with rights to rule in Boina. Ndramandiso's older son, Ndramamonjy, half-brother to Ndramboeniarivo, produced his own line of lower-ranking royalty. As his descendant, Mady Simba related, "Arriving from Menabe to find that their father had given the new kingdom to his younger sibling, Ndramamonjy petitioned for something for himself. Everything had already been divided up except the 'gate' (*varavaraña*). So he and his descendants were given power over the gate; whoever they say can enter, enters; whoever they say can't enter, cannot." This means that the descendants of Ndramamonjy, the Maromadiniky, have charge over the royal tomb enclosures.[16] One of their number lives at Bezavodoany, although he in turn is under the authority of the ruler in the senior line (that is, Ndramboeny's), namely Ampanjaka Amina, in whose domain the cemetery lies. Amina could dismiss him, but she must replace him with someone of his descent group. Maromadiniky have their own burial enclosure at Anjiamañitry and produce *tromba,* among whom

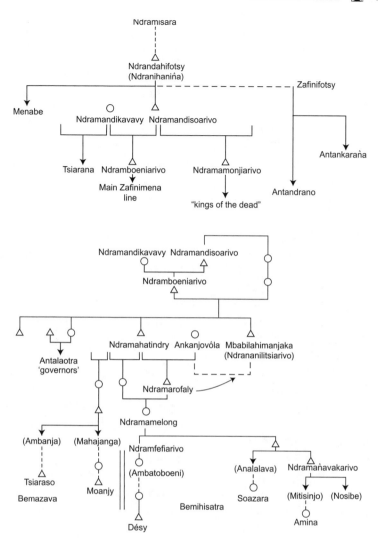

4.3. Simplified Royal Genealogy Indicating Major Branches

Ndransinint' was very popular in the 1990s (chapter 9). There is thus an organic and hierarchical division of labor between the line of the rulers of the living and the shadowy rulers of the dead.

Ndramboeniarivo (Lord of the People a Thousandfold; known as Andriantoakafo while he ruled) is the next of the "four men."[17] His reign, from about 1710 to 1733 (Vérin 1986: 110–11), is recalled as a period of unity, and he symbolizes the importance of exclusive filial succession in maintaining that unity. Ndramboeny is a powerful and intimidating *tromba,* known for his propensity for rage; yet he is also a loyal son and solicitous of his mother.

From Ndramboeniarivo to the Present: A Brief History of the Monarchy and Relics

After Ndramboeny, the kingdom passed into the hands of his son, Ndramahatin-driarivo, who moved the capital to Marovoay (Vérin 1986: 310) to be in better communication with foreign slaving vessels while remaining at a safe distance. He also encouraged Antalaotra from the older port of Boeni to settle in Mahajanga, where they would be safely beyond the influence of his competing half-siblings (Guillain 1845; Vérin 1986: 312–14). Foreign traders included Americans, Danes, Dutch, English, French, and Portuguese. Although goods such as cloth, jewelry, and knives were exchanged for skins, incense, benzoin, amber, wax, wood, and turtle shell with the Antalaotra and Arabs living in Mahajanga (Vérin 1986: 315), Europeans had to go upstream to ask the king's permission to trade in slaves and to discover the price. A Dutch expedition of 1741 recorded that each slave cost "2 muskets, 3 pounds of powder, a pound of bullets, and 10 flintstones, or else 45 piastres per man" (ibid.: 311, citing O. Hemmy). In addition, the Dutch gave the king presents that included cloth and alcohol. The Sakalava had a shortage of slaves to sell at the time—a campaign in the highlands having failed—and wanted to trade rice and oxen. They were interested in acquiring light rifles, flints, powder, lead, iron vessels, glass beads, and Spanish piasters (ibid.: 311).

The Dutch visitor, Hemmy, described Marovoay in 1741 as having thousands of houses and a palace "surrounded by four or five concentric rows of fences made of great stones sharpened at the ends" (Vérin 1986: 312). Inside were warehouses full of merchandise. Hemmy saw over 100 muskets, chests said to be full of vases and silver objects, and a lacquered throne brought by the French from China. The royal reliquary already contained fragments of the four ancestors presently venerated at the shrine in Mahajanga. These were described as gold and silver "escutcheons," each fastened on a stake with four big gold teeth "like those of the sea-cow (or the crocodile) and covered with material sewn with piastres and a large silver chain" (ibid.).

Ndramahatindry is remembered today as violent and ready to kill potential rivals. In part, this refers to problems he had controlling the many children Ndramboeny appears to have fathered out of wedlock. Members of the Bemihisatra royal faction depict complex and incestuous family relations among Ndramboeny's and Ndramahatindry's legitimate offspring and point to this period as the source of the split between them and the Bemazava. The factions eventually divided the territory in and around Mahajanga and developed separate tomb precincts on either side of the Betsiboka near Marovoay (figure 4.2).[18] Overlooking the Betsiboka from the west is the Bemazava precinct of Mahabo in which Ndramahatindry lies, whereas his younger brother, Mbabilahy (Mbabilahimanjaka, formally known as Ndrananilitsiarivo), is the senior resident at the Bemihisatra cemetery of Betsioko on the east side of the river. Significantly, this side is closer to Antananarivo. I have not knowingly seen the *tromba* of Ndramahatindry rise among Bemihisatra; conversely, he is critical for Bemazava. Both groups describe him as extremely fierce.

It is clear that dynastic struggle and the maintenance of centralized power was always an issue. Ndramboeny had to deal with competing half-siblings. Ndramboeny's male offspring quarreled, whereas a daughter, Ndranantanarivo (Samoa in life [Vérin

1986: 310]) was married by her grandfather to the leader of the Antalaotra merchant community whose activities Sakalava both depended upon and wished to control. She was probably the first member of the royal family to accept Islam, and she married husbands successively from Surat and Patta (Guillain 1845: 21–22) but was buried at Bezavodoany. The ideology of ancestral devotion kept the descendants of the siblings in relation with one another.

Despite problems over succession following Ndramahatindry's death, Boina remained reasonably unified and very prosperous, notably under Ravahiny (Ndramamelong), who ruled from about 1785 to 1819.[19] Boina was then "at the height of its power, even though it was constantly torn by palace revolutions and risings among the peoples who had been less completely subjugated" (Vérin 1986: 111). Ravahiny had thousands of oxen at her disposal and engaged extensively in sea trade, mediated by the Arabs and Antalaotra whom she allowed to build houses in stone as they had done prior to Sakalava arrival.[20] She received tribute from the Antankaraña in 1792, which included forty to fifty canoes full of wax, tallow, and especially turtleshell (Du Maine 1810: 23).

Guillain wrote of Ravahiny that she "has remained surrounded by a veneration among Sakalava so much the deeper as her reign continues to evoke the liveliest memories of their past grandeur" (1845: 35; my translation). Yet her *tromba,* Ndramamelong, rarely appeared in Mahajanga in the 1990s, and the power and wealth that her reign represented were not particularly salient. Her relative absence is partly a function of a general conflation of this period of Sakalava history, the fact that female *tromba* are more discreet than male ones, and that she is portrayed as so old as to be virtually immobile; but more, I think, because she was buried at Mahabo, the burial ground that retroactively became specifically Bemazava.

In all of these respects, Ndramamelong can be contrasted with her father, Ndramarofaly, one of the most popular *tromba* in present-day Mahajanga. Ndramarofaly is present at virtually every occasion when more than one spirit is called (chapters 3 and 7), though he never ruled. He is remembered as a warrior, unfamiliar with Europeans, who died hunting in the forest with his dogs, whereas earlier traditions suggest either that he was poisoned or killed in warfare between his father and uncle. Some say he was discovered in a tree by a faithful dog and returned for burial at Betsioko. More likely is Rusillon's version (1922–1923) that his body was never found and Betsioko contains only a cenotaph in his honor rather than a full tomb. This is a matter of uncomfortable silence, or denial, today.

A reason for Ndramarofaly's ubiquity is the fact that his mother was well known as a Zafinifotsy queen; thus, he conjoins both Zafinifotsy and Zafinimena lines in his person and is present at events pertaining to both sets of spirits. At Zafinifotsy gatherings he stands out as the one person dressed in red. His name means Lord of Many Taboos, but he is remembered more for not following taboos and especially for opening women's bellies to discover the secrets of childbirth. This may be linked to questions of his own maternity and birth. Despite or because of his evident unruliness and "rude" manner, he is viewed very fondly in present-day Mahajanga and in return is solicitous of his many clients.

Toward the end of the eighteenth century Ravahiny exchanged gifts with Andrianampoinimerina, the king of the highland Merina (or Hova), who was nominally her vassal. The Hova sent red cloth and money for her tomb and 400 oxen for sacrifice at her funeral (Guillain 1845: 33). As Sibree remarked (1885: 462), "This was the last act of homage rendered by the Hova chiefs to the sovereigns of Iboina." By 1824, the tables had turned, and the Merina king, Radama, was demanding that Ravahiny's eventual successor, her grandson, Andriantsoly, declare himself Radama's "son." When Andriantsoly refused, Radama captured Mahajanga and placed Andriantsoly under house arrest in Marovoay. An uprising the next year failed and Sakalava royalty then dispersed, the senior line moving northward along the coast and Andriantsoly eventually settling in Mayotte where he was buried, becoming (under his posthumous name, Ndramañavakarivo) the leading *tromba* there (Lambek 1981, 1993). When Hastie, who was aiding the Merina, entered the young king's hastily abandoned royal residence at Mahitapanzava, a town of some 740 houses, he found nice furniture, mirrors and in the largest room—"the unimpeachable proofs of the royalty of the chief, the remains of his ancestors preserved with great care in a sort of sarcophagus surrounded by hangings of white cotton" (Guillain 1845: 77; my translation).

Sakalava who remained in Mahajanga were treated harshly by the Hova governor, and those with any wealth were accused of sorcery and their possessions appropriated (Allibert 1999). Antalaotra, formerly the main inhabitants of the port, fled elsewhere, and their role as middlemen was taken over by South Asians, notably Muslims from Surat, and Europeans. Although the Merina had captured Mahajanga to have direct access to external trade (as well as to restrict Sakalava access to firearms), maritime traffic shifted to Toamasina on the east coast. In addition, the interests of the Europeans were changing from mercantile relations to incorporation; this global shift accounted in part for the inversion in the balance of power between Sakalava and Merina (Lombard 1988).

The decline in Mahajanga is evident. Recall that in 1792, Du Maine observed "more than 6,000 Arabs and Indians" as well as mosques of various sects, schools, and workshops of all kinds. Boat building was active, food plentiful, and commerce of an "astonishing activity" (1810: 27–8). In 1869, A. Grandidier estimated the population of the city to include some 700–800 Antalaotra, 150–200 Karana (South Asians), 2 Arabs from Muscat, 10 Comorians, 150 Merina traders, 300–400 Sakalava, 1,500 African slaves, and possibly 3,000 men in the Merina garrison (Vérin 1986: 335). In 1883, when Mahajanga was first occupied by French troops, there were probably between 8,000 and 10,000 inhabitants; Sakalava, Makoa (former Africans), and Comorians lived in simple huts; Europeans, Arabs, and Indians in stone houses (ibid.: 339). Outside the city were a number of small independent polities led by Sakalava kings or queens supported by Muslim advisors (Catat 1895: 246).

The traumatic experiences of the period of Merina conquest and rule are not directly displayed by the Zafinimena *tromba* in Mahajanga but are instead displaced onto the Antandrano (Water-Dwellers), a group of Zafinifotsy spirits who drowned in the Loza River while attempting to escape the Merina (Lambek 1996). Andriantsoly is remembered for his conversion to Islam; his mediums sponsor Maulidas, musical performances in which he joins in honoring the Prophet.

The direct descendant of the last official ruler before French conquest is Ampanjaka Soazara in Analalava, whose immediate ancestors are buried at Nosy Lava (Feeley-Harnik 1991a). Soazara's line is identified as Bemihisatra, though she also has Bemazava antecedents; Bemihisatra descendants of Andriantsoly rule at Nosy Be (Baré 1977, 1980) and at Mitsinjo. Bemazava polities were established at Ambanja (Sharp 1999a) and at Boeny. As noted, Merina conquest separated the rulers from the ancestral relics. The Merina placed the relics in Mahajanga where they eventually came under the control of those branches of the royal family who had stayed in the area or returned to it during the colonial period. These include the Bemihisatra of Ambatoboeni, on the east bank of the Betsiboka upstream from Mahajanga, and the Bemazava of Boeny on the west bank by the coast.

In chapter 2, I cited the observations of the British trader, Leigh, who saw the relics in the Merina fort in the period 1836–1840. These are supported by British missionary, Pickersgill, who on a visit to Mahajanga near the close of the Merina period noted that "the Sakalava . . . continued to regard the town as their head-quarters, their chief being resident there and holding an influential position in the government of the province; but a large number of them had gone up the river to live near the rice-fields and cattle-pastures. Many of those who remained were owners of outrigger canoes and earned an easy livelihood by ferrying, fishing, and cutting wood. There were likewise some thousands of lately-freed Africans." (1893: 34).

Pickersgill writes in florid prose: "The fetish itself was honored with a house inside the Governor's stockade, and is guarded to the present day by Hova [Merina] muskets. . . . These relics are contained in a silver vessel, which is so swathed about with strings of beads and crocodiles' teeth and horns of fat, and sacred honey, and other potent charms, as to be quite a strong man's load to carry. Such is the only bond of Iboina union, and to this the tribe gathers once a year with rum and drums and wild singing: and the priests of their uncouth religion agonise with inspiration from the fragments of mouldered mortality, and Sakalavadom holds high festival" (1893: 39; cf. Catat 1895).

He adds, "besides the big yearly festival there is a minor gathering once a week, on Friday, which for the most part is attended only by women and children; and another on the Mondays immediately following the appearance of a new moon. . . . These monthly celebrations are generally wound up with the slaughter of a bullock and signs of a general feast. . . . The more devout of the Sakalava visit the 'Zomba,' as their place of worship is called, on every occasion of private importance, to make offerings of fowls and money and ask for intercession, but their numbers are decreasing every year" (1893: 40).

Pickersgill argued, "To the Sakalava the sovereign never dies. Departing the earthly life, he 'bends' or 'submits,' and passes to another existence, in which, by right of rank, he abides with the deity who rules the world. Thus, he becomes a mediator for his people, and speaks to them through the priests who have charge of his mortal remains. 'Ask of me, and I will ask of God,' is the oracle's injunction to the multitude when the relics are brought forth at the yearly festival and carried in procession through the streets" (1893: 39).

Contemporary Bemihisatra and Bemazava in Mahajanga differ in their respective accounts of the relics during the Merina period. Bemihisatra argue that Ndramfefiarivo, a daughter of Ndramamelong (Ravahiny), followed her ancestors (the relics) to Antananarivo and even married King Radama to better protect them.[21] Rusillon identifies her as Ratarabolamena and says that she reigned at Mahetsapanjava (presumably after the escape of her nephew, Andriantsoly) but was taken to Antananarivo in 1825. Subsequent rulers in the Bemihisatra line are descended from this union and thus part Merina themselves; the *tromba* are portrayed with Merina cultural traits, and several are Christian. Eventually, Sakalava were able to bring home the relics, as well as the body of Ndramfefiarivo.[22] The former queen declared, via her *tromba,* her fear of capsizing while crossing the Betsiboka; and so, rather than being buried with her mother at Mahabo on the west side, she was buried at Betsioko, the cemetery that thenceforward was identified as specifically Bemihisatra and home to the part-Merina line.[23] The oldest ancestor buried at Betsioko is Mbabilahy, the youngest son of Ndramboeniarivo, who became, somewhat retroactively, senior ancestor of the Bemihisatra (figure 4.4).[24] Mbabilahy plays a critical role in representing Bemihisatra interests today. He appears widely and has senior mediums resident at both Betsioko and Mahajanga. He is portrayed as the father of Ndramarofaly, who is thus also identified as Bemihisatra.

The Bemazava record of this period is less clear to me, but some argue that they kept the relics in hiding from the Merina and never engaged in dubious marriage relations with the enemy. Pickersgill confirms that Radama's search for the relics "was fruitless: the fetish had been hidden in the forest" (1893: 38). Eventually, "a heavy bribe induced a Sakalava to tell of its whereabouts" (ibid.: 40). Vérin also suggests that some of the relics may have remained hidden (1986: 330, no. 4).

Feeley-Harnik places the movement of the relics to Antananarivo much later than the contemporary Bemihisatra account and leaves them there for a much shorter period. She cites "an interpreter for the French colonial administration in Majunga, writing shortly after the conquest, [who] said that when the Hova fled Mahajanga in 1895, they took the relics to force 'the queen of Boeni, Ramboatofa' and her followers to follow them to Antananarivo, where they were all captured by the French, who eventually returned the relics to Mahajanga where they also used them to control the local population" (1991a: 505, no. 60).[25] It would appear that having come from Antananarivo in 1861,[26] Ramboatofa had to return there in 1895 or the Bemihisatra transposed this visit to Ndramfefiarivo seventy years earlier.

Whatever the case, the two Sakalava factions worked out complementary roles in maintaining the relics but then spent much of the twentieth century quarreling over them. A fire in 1958 destroyed the shrine just after it had been transferred by the government into Bemazava hands. Bemazava constructed a new shrine, known as Mandresiarivo, or Doany Avaratra (the Northern Shrine), which housed the relics (replenished from Menabe) from September 15, 1961 until Bemihisatra took them by force the night of June 8, 1973. The Bemazava temple and its palisade were bulldozed

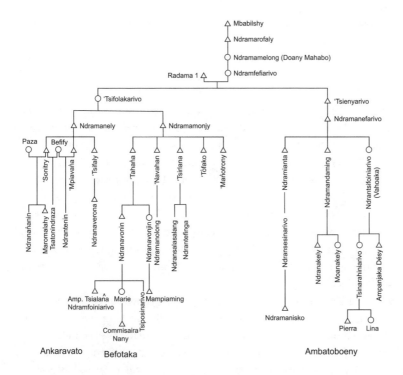

4.4. Version of the Bemihisatra Branch at Betsioko as provided by Amady bin Ali

October 19, 1979 at the instructions of a government official associated with Bemi-hisatra. These were traumatic events; the spot remains a sacred one, and the despoiled Northern Shrine is still inhabited and frequented by Bemazava.[27]

The dispute over the relics has been long and bitter, and both sides are capable of arguing the justice of their respective causes with great passion. Bemihisatra relodged the relics in the Doany Miarinarivo in 1973 but have been worried ever since about having them taken back by Bemazava (chapter 10).

Several members of the Bemihisatra line served minor roles in the colonial admin-istration, and their *tromba* are portrayed with French consumption habits and know-how. These *tromba* are referred to as "*zafy,*" grandchildren or descendants, even though some of them are at an advanced stage of the personal life cycle. They are among the spirits most frequently visible at possession ceremonies in Mahajanga today. They in-clude Ndramandaming (Andriamandaminarivo, Lord Who Arranges a Thousandfold), who was "governor" of Ambatoboeny in the indigenous administration reporting to the French Chief of District. He may have been the first Sakalava monarch to travel to France, visiting the Colonial Exposition of 1931. His *tromba* is portrayed exactly as Delorme and Grellier describe the man himself, "indulgent and debonair" (1948: 4). His older brother, Ndramiantarivo, appears much more taciturn and wearing a red fez. Their father, Ndramanefiarivo, a grandson of Ndramfefiarivo, is portrayed as more Merina in appearance and habits. They are all buried at Betsioko. Bemazava were less involved in the colonial administration and did not convert to Christianity;[28] many are Muslim.

The Senior Relics

I now go back some 350 years and return to the third relic in the shrine. Originally, Ndranihaniña (Lord Missed by Thousands) almost certainly represented Ndramandiso's father (ca. 1600–1680), the conqueror of Menabe in 1649 (Guillain 1845: 8; Rusillon 1922–1923). This man's name during life was Ndrandahifotsy (White Lord), and the spirit has come to be identified primarily with the Zafinifotsy (White/Silver Clan), who conquered the north ahead of their brothers, the Zafinimena (Red/Gold Clan).[29] Zafinifotsy are possibly descended from Ndrandahifotsy, whereas other associations suggest deeper roots in the North. Whatever the case, they also intermarried with Zafinimena and thus form the mothers and mothers' brothers of several important members of the dominant Zafinimena line. So Ndranihaniña is also an important affine to the early rulers.

Ndranihaniña qua Zafinifotsy saw Boina as originally his own territory and is reputed to have said to the Zafinimena, "Since we now all live here together, let us 'serve' (manompo) each other. So they put the relics together; that was his condition for peaceful cohabitation." However, Zafinimena accounts view Zafinifotsy as secondary; Rusillon records them as offspring of Ndrandahifotsy's commoner wives.[30] Whatever the emphasis, Ndrandahifotsy/ Ndranihaniña serves as a link between the rulers of Boina and other royal descent groups in Madagascar, epitomizing their relations through both descent and affinity. In effect, as one person expressed it, members of the royal clan are Menafotsy (Red-White).

At present there are a number of people in Mahajanga who quietly affirm Zafinifotsy ancestry. They have no special privileges since Mahajanga is not within their jurisdiction. However, Zafinifotsy tromba are widespread and their mediums numerous. They are well represented at ancestral events, especially by Antandrano (Water-Dweller) spirits, so named because they died by drowning. Their popularity is due in part to the fact that the majority are youthful siblings who like to drink and party and who generally appear as a cohort. Their predominant kinship mode aside from siblingship is that of nephew to mother's brother (zama) or father's sister (angivavy) rather than parent to child—thus nonauthoritarian. They take a secondary place to the Zafinimena spirits who are proprietors (tômpin) of Mahajanga. For all of these reasons they are more approachable and less fearsome than the older generations of Zafinimena spirits and much sought after by clients.

Another reason for the popularity of the Water-Dwellers is that they evoke a dramatic and sentimental story with strong political implications. As recounted today, they were pushed to the banks of the Loza River and jumped in, preferring to drown rather than submit to their Hova (Merina) pursuers. Indeed, several who lingered were shot. Merina still play the dominant role in national life, and Mahajanga today has more Merina than Sakalava inhabitants. Antandrano are recognized across northwestern Madagascar, from Mahajanga through Antsiranana (Diego Suarez), that is, across an area where people bitterly resented Merina conquest. The story unequivocally evokes that trauma and the courage displayed in defeat. Hence, the Antandrano stand in juxtaposition to the Bemihisatra spirits buried at Betsioko who descend from the conquering Merina king and who served in successive administrations. The opposition is heightened by the anti-Christian sentiments of the Antandrano. The youthful, gentle, playful, and frequently drunk Antandrano spirits provide an earlier layer of the historical palimpsest but one that speaks directly to the fate and contemporary condition of Sakalava.

I asked Kôt' Mena, an Antandrano spirit with whom I had become friendly in a male medium, why Zafinifotsy spirits were so evident at the Great Service when it concerned Zafinimena ancestors. He corrected me, explaining that the Zafinifotsy served (manompo) their own ancestor, Ndranihaniña, who, after all, was one of the "four men." His presence provides a cover for both their submission and multilayered resistance.

When he had drunk a good deal, Kôt' Mena went so far as to suggest that Ndranihaniña may have been the illicit lover of his "brother" Ndramandiso's wife (Ndramandikavavy). "It is notable," pointed out Kôt' Mena, "that when they rise together she doesn't harangue or chase him away." Whatever the truth of this (in the genealogy presented above, he is her father-in-law), it is an interesting illustration of how people in Mahajanga (including other tromba, themselves!) draw historical conclusions from their observation of the bodily comportment and practices of tromba. Performances are thus understood to encode historical memory that may no longer be directly accessible in discursive form (cf. Connerton 1989). As Kôt' Mena also noted, when the senior spirits rise together, they arrange themselves from north to south such that Ndranihaniña has the most senior, northern position. This suggests that he is indeed Ndramandiso's father.

The fourth, and oldest, relic is that of Ndramisara, after whom the shrine is informally named (Doany Ndramisara, or Ndramisara Efadahy). Unlike the other three personages, there is some debate as to whether Ndramisara was an ancestor of the royal line. Opinion today is divided, and even Rusillon (1922–23: 171) could not get a clear answer. Although some people concur with older genealogies that place him as father (or earlier ancestor) of Ndranihaniña and thus paternal grandfather of Ndramandisoarivo, most consider him an outsider. Some go so far as to locate his origins in East Africa and, thus, tacitly of slave background.[31] This makes him neither "white" (fotsy) nor "red" (mena), and in this version no children or descendants are mentioned.[32] His lesser status is implicit in the way his tromba is treated relative to the other personages at the shrine. Ndramisara is an instance of the common symbolic principle (Turner 1969) of the "powers of the low." He differs from other tromba in that he never participates at possession ceremonies held by private individuals, rises only at the shrine, and only in male mediums.

Whatever his status and origins, Ndramisara is a critical figure in the Sakalava version of the creation of their kingdom—a version that has nothing to do with the exchange relations established with American slavers described above and everything to do with destiny, sacred potency, and the intimate relations of family rule. Ndramisara was a diviner of great power (*moas*) and consultant to the early kings, who was able to foresee and meet the conditions necessary for the establishment of Boina. It was he who advised Ndramandiso that the success and perdurance of his kingdom (and line) was predicated on sacrificing what was dearest to him. It is to Ndramisara that the organic division of labor claimed to be the basis for political cohesion is attributed (chapter 5). In both his nonroyal origins and as architect of the political order in which they play a central part, Ndramisara is patron and symbolic ancestor to the clans of Ancestor People who most closely serve royalty. He is also patron to diviners. He links the political and social order to wider cosmic processes accessible through divination.

Although oral histories (*tantara*) describe Ndramisara accompanying the conquering Ndramandisoarivo and advising him en route, what is meant is that Ndramisara spoke through spirit mediums or by means of signs accessible to diviners. Instead of representing an individual ruler with distinct personal habits, family life, and descendants, he indexes structure—principles of hierarchy, mechanisms for the circulation and reproduction of power, wealth, and life in the face of destiny, as well as the sense of a centralized yet encapsulating whole. It is he who established the shrine, and it is referred to as his. Ndramisara designed the implantation of the monarchy and ensured the line of succession that has continued to this day. He is sometimes likened to Moses, who was able to part the Red Sea with his staff and perform other miracles and who saw the Israelites through their migration to the establishment of a new kingdom. What Ndramisara represents is less the kingship or the kingdom than their very possibility. He is their guarantor and, as such, more sacred and less human than the early monarchs.

Conclusion: Division and Unity

The Sakalava polity always had fissionary tendencies. Boina, itself, was produced by a segment of the royal lineage emerging from Menabe. Merina incursions led to the establishment of a number of smaller polities across northwestern Madagascar by descendants of the original rulers. These polities each replicate, more or less, the structure of the unified kingdom, and none is seen as supreme. Each established its own line of succession and burial ground, reproducing both itself and the vision of the whole. Their common historical experience is symbolized and materialized in the ancestral cult and especially in the relics at Mahajanga.[33] The relics signify, both in themselves and in their mutual relations, speaking variously to diverse groups about their relative status and order of precedence but also affirming unity.

The relics form a source of symbolic capital for the living rulers in whose domains they happen to fall. But the power of living rulers and the immediate political differences among segments are understood as superficial and transient relative to the concerns of the more senior ancestors. As epitomized by the fact that they owe their power to Ndramisara, the diviner of lower and external origins, members of the royal clan are located within an encompassing order, not above it.

Finally, it should be evident that I have intertwined two versions of Sakalava history in this chapter: one based primarily on European sources, sometimes drawing on earlier Sakalava accounts, and one derived from contemporary Sakalava narrative and performance. I have not attempted to provide full versions of either, nor to adjudicate which is "correct." Each pursues a dialectical course using "structure" to illuminate "events" and portraying "events" to elucidate what is understood as "structure." I would emphasize (along with Lombard 1988) that although the mercantile period and especially the European role in the firearms–slave trade provided the conditions for the development and expansion of Sakalava power, the polities evolved according to their own dynamic and understood themselves according to their own logic. Similarly, the shift from mercantilism to colonialism that triggered the collapse of Sakalava auton-omy did not cause the disappearance of the politico–ritual structure (Lambek 2001).

As early as 1845, Guillain recognized that the system that had once been charac-terized by "*monarchie absolue pour le gouvernement, féodalité pour la constitution politique et territioriale*" was shifting so that "royal authority was tempered by oligarchic and de-mocratic institutions" (1845: 38, my translation). It is evident that despite the radical decline of Sakalava autonomy, "feudal" tribute, and royal absolutism consequent to successive waves of Merina, French, and Malagasy incorporation, the relics have only gained in importance. It is as if material power has etherealized into sacred potency and recondensed in the relics. If it is structure, as expressed in the relations of sacred objects, places, and spirits, that has afforded Sakalava a common and continuous his-torical practice and consciousness, so it is successive historical experience that has pro-vided structure its substance.

5

The Legacy of Lord Diviner, Ndramisara: Organic Division, "Kindedness," and Sakalava Subjects

The main shrine at Mahajanga is known as Ndramisara's (Doany Ndramisara). Ndramisara is remembered not as a conqueror or an administrator but as a kind of political philosopher who laid out the basis of the kingdom. He established a balance of powers, rights, responsibilities, and obligations that are described as matters of *anjara*. *Anjara* refers to something that is at once a responsibility and a right, a function or portion, a lot, or destiny. Ndramisara was a powerful seer (*moas*). His name means Lord Diviner and what is divined (*misara*) are *anjara* (lots, destiny). Thus, people's roles or obligations, their lot in life, and the entire division of labor so composed are simultaneously political, moral, and cosmological. Ndramisara was able to discover, recognize, or determine people's place in the larger scheme of things.

It is of interest that the political order is attributed neither directly to cosmological principles nor to the dictates of a powerful ruler but to the knowledge, wisdom, and intervention of a particular diviner (*moas*). Ndramisara's constitution, however, is not a document or rationalized blueprint to which one can turn; it is to be found not in discursive accounts but in the distribution of kinds of persons and their practices. Although it underlies Sakalava thought and practice, the scheme is no longer explicit or fully systematic (if it ever was). Thus, although it is described as a "model for" the Sakalava polity, I can only attempt to assemble pieces of a secondhand "model of" the polity.

The genius of Ndramisara, knowledgeable consultants say, was to build organic links among component groups around the cult of the royal ancestors.[1] Although Sakalava would not put it in such a blunt way, the care and reproduction of the ancestors involves three sorts of people. First are *ampanjaka,* members of the royal clan, who are responsible for becoming ancestors, and for overseeing the maintenance of their ancestors and ensuring that they receive proper respect. *Ampanjaka* represent the principle of ancestry in its fleshly, living form. Second are *razan'olo,* the "Ancestor People," who are responsible for curating the shrines, relics, bones, tombs, and artifacts of ancestors and for managing also the purity of living royals, the mortuary transitions of royalty, and the general separation of life from death. *Razan'olo* thus represent ancestry in its form as death and enduring material remains. Third are spirit mediums (*saha*), who incarnate the ancestors as *tromba* (spirits). Mediums represent ancestry in its enduring spiritual form through the embodiment and voicing of ancestors and mediate the division between life and death.

Today, all of the people described here—living rulers, *razan'olo,* and spirit mediums—can be described in a certain sense as "scapegoats," of Sakalava society—those people devoted to pleasing and reproducing the ancestors so that everyone else may get on with their lives. Lay people can take for granted that they provide ancestral care and will serve as the instruments of ancestral blessing should they need it.

Since descent classification is nonexclusive, a given individual may partake of all these statuses and roles. Some spirit mediums trace descent from both royalty and Ancestor People. It is an attractive feature of the Sakalava system that although it is an encompassing hierarchy in the abstract, ascription is multiple and recruitment frequently fluid. To elucidate this I begin with the way in which descent provides a means to locate and classify people.

"Kindedness" and Kinship

The Sakalava state was built on Ndramisara's ideology of "kindedness," applying distinctions among people found in, or brought to, the North to compose an organic division of labor in which different "kinds" of people held distinctive obligations and responsibilities. Clarifying the significance of "kindedness" is difficult for several reasons. First, kindedness has had significance in both the kinship (domestic) sphere and the political one; the idioms transect these spheres, and the boundaries between them are not always clear (cf. Evans-Pritchard 1940; Fortes 1969). Second, Sakalava appropriated categories that operated in one political system for use in quite a different one (cf. Leach 1954). Third, this latter system itself has been eroding and becoming replaced by the curious mix of individual rights and ethnic claims that characterizes the modern African state (Comaroff and Comaroff 1997; Solway 2002). Hence, kindedness that provides autonomy in one context signifies ranked embeddedness in another, and competing rights claims in yet a third. Yet these distinctions are not clearly articulated.

Sakalava distinguish one another as different "kinds" of people. The most salient criterion for being a particular "kind" (*karazaña*) of person is descent from a particular ancestor (*razaña*). Descent is thus a primary means of recruitment. Yet to speak of a person as kinded is not necessarily to imply that the kindedness is based on descent. Kindedness, itself, does not specify the criteria by means of which people are recruited or distinguished from each other. That is, there are different kinds of kindedness, and they are not fully commensurable with one another. This means, in turn, both that "kinds" are not, for the most, part nested within one another or articulated within a comprehensive system of ranking and that the designated categories of people can overlap. In some instances, the salient criterion may be political; in others, locality or religion. Descent is often implicit in kindedness, but one and the same name can designate both a descent-based unit and a political or territorial unit, much as among the Nuer. Given the nature of bilateral descent, however, kin groupings hardly approximate discrete, relative, and nested units in the way that the Nuer were said to do. And because marriage is exogamous, there are no clearly demarcated demes and, hence, no clear mapping of descent group with territory, as in highland Madagascar (Bloch 1971).

Sakalava

Descent is not the most critical feature in constituting the category of "Sakalava" itself. The clans to which Sakalava belong readily traverse Sakalava borders and are found in neighboring groups. Moreover, many Sakalava include non-Sakalava among their ascending kin. The following conversation is within the range of common responses to questions about personal kindedness.

I ask M. about her *karazaña*. "Very mixed," she says, "part Betsileo, part Tsimihety."

" Not Sakalava?" I ask.

"Yes, Sakalava, because I was born in Majunga!" In addition, one parent was "métisse French/Antalaotra."

What exactly is comprehended by "Sakalava" is complex and has undoubtedly shifted over time. Northwesterners became Sakalava with conquest by Ndramandisoarivo. It is thus probably most accurate to understand Sakalava as originally a political term and form of political subjectivity. Identification as Sakalava has immediate political connotations—Sakalava are those who are subject to Zafinimena royal ancestors and, hence, those who bear the history described in this book. Tsimihety are so named and in part constituted by the act of having once successfully refused this political subjectivity.[2] But considerable ambiguities remain.

Although the political categories are not grounded in descent, recruitment draws on descent. The offspring of Sakalava are Sakalava, and the offspring of Tsimihety are Tsimihety; the offspring of unions between Sakalava and Tsimihety are both. Over time, the political categories become somewhat clan-like and also, in the context of the colonial and postcolonial state, ethnicized and subject to ethnic stereotyping.[3] Many people in Mahajanga suggest that Tsimihety provide the majority of new spirit mediums at the present time. In every respect, beginning with the most embodied and literal, mediumship is itself a form of political subjection, and insofar as this is so, Sakalava may be said to be recruiting new subjects from the very category of people whose ancestors once refused them. Yet in becoming the mediums of Sakalava spirits, such people today do not lose such prior identifications as Tsimihety (or Betsimisaraka, Betsileo, French, etc.).[4]

Ethnic stereotypes can be revealing of those who produce them. Tsimihety have been a salient, because fundamentally similar, "Other" to Sakalava. Some Sakalava say disdainfully that Tsimihety eat lemurs and, indeed, that they will *eat anything*. They say that there are a disproportionate number of Tsimihety women among Mahajanga's prostitutes. If Tsimihety are thus depicted as *holding no taboos,* the point is that to be

Sakalava is to be constricted or constrained in certain ways. The dignity entailed in being Sakalava is grounded in adherence to an ancestral order of constraints. In contrast to Tsimihety, Sakalava sometimes describe themselves as *miavong*—reserved, standoffish, or proud. This implies a degree of conservatism.

I have also heard Sakalava describe themselves as "lazy," preferring fishing and collecting shellfish to the labor of growing rice. This plays off a prevalent symbolic contrast between people of the coasts and those who live inland (as well, no doubt, as a colonial stereotype). Together, the work ethic, reserve, and ancestral constraints have meant that Sakalava were slower than others to take up French education or capital accumulation. As a result, there are proportionately fewer Sakalava in administrative and managerial positions or among the relatively wealthy and influential modern elite and middle classes. The president of the First Republic, Philibert Tsiranana, was Tsimihety. His widow resides in Mahajanga. Insofar as patronage is both critical for upward mobility in the modern sector and tied to "kindedness," this has had long-term repercussions for Sakalava—or at least, so it has appeared to them. This perception serves to build further "reserve." On the other hand, Mme Tsiranana, herself, was rumored to have taken a "reserved" attitude to the appearance of her late husband as a Sakalava *tromba*.

Clanship and Descent

By clans, I refer to named groups whose recruitment is by means of bilateral descent. Not all such groups posit an apical ancestor. In Astuti's useful terminology (1995), members form a specific "kind" (*karazaña*) of people rather than necessarily each other's kin.[5] Clans are not bounded and never (or rarely) act as a collectivity. With the exception of the royal clan, descent is relevant only for recruitment; so far as I know, internal alignment of clan branches is irrelevant, and clans are not segmentary.[6] One knows one is a particular "kind" of person by calculating up to a particular ancestor who was identified as that *karazaña,* and one realizes it by one's actions, especially by following particular taboos (Lambek 1992). Hence, the "clan" is more appropriately thought of as a descent-based category or class of people than as a group. This is perhaps implicit in the word *karazaña*.

The relative degree of kindedness has, in part, to do with the way Sakalava conquered the North and the degree to which they assimilated people living there at the time, as well as those peoples' categories of discrimination. Prior to conquest, when inhabitants were "not yet" Sakalava, the North was divided into named political regions, each with its *ampanjaka* and its dominant local clans. Some clans, such as the Antifañahy (famous for their skill as bone-setters) were found across the North. Other clans who are well-recognized in the North today arrived with the Sakalava conquerors.[7]

Not all clans were subordinated or incorporated in the same way. As one servitor at the shrine remarked,

> Antifañahy do not have to work here. One of them once married an *ampanjaka*. They had a son. Antifañahy had a special practice (*fômba*) for carrying out circumcision which the mother wished for their son, but her husband, the *ampanjaka* over-ruled her, saying

that *ampanjaka* were higher ranked. They followed the *ampanjaka*'s practice and the boy died. "From that point," said the king, "the Antifañahy are our equals (*mitovy*)."

The story shows that royalty could not simply absorb their children with Antifañahy partners as they could offspring conceived with people from lesser clans. Quite typically, the fact is recognized through evidence provided by the disregard of a taboo, thereby validating the significance of descent-based transmission of custom.

As this story implies, clanship is normally acquired bilaterally and, hence, clans are nondiscrete; a person is a member of both their mother's and father's clans. Sometimes this is said to include eight ancestors (hence, potentially eight clans) on each side. Some kinds must drop out at each generation, and in practice most people "remember" or activate far fewer. Some people prefer or are constrained by circumstance to play on multiple clan identities; others emphasize a single one.

When asked the overly blunt question of what kind of person they are, most people in Mahajanga responded at first with a single category. There is a patrilineal bias; if other things do not intervene, children are most likely to take their father's identity as primary.[8] But this identity is never exclusive, and other factors often do intervene. As Amana Ibrahim put it, those who are interested follow a particular parent or grandparent. Thus, a well-known Antalaotra leader in Mahajanga took on this identity by following his mother's mother. At times, parents direct individual children to emphasize connections with specific ancestors, dividing their kinship "inheritance" among their offspring. This comes about through naming practices, property devolution and land use, postmarital residence, fosterage or temporary visiting patterns, religious education, interpretations of illness, and the like. Thus, as Feeley-Harnik notes, kindedness in fact often distinguishes among very close kin, emphasizing individual uniqueness (1991a: 171).

With the partial exception of the *ampanjaka,* clans have no rules or preferences for endogamy; indeed, quite the reverse. Most Sakalava demand wide exogamy. There are no prescriptive marriage rules and no *karazaña* defined with reference to positions in wife-exchanging circuits.[9] The general rule for Sakalava is that one cannot marry anyone to whom one knows one is related. This holds whether one knows a wide spectrum of kin or just a narrow one. It has the effect of maximizing the *karazaña* from which any given individual is descended. Indeed, it leads to frequent marriages with non-Sakalava. In a port city like Mahajanga, unions with non-Malagasy have been common and it is not difficult to find people whose four immediate grandparents hail from four continents.[10] Tsimihety say that is why the Sakalava are dying out. They mean by this not that there are fewer biological individuals but that Sakalava identity becomes more diluted at each generation. Of course, this is also a way of incorporating outsiders.

Kinds and Persons

It is critical to distinguish kinds of people, which are abstract categories, from actual persons. Kinds are clearly distinct from each other, but individual persons are each multiply kinded; and each way in which they are kinded may make a different sort of difference. A chief reason why kindedness is not in and of itself constraining is that everyone is simultaneously many "kinds" at once because of bilateral descent. In practice, only a few of these are salient for any given individual, and even fewer may be exercised at any given time. But in principle, they provide multiple avenues along which people choose how to live, divide up their time, orient their interests, and so on. Different elements can emerge into prominence at different points of the life cycle. As reigning monarch, Ampanjaka Moandjy avoided association with death; but prior to his accession, he served as a washer of the dead within the Muslim community.

The organic division of labor is complemented, reinforced, and indeed partially constituted by the taboos adhering to each kind of person—by the work they cannot do as much as by the work they can. Insofar as people are multiply kinded, taboos derived from one of their identities can interfere with the performance of another. That royalty cannot approach the dead is particularly embarrassing to Muslim *ampanjaka* and senior spirit mediums, who cannot attend funerals nor visit the houses of mourners. Not only are such actions of great moral significance to Muslims, but people worry that if they do not participate, there will be no one to bury them when they die.

Nothing can illustrate more clearly the contrast between these two moral-political systems. Islam is based on equality, equivalence, balanced reciprocity, and mechanical solidarity among individual adherents; every Muslim is fundamentally and fully the same kind of person, united in faith. Sakalava order is based on categorical distinctions among complementary ritual tasks, obligations, and differentially kinded identities. The point is reinforced by the story that when the coastal Antalaotra of the eighteenth- and nineteenth-century ports of trade tried to convert Sakalava to Islam, the monarchs permitted virtually anyone to do so with the exception of those "kinds of people" whose death services were important to them.

All of this suggests that being Sakalava and being a particular kind of Sakalava are largely moral choices, performatively established. However, due especially to the influence of parents and grandparents (alive and ancestral), they are not necessarily individual choices. Just as each of a person's ancestries provides options for an individual, so they may provide sites and sources of more or less compelling interpellation.

Everyone has multiple connections acquired through descent that they can call upon or that call upon them. Such identities are not merely descriptive but index various forms of practice—sometimes mutually incompatible and necessitating positive, but rarely exclusive, adherence and commitment. Choice, interpellation, polyvalence, and multiplicity extend to religious affiliation. Ampanjaka Moandjy could continue to pray at the mosque, but as *ampanjaka,* he had to take care that no one prayed in front of him. Offspring of marriages between a Muslim and a Christian or a person outside these religions who "does not pray" (*tsy mivavaka*) must come to realize which of these identities is salient for them and follow through in practice. The degrees to which people practice particular food and body taboos, dress codes, and rites of passage or suffer the effects of ignoring them (remember the story of the Antifañahy circumcision) vary. Offspring of mixed marriages might follow their Muslim fathers—a path rein-forced both by Sakalava patrilateral bias and perhaps by Islam itself. Bilaterality may enable one to avoid unwelcome obligations by emphasizing a *karazaña* that, for ex-ample, has less onerous tasks within the royal service. In the end it is a matter both of competing obligations with respect to one's *anjara*—load, share, destiny—and of what one is prepared to do or dares (*mahasaky*) to get away with.

Discrete identity may only be finalized, as Bloch (1971) and Astuti (1995) argue for other regions of Madagascar, at death and is indicated by the place and manner of the disposal of the body. But even this last move on the path "from womb to tomb" may not be unambiguous. Sakalava commoners say that they simply take the body to whichever family tomb is nearest, as long as people are buried with their ascendants rather than their spouses. Revealing the patrilateral bias, it is said that the reason many Sakalava women raise a brother's son is so that there will be someone to bury them. But there is often some conflict about the manner and locus of burial. The coffins of Muslim *ampanjaka* may be seized by traditionalists en route to the Muslim cemetery. Should their *tromba* subsequently appear, they may continue to exemplify the hybrid-ity, mediation, and indeterminacy of identity. Ndrankehindraza, a *tromba* spirit who wears a Muslim red fez yet drinks alcohol, told me that he was *kosilim* rather than *kisilam,* meaning by this wordplay that he was a "mimetic Muslim," a Muslim in out-ward appearance rather than in religious observance.[11] The multiplicity of custom and tradition serves as both the ground for serious moral commitment and as the source of playful signifiers in meta-performances of identity.

The prevalence of a birthing taboo illustrates another dimension of the Sakalava understanding of the origins and transmission of identity via kinship, although in this case the distinctions that it makes are not reified as distinct *karazaña*. People who are *ranginalo* must use cold water for postpartum care rather than the more common ap-plication of heat (*mañapataña*). In the past, they might have bathed in the sea; nowa-days in the clinic, a few drops of water are placed on the baby's head. *Ranginalo* women also abstain from "hot" foods (ginger, chili, etc.) during pregnancy. *Ranginalo* is a "dominant trait" inherited from either parent or followed by a woman who has con-ceived with a *ranginalo* man. It makes no difference whether either parent was royalty.

A *ranginalo* woman told me that they are forbidden to wear tubular wrappers (*salo-vaña*) during pregnancy but have to slit open the seam. She forgot this once and nearly died. She remarked how rapidly these taboos spread. Once a woman takes on her part-ner's condition of *ranginalo,* she has to keep it with subsequent partners, whether they are *ranginalo* or not, and she passes it on to all of her offspring. *Ranginalo* is widespread among "coastal" people but absent among the more "distant" Borzan (Merina) and Antandroy.

Most interestingly, the source of the taboo is traced to a single individual, a mer-maid (*zavavindrano, fée*), who married a human fisherman. Thus, all people holding the taboo can claim connection with a single ancestress, whether through direct de-scent, affinity, or in some cases nursing from a *ranginalo* woman. *Ranginalo* makes the point that a single ancestor somewhere way back can have a lasting influence on masses of descendants. In principle, any and every ancestor can leave a mark on their descendants, and thus descendants are marked by multiple ancestors. Hence, accounts frequently heard along the coast that people originated from "Arabia" or elsewhere need imply only that a single ancestor out of many came from such a place and married into local society, a point that is often missed by migration theorists. Neglect of the *ranginalo* taboo was said to risk paralysis or death for the mother; clearly, any one of these "ancestries" is relevant only as long and to the degree that it is associated with a salient practice or ideology, often legitimated by embodied response. But such practices need not produce bounded groups with shared interests.

Although kindedness is frequently based on descent and that is how members are usually recruited, it is realized in practice. As indicated, traits that are useful for distinguishing kinds in the abstract become problematic in practice given that individuals are multiply kinded. In fact, then, what different kinds of people can do changes over time. In Menabe apparently only the reigning monarch kept strict taboos against association with death. In Boina, there was a time when members of the *ampanjaka* category would remove themselves from a dying spouse. In sum, there is historically situated mediation as to what is "essential" and what is contingent in peoples' practice and, hence, in the substance of their kindedness.

Kindedness and Service

The Vezo studied by Astuti (1995) are adamant that they are not a kind of people. I take this to be primarily a political statement—an attempt to avoid being classified within the (southern) Sakalava royal system and subjected to certain forms of tribute. The very fact of being kinded—whatever its basis—made one susceptible to Sakalava incorporation. Vezo make their claim by asserting a clear distinction between descent and performance-based identities. However, this does not hold absolutely since in the end, tombs provide their members with a categorical identity. Just as performance-based identities like Vezo retain a descent component or residue, so does descent among Sakalava retain a performative dimension as choices of emphasis are realized through acts of acceptance and commitment to some of one's several "kinds" over others. Kindedness is performed and realized performatively. Sakalava, therefore, would see Vezo as a kind of people.

Vezo in the area studied by Astuti distinguish themselves from the Masikoro living inland. This is an instance of a more general division reminiscent of other parts of the Austronesian world between land and sea. Around Mahajanga, people associated most directly with the sea consider themselves less firmly embedded in the Sakalava domain than others. This may reflect the fact that Sakalava royalty in the past lived upstream.

Coastal people say that ferrying was their *anjara*—their lot or share; *anjara* can refer to what one owes or contributes to royal service. Thus, the *anjara* of the state sugar plantation has been to repair the road before the annual service at Bezavodoany. Any group that takes on a specific *anjara* receives recognition as a result, and its identity is reaffirmed each time it fulfills the obligation. As people say, "Each *karazaña* has its *anjara*."

Not all forms of kindedness are equally salient, encompassing, or realized in practice; and no two people may be kinded to the same degree. That is, some people and some kinds of people find their ascribed attributes more clearly specified and more compelling than do others. In Astuti's terms (1995), there are differences in degree of determinacy versus indeterminacy, differentiation versus undifferentiation, and fixity versus fluidity. The greater a person's kindedness, in a sense, the deeper their embeddedness in the monarchy and the stronger and less negotiable their political obligation. One can turn this around and say the deeper a person's political obligation or commitment is, the greater their specific kindedness. Degree of kindedness and interpellation are products both of external, material and social factors (stemming both from royal power and traits attributed to specific ancestries) and internal, subjective ones.

In sum, although Sakalava have long been subject to various forms of categorization, the different kinds of distinctions that they produced were not fully rationalized: Distinguishing people in terms of where they are from, where they live, or how they live is incommensurable with distinguishing them by descent or by what their ritual function is. These are neither exclusive nor contradictory modes of classification. Rather, in Kuhn's terms (1970), these discriminations fail "to make *complete* contact." It is a mistake to look for a rationalization or logic that is not there; instead, one needs to be attuned to the overlay of diverse historical models and cultural idioms that do not succeed or replace each other so much as coexist. Their mutual interference thickens the social context in which people live; and it provides ambiguity, escape, prestige, play, and the avoidance of overdetermination or excessive subjection. As Foucault would have noted, the ambiguity and polyvalence generated through under-rationalized Sakalava discursive practices is not the subjectification characteristic of the apparatuses of modernity, state bureaucracies, and expert discourses nor that of the labor market. The state did not invent ethnicity nor colonialism invent the tribe so much as they attempted to systematize and rationalize subject populations and to fix people permanently and exclusively into existing categories.

Bilateral descent and exogamy meant that individuals could often navigate among their kinded identities. It was not directly obvious which "kind" would be primary, and secondary identities were available to modulate primary ones. But if some people could spend their lives articulating multiple dimensions of their identities or avoid unpleasant or unremunerative duties associated with a particular form of kindedness, this does not explain why others found themselves committed to specific subject positions.[12]

Why do some people appear more specifically kinded and more constrained by their kindedness than others? Sometimes parents select particular offspring to follow them in carrying out specific roles. Sometimes a *tromba* unexpectedly singles someone out. In part, the degree of subjection depends on the task to which persons or communities were apportioned by the Sakalava state, and this was sometimes accomplished by wrenching people from ancestral ties. An annual levy of honey or a ferry ride on demand are different in their consequences from the obligation to guard tombs. As some tasks become consuming, so do the identities associated with them. A positive feedback loop is produced such that some kinds of lives come to seem increasingly overdetermined.

In effect, hierarchy itself may be understood with respect to freedom of movement. It is the people at the top and bottom of the system—the ruling *ampanjaka* and the "Ancestor People" (*razan'olo*) most closely associated with the intimate service of living and deceased royalty—who are, or remain, most determined by their kindedness. The extremes begin to resemble each other: each specifically kinded; each descent-based yet emphasizing discrete kinds of work within the overall division of labor; each more focused on the ancestors than the remaining population. Even members of the royal clan view their status as a mixed blessing. I turn now to a closer look at these "specific" kinds of people.

RULERS AND ROYALTY

Ampanjaka and *razan'olo* refer not to discrete descent groups but represent functions in the social totality. Yet in effect, descent is a critical and necessary criterion by which one accedes to either position. *Ampanjaka*—royalty, rulers—are members of a set of related clans. The clan names—Zafinimena, Zafinifotsy—are rarely used in Mahajanga; to call people *ampanjaka* is to imply that they are Zafinimena. All and only members of royal clans are acknowledged to be *ampanjaka;* however, only particularly active persons or those closely related to the present or former reigning monarchs are likely to be so addressed or referred to specifically. The royal genealogy includes every reigning monarch, their legitimate children, and collateral lines. Members of peripheral branches remember additional persons. There are also accessory clans composed of descendants of illegitimate offspring. "Illegitimate" refers to children born out of wedlock who were not taken to royal ancestors to have their status officially acknowledged.

Because descent is reckoned through both men and women, because many royal unions are not clan endogamous, and because royal ancestry takes precedence, the number of *ampanjaka* is large. When a commoner marries an *ampanjaka,* the ancestors must be informed and asked to accept their new "descendants;" it is forbidden (*fady*) to bury any legitimate offspring of a mixed union in a commoner tomb. "*Dangereux!*" exclaimed one person. However, it is not the primary referent for everyone who could assert royal status. Many people do not readily admit to being *ampanjaka;* it is considered arrogant or pretentious "rattling" (chapter 2). And, despite the prestige ostensibly associated with royalty, it brings with it many unwelcome taboos.

Some people in Mahajanga think every group in Madagascar had its own *ampanjaka*. The category does seem widespread across the island, but the functions and power associated with it vary widely. The most important thing about contemporary Sakalava *ampanjaka* is as one person put it, "Whether ruling or not, royals are different from other people because they produce sacred ancestors (*raza masing*)." They are respected for their status as descendants of revered ancestors and for their prospect of becoming ancestors. These past and future-oriented dimensions of identity overshadow present attributes. Although reigning monarchs are prominent figures, many *ampanjaka* living in Mahajanga are quite poor and their lifestyle is virtually indistinguishable from that of their neighbors.

Thus, although *ampanjaka* is often translated as royalty, as I have done here, this specifically political function is overshadowed by their place in a larger whole—a social cosmology or hierarchy that encompasses and mediates between life and death. This is evident in the fact that every member of the royal clan is referred to as *ampanjaka,* although only one among their number (within any given political domain) has the political function of ruler. The ruler is referred to as *ampanjaka be,* the great or senior *ampanjaka* (or sometimes as *ampanjaka manjaka* or *amp. manoro,* the ruling/reigning monarch).[13] As noted in chapter 1, the root *jaka* means "burden."

The *ampanjaka be* today is responsible for maintaining ancestral devotion. Monarchs officiate over ancestral sacrifice and receive it. Ancestors address living monarchs through their mediums to express support or camaraderie, offer advice, or register disapproval. Monarchs oversee ancestral tombs, represent the interests of the monarchy vis-à-vis the Malagasy state, and, in the Mahajanga area, ensure the management of the shrine and the success of the annual Great Service. These tasks are carried out in consultation and with the assistance of key officials, but it is *Ampanjaka* Désy who sends out invitations, now written by computer, to the annual Service.[14] During the festival he serves as primary witness to the treatment of his ancestors. Whether he has responsibility for dividing up the proceeds of the event and how much he keeps to cover expenses are matters of debate, but he should see that some of the money is distributed among the people who live and work at the shrine. Direct power over subjects has been reduced but is still evident (chapter 10).

In his discussion of the Shilluk of southern Sudan Evans-Pritchard remarked, "Kingship everywhere and at all times has been in some degree a sacred office. *Rex est mixta persona cum sacerdote.* This is because a king symbolizes a whole society and must not be identified with any part of it. He must be in the society and yet stand outside it and this is only possible if his office is raised to a mystical plane. It is the kingship and not the king who is divine" (1962: 210). This is certainly true in Mahajanga. Contemporary rulers are neither omnipotent nor divine. Reigning monarchs mediate between their ancestors and the people, but they are not the only ones to play such a role. They will become ancestors themselves, but so will other members of the royal clan. During the period of Sakalava autonomy, the ruling monarch had a good deal of earthly power and served as a leader in war and trade negotiations. Today, rulers are treated with respect and formality but can be readily avoided. Rulers have economic privileges, but their rights over cattle, money, and labor are, in principle, custodial; ideally, resources are used for serving ancestors (for example, for renovating their tombs) rather than for the personal use or investment of the living.[15] Rulers should offer hospitality to visitors, and they are also entitled to consume part of what they receive.

Of the Shilluk, Evans-Pritchard said, "The king . . . reigns but does not govern" (1962: 200). Sakalava monarchs did govern in the past. Today, they serve as chief figures in decision-making concerning those aspects of the cult that fall within their domain. Rulers are leaders of the interests of their factions and domains. They serve as lobbyists rather than holding high official positions within the state government. They may hold minor positions within the bureaucracy (following a practice of the French colonial regime) that provide small stipends; but there is nothing equivalent to the situation in Botswana where chiefs oversee customary law and sit in a House of Chiefs in the capital, which forms a kind of upper house of parliament (Solway in press). The Malagasy state treats rulers with the respect due people who can influence local public opinion and deliver votes but does not defer to them. Local rulers are well aware where the balance of power lies.

So are their subjects. The choice of royal successor is ideally determined by the collectivity (*fokon'olon*), including royal kin, advisors, and Ancestor People. At the time of Ampanjaka Désy's selection in 1970, it was considered more important that the candidate be literate and have some prior experience in the civil service than that he or she be well versed or interested in ancestral knowledge. The kingship often co-opts people who are not especially seeking it, and the community persuades them to redirect their energies and means towards its ends.

Although the *ampanjaka be* may expect obedience and respect, all *ampanjaka* are subjected to the larger ancestral order. Their particular social identities are products of the kinds of discursive, disciplinary, and ethical practices that create all Sakalava subjects. In some respects, *ampanjaka* are more constrained than ordinary people; it is for this reason that many Sakalava decline to emphasize their royal ancestry and might regret selection as *ampanjaka be*. For example, one middle-aged and part-European man who owned a successful video parlor refused the offer to reign in an inland town, explaining, "As an *ampanjaka* you depend on others. You can't work or earn your own money, but have to stay put."

Ampanjaka do not automatically inherit a material estate; they are not landed in the sense of the European aristocracy. Only those closest to the kingship partake in the privileges and rewards of the office, whereas even those who do not bear authority are subject to respect and pollution rules. Gradations in rank and distinctive sumptuary rules have been reduced. Only *ampanjaka* have the right to wear new clothing that is red. Royal women pierce the right nostril for a jeweled ornament (*zoping*); all others pierce the left.

One person said that where he came from people were reluctant to marry royalty because of the trouble entailed by the restrictions. To be able to marry freely, *ampanjaka* sometimes even tried to conceal this aspect of their identity. (In contrast, it was said to be unseemly that so many people laid claim to royal status in Mahajanga.) When an *ampanjaka* marries a commoner, their offspring (who are *ampanjaka* as well) cannot nurse or bathe the aging or ailing commoner parent. A man who marries an *ampanjaka* cannot ask her to wash his underwear or socks or wash his feet with water that she has carried on her head. A woman who marries an *ampanjaka* cannot climb over him to get in or out of bed. A commoner spouse has to observe mourning for a month; conversely an *ampanjaka* should leave immediately upon the death of a commoner spouse, if not before.

There is a story about a thirsty traveler who asked a woman for water and was directed to her neighbor. As an *ampanjaka,* she could not serve him; royalty do not work for others. This illustrates how the "privileges" of royalty are often a nuisance and also how identities are sometimes revealed through trivial circumstances. When *ampanjaka* share an urban courtyard with other families they cannot take their turn cleaning the toilet and have to hire someone to do it, as one person said, "even though they shit like everyone else!" And women hired to clean for them have to provide replacements during their menses.

If royals are distinguished from commoners and others by the fact that only they have history—in the sense that it is only their ancestors who are known and commemorated (Feeley-Harnik 1978)[16]—they are also distinguished by being charged with being fully subjected *to* history—becoming the subjects and objects of history. In them, the subject/object distinction is most fully transcended; the history that they "have" is who they "are."

However, this becomes everyone's history. If royals are virtually the only people who become public "ancestors," they do this on behalf of everyone. The critical point is not only that the possession of such history distinguishes Sakalava royalty from commoners but that the acceptance of royals *as* history is what has distinguished Sakalava from their neighbors. Sakalava are the people who acknowledge the historical burden of their *ampanjaka* and who help them carry it. If the royals become history itself, their subjects are literally subject to history. But these subjects are also, in a variety of ways, its enablers, producers, performers, and appreciative audience (cf. Walsh 2001).

Monarchs' activities are constrained by the fact that, as discussed in chapter 2, their power should emanate from them without advertisement. Service (*fanompoa*) should find its way to them without them having to ask for it. Subjects should approach rulers; not rulers their subjects. For all of their prestige, and part and parcel of it, rulers are constrained in their movements. For example, Ampanjaka Soazara of Analalava cannot enter Mahajanga.[17] From this perspective, ruling monarchs are more "pawns" of the system of invisible potency than they are powerful agents in their own right. Potency is achieved by sacrificing the living monarch's personal autonomy on behalf of the office, as Evans-Pritchard implies. Thus, the monarch too "bears" the kingdom.

The potency that attracts without advertising is most fully realized at death when abstract agency entirely overcomes individualized self-representation. Deathly power may be a kind of paradox to Westerners but is a basic assumption of Malagasy thought. The death in question is distinguished from the immediate contingency of ordinary death. Indeed, despite their role in the upkeep of the tombs, living rulers should be kept as far from death as possible and must be purified if they come into contact with it. Members of the royal clan are said not to die but to "collapse" or "lean precipitously" (*mihilaña*).

Indeed, there is a special vocabulary associated with royalty and royal bodies. One reason for it is that words associated with royal death become taboo. A collapsed royal is given a new name (that is, that of the *tromba*), and the old one is forbidden. If the former name contained morphemes with other referents, these usages too will be replaced by synonyms. This is one of the ways in which the contingencies of historical events are incorporated into the structure. However, this process is characterized by the opacity of the "maze" insofar as the event or person is inscribed in the form of an absence—a dropped word—and a substitution—the new one.

Despite attention to their bodies and the importance of descent in establishing their status, being *ampanjaka* is not a biologically essentialized condition. Virtually all reigning monarchs are associated with rumors concerning their parentage, and this does not appear to undermine their ability to act as royalty. Whereas the blood of children comes from the birth parents, their manners or customs (*fômba*) depend on the context in which they are raised. Indeed people sometimes distinguish between a person's "kind" (*karazaña*) by descent and "kind" by the practices or habits they follow.

ANCESTOR PEOPLE

Razan'olo means literally "Ancestor People." One connotation is "people with ancestors," that is, commoners, not slaves. But unlike most commoners, *razan'olo* are Ancestor People in a second sense, namely, that they work for, on behalf of, and in ways determined by, royal ancestors. They are the people at once most fully integrated into ancestral service and most specifically "kinded" insofar as their social kindedness determines the service they perform. "*Razan'olo*" might thus also be a shortened form of *karazañ'olo*, "kinded people." Those commoners who remain at arm's length from royal work are not specifically referred to and do not refer to themselves as *razan'olo*, though in the first, more general, sense they are. "Ancestor People" is a category not a group. Active members are drawn from a series of named clans (*karazaña*), but unlike royalty, these clans do not sustain elaborate genealogies.

Although everyone should perform service on behalf of the ancestors and participate in "bearing the load" of Sakalava history, *razan'olo* are specifically defined with

Servants entering through the main gate of the sacred enclosure, Doany Ndramisara, Mahajanga.

A group of shrine servants, Doany Ndramisara, Mahajanga. Looking northwest across the plaza, the house of Ampanjaka Désy is visible and beyond it, that of Mbabilahy.

reference to their portion of labor. The obligations of Ancestor People have been at once among the most specific and the most intimate. Insofar as they demand relatively continuous presence, they also undermine autonomy. Notable among the tasks of *razan'olo* are managing shrines, disposing of noble corpses, guarding tombs, shrines and material remains of royalty, purifying living royalty, serving as messengers, and escorting supplicants into the shrine and intoning prayers on their behalf. The most intimate servants have a joking rather than a reserved relationship with royalty. As one person said, "They would not be afraid of sleeping on the same bed."

Some clans of *razan'olo* are closely associated with death. It is they who transform and preside over the precious physical remains of the ancestors—the ultimate source of sacred potency (*hasing*). They are the midwives in the separation of fallen rulers into *tromba*, relics, and buried remains. They guard access to the material remains very strictly, forbidding even spirit mediums entry into the inner precincts of the tomb enclosures and fining those who violate any of a series of taboos associated with places of death. They test and authenticate spirits risen in particular mediums. They remove the pollution of death from the living while embracing or absorbing it themselves.

The *razan'olo* might be described as a kind of priestly estate. In effect, they are the gatekeepers, guides, and guardians of the maze and the custodians of the ancestors. At the same time, they are lodged within the chronotopic maze they maintain. With the exception of the most senior managers, they are generally able to see only their own stretch of it and, by the same token, are unable to provide an overall picture.

In the past, most *razan'olo* would have been dependent on rulers for access to foreign material goods and limited as to what they could use. Segments of the *razan'olo* remain the most conservative element of the population with respect to their sumptuary practices. Insofar as they fulfill their duties by living at sacred places, their material lives must remain rooted in past practices. Thus, not only do shrine dwellers live from service rather than wage labor, they reside in houses constructed of traditional materials and forego such domestic amenities as latrines and electricity.[18] This is their function with respect to the master chronotope. However, some perform wage labor concurrently and maintain more modern homes outside the shrine precincts.

In the mid-1990s there were some ten guardians (*mpiambing*), evenly split between men and women, who took turns sleeping inside the temple next to the relics at the Doany Ndramisara. Each was responsible for about four nights a fortnight. Along with other shrine servants who only performed day work, they also swept the grounds and were available for various duties whenever the shrine was open. Custodians were not well reimbursed for their labor. They could build a house in the shrine village and live rent free and they could cultivate a few vegetables on the shrine grounds, pick the mangos and sell mango pickle, but all of this was not really enough for them to subsist. Payment was somewhat irregular and cast in a gift mode. Guardians could expect to receive a lump sum after the Great Service; this ranged from fmg10,000 in 1993 to fmg50,000 in 1998. They also received a monthly stipend, which, in 1998, was fmg5,000. Those who did only day work received fmg2,000 per month in the same year.[19] Finally, they could expect to receive a small portion of the money taken in by the *lambantañana*, the "palm of the hand," that is, the man who escorted supplicants and recited on their behalf.

Ancestor People showed deference to royalty. Hierarchy was evident in spatial arrangements. In the shrine village, houses of Ancestor People were to the west and south of those of royalty. Similarly, they placed themselves along the south and west sides of a room. They carried out tasks deemed polluting to royalty.

Mead Cooking

Ancestor People are the central figures in the Service that takes place at the Doany Ndramisara in June, a month before the Great Service, to prepare the mead (*tôa mainty*) used to bathe the relics. This is honey cooked with ground wood from the *katrafay* plant.[20] Mead is prepared by Ancestor People within the sacred enclosure (*valamena*) and then formally offered by them to the ancestors inside the temple. The event I observed on Monday, June 8, 1998, was witnessed by mediums and others and greeted with joy by several *tromba*. Eight pots of mead were cooked, representing eight clans/kinds (*karazaña*) of *razan'olo*. According to one of the workers, four pots were destined for the four ancestors (*efadahy*), two for the Ancestor People, and two for people at large (*bevahoaka*, Bemihisatra); of the latter, one each for the mediums and the general public. The mead, which ferments to the strength of beer, is reserved for drinking and anointing at the Great Service as a form of blessing. While the mead was cooking, about one-quarter of the raw honey was passed around and enjoyed by the celebrants.

That this is the Service (*Fanompoa*) of the Ancestor People, as some referred to it, is to say that this is the occasion when their "free" offering of service to the ancestors is rendered explicit. As one shrine servant explained, "If a celebration were being held on my behalf, my servants would want to make their personal contribution as well." But this is also an occasion at which the contribution of their servants is acknowledged by the ancestors. This acknowledgment comes primarily in the gift of the meat of a slaughtered cow.

The beast, which was sacrificed at the shrine early in the morning, was understood to belong to the *razan'olo*, supplied them by the mediums of the four shrine ancestors. Earlier, a letter had gone out from the shrine manager to all relevant mediums requesting financial contributions toward purchase of the beast.[21] The animal was referred to as *paré*, a contribution to royal service, and as the lot or due (*anjara*) of the mediums; but unlike other contributions, its recipients were not the ancestors, but specifically the Ancestor People (though in fact they just received the hind legs).

Mediums of more junior spirits observed the cooking and tasted the mead to ensure that the proportions of sweet honey and bitter *katrafay* were correct. Once the mead was presented to the ancestors, the Ancestor People asked for gifts of money. Senior officials and *tromba* (and anthropologist) collectively gave fmg41,000 (carefully divided among twenty-four persons) plus two bottles of liquor.

One of the honey cooks explained very clearly that Ndramisara brought eight *razan'olo* with him from Menabe and that each kind must be represented among them to commemorate the fact.[22] Each clan provided two women and one man for the work. They were assigned to the pots of particular ancestors, which the women fanned—a sign of respect to the recipient.

In sum, the event recognizes the attachment of the Ancestor People to Ndramisara and, hence, indirectly to royalty, and articulates the relations of dependency that link the various constituents as ones of mutual responsibility and respect.

Varieties of Ancestor People

There may have been a time when each of eight *karazaña* had its specific place in Ndramisara's model of ancestral service. Only vestiges of such an order appear today; information is rather fragmentary, at times contradictory, and in part secret. I do not pretend to know it all and will only mention certain points.

Consultants affirm that Ancestor People do not include slaves (*andevo*). Yet *Sambarivo*, whom Feeley-Harnik (1991a) explicitly contrasts to *razan'olo* as slaves, unkinded, and separated from kin, were identified alongside Antankoala and the rest as but another *karazaña*. Following the abolition of slavery, recruitment to Sambarivo status has occurred by means of descent, thereby transforming the category into a descent-based *karazaña*; by referring to themselves as Ancestor People, Sambarivo confirm their place in the system while dropping the attribution of slavery.[23]

Ancestor People do differ among themselves in prestige; *karazaña* who handle most directly the most polluting but also the most sacred royal remains, are of lower or more liminal status. Thus, Jingô and Sambarivo (with some notable exceptions) act more deferential, stay farther to the west and south, and hold more menial jobs at the shrine than do Tsiarana and Vôrong Mahery.

Most powerful and prestigious among Ancestor People are those who become royal advisors and, in Mahajanga, who manage the shrine. Asindraza Doara (Edouard), the *fahatelo* (third) maintained virtually complete control over the shrine for a number of years. Working closely with Ampanjaka Désy, Doara managed the Great Service and publicly acknowledged the gifts and money brought in. He also presided over the "testing" (*fitsaraña*) of new mediums at the end of each festival.

Doara's *karazaña* is Vôrong Mahery (Birds of Prey). "We were Ndramisara's warriors," he explained to me. Others suggested that Vôrong Mahery were recruited as royal messengers, or *service postale*. Doara asserted that he, not Ampanjaka Désy, was accountable for the money brought in during the Great Service. The money, he said, was given to maintain the shrine, not to support the ruler, and it was he, Doara who decided how to dispense it (including, presumably, how much to give Désy).

It is illustrative of the way that "kindedness" works that one of Doara's full brothers, also resident at the shrine, accepted as his primary identity the *karazaña* of the opposite parent to that of Doara. His clan is Tsiarana, and as such he played a leading role in ritual associated with the founding queen of Boina, Ndramandikavavy, whose descendants via a first, non-royal union the Tsiarana are.[24] Tsiarana have rights and obligations regarding her commemoration, which today includes an annual bovine sacrifice before the Great Service can begin (chapter 8). This man also bore the saber (*viarara*) belonging to Ndramandikavavy during the annual circumambulation of the relics. Used in ancestral sacrifice, it must be wielded by a Tsiarana. There have been many unions between Tsiarana and *ampanjaka* who, after all, are descendants of Ndramandikavavy as well. Unlike those of other royal unions, offspring are able to keep the identity of the Tsiarana parent.

The specific duties (*anjara*) assigned to Ancestor People can also be considered privileged access to royalty and royal ancestors. Certain clans are described as "ready" or daring (*mahasaky*) to perform tasks forbidden to others. Antankoala have the right to stand above royalty and so are engaged in erecting the roof of the temple. Zafindramahavita (Descendants of those who finish things off) were said to hold the right to kill a seriously ailing monarch.

A "collapsed" monarch was apparently treated by members of four distinct *karazaña*, each of which had responsibility for a certain portion of the royal body and stood at a particular cardinal direction while doing the mortuary work. A good deal of this specificity has lost its saliency in Mahajanga though it may remain relevant at mortuary villages.[25] But the ideological principle of organic solidarity remains. It is found in the comments of a spirit medium critical of the way wealth and decision-making appeared to have become centralized under Doara: "The work of Ndramisara is exactly opposed to this. That is precisely why there are so many *karazaña*. Every kind of work has someone responsible for it. It is not good for one person to take responsibility for everything." This model of Ndramisara is what is displayed and enacted at the cooking of the mead.

Irrespective of the former practice of slavery, today ancestral work at the shrine is allocated according to the principle of kindedness. One point of similarity between Ancestor People and slaves is that neither could quit (*niala*) their work.

The Jingô and Subjection

There is a saying: There cannot be royalty without Jingô.[26] Jingô is the most salient category of Ancestor People, and one with which more people living at the shrine identify than any other. A Jingô woman active in royal service asserted that Ndramisara appointed the Jingô, saying, "You people will see to all my descendants (*mañajary zafiko jiaby*). You will always be there for us."

I first met someone referred to as Jingô during a visit to the Antankaraña royal capital in 1992. Folo stood out from those around him. Gaunt and silent, he carried himself with an austere dignity and reserve. His house was at some distance from the village, and he had no wife or children. Folo acknowledged his difference. He was from Analalava, more than 250 km away, one of six children of a couple who were Jingô and Sakalava Mañoroñomby. He succeeded to his father's position (and his father's father before him) as a *mpanasa,* a purifier of living royals. When he dies one of his siblings [or children] will take over. He explained that a *tromba* had sent him north. The *tromba* Ndramarofaly looked after him and was his patron (*olo be*) in the Analalava region. But it was Dady ny Kôto (an Antankaraña *tromba*) who sent for him because they had no one to do the job. Once he arrived in the North he settled in a fishing village where he was discovered by the ruling monarch and invited to live in the capital and perform "custom" (*fômba*). The king offered to supply his needs, and his successor has continued to provide Folo with gifts of clothing and money. Folo also worked as a carpenter.

Folo played an important role. He accompanied the monarch on his ritual pilgrimage to ancestral graves and prepared the sacred mead that is deposited there (Walsh 1998). Antankaraña monarchs are Zafinifotsy, and Folo explained that Jingô Malandy (White Jingo, a term I heard only here) had a special relationship (*lohateny*) with them of mutual pardon and support.[27] It was his job to purify the king whenever he committed an error or suffered an insult. Folo could remove the effects of anger directed at the king, whether it came from the ancestors or the public. His abilities had been successfully tested when the reigning monarch, Ampanjaka Isa, had been insulted by a few of his constituents who opposed his position on the politics of national reform. Folo gave the king an elaborate bath using thirty-two pots of honey (in multiples of eight), and the king's dignity, well-being, and, in effect, his sanctity were restored.

Folo felt the presence of ancestors (*razaña*) strongly. He would not speak about them without checking with them first and making an offering of rum and money. His Jingô ancestors demanded that he live apart; he often spoke with them at night, and the noise would have disturbed neighbors. He said forcefully, "A person who doesn't take on the work requested of him by the ancestors will be abandoned by them and have no work or luck. However, many people have themselves abandoned the ancestors because their path is too hard; if you don't do the work properly you go crazy because the ancestors are angry with you."

Folo also spoke about the custom of sacrificing a Jingô on the death of an *ampanjaka:* "Incredible things happened at a royal death; a series of obstacles emerged to prevent the burial of the monarch. A Jingô was brought in to say a prayer (*jôro*), then the Jingô was slaughtered and his blood used to mark the monarch's grave. Once the Jingô was buried, the king could be buried easily. Any Jingô could be chosen; however, it was taboo to kill one whose tears were so plentiful they reached the ground." Folo acknowledged that the custom had been abandoned but asserted, "Every time a monarch dies, so does a Jingô; indeed, now that Jingô are no longer deliberately slaughtered, it is not uncommon for two or three to die. When they carried out the ritual (*fômba*) properly, only one died."

Jingô as Victims

Although Folo's case is undoubtedly unique for reasons of personality, isolation, and the particular nature of Antankaraña royal ritual, many of the themes he raised resonate with what follows. In Mahajanga, Jingô purify living *ampanjaka* and their houses that have been polluted by death. Jingô cannot visit *ampanjaka* who are sick; they only show up once someone has died. Although some Jingô attend the needs of living monarchs and others serve as guardians at the shrine, still others conduct the royal mortuary rites and process the flesh.[28]

Jingô are the clan most closely tied to royalty and most closely identified with Ndramisara. Both Ndramisara and Jingô are sometimes referred to as Makoa (of African origin). Jingô were said to share the most intimate secrets of royalty. However, as one person exclaimed, "Jingô are now free, *fa civilisation!*"

Ndramisara designated the Jingô as sacrificial victims for royal transcendence. A primary obligation, exclusive to Jingô and no longer enforced, was to supply a human sacrifice on the demise of each reigning monarch. These were referred to as *lafiky antany,* "ground sheets," as though the Jingô lay between the monarch and the ground, perhaps in the tomb itself. They may have been killed at the *varavara mena lio,* the red blood gate, also known as *varavara faly,* the taboo/sacred gate. A conjunction of portal and mortal, the site is a critical configuration within the chronotope and one that would not have surprised van Gennep (1960), who understood that transitions in time are marked by crossings in space and that transition is often symbolized by cutting or death.

Here the death was not simply symbolic. Human sacrifice is said to have been a privilege of royals, distinguishing them from ordinary people who had to make do with cattle. But the significance was more profound. Essentially, the Jingô deaths displaced those of monarchs (who cannot be said explicitly to "die"), leaving the latter to become potent ancestral *tromba* and relics. Just as the bodies of "collapsed" *ampanjaka* were treated to draw off the fluids and rotting soft parts from the enduring bones, so death itself was divided into the immediate and terminal, represented by the body and blood of Jingô, and the transcendental and enduring, represented by the bones of the royal body. However, the transformation of *ampanjaka* into ancestors was not mechanically dependent on human sacrifice; many *tromba* have emerged since human sacrifice has stopped.

Jingô selected as victims appear to have been strong and healthy. Some people said that they were of the same sex as the monarch, implying a kind of younger "double;" others, that they were attractive people of the opposite sex, implying a sexual partner. More abstractly, the sacrifice can be seen as an instance of the conquest of youthful vitality in the constitution of ancestral transcendence (Bloch 1986).

Human sacrifice evokes and may have been designed to emulate the voluntary sacrifice of the first queen that originated the monarchy of Boina. But the Queen's descendants are Tsiarana, not Jingô, and they have inherited the right to perform cattle sacrifice and to curate the weapons for it, not to be its objects. The voluntary quality of Jingô sacrifice is affirmed in some accounts; its coerciveness, in others. Voluntariness is a theme that remains with the cattle sacrifice. A beast that cries or bellows is released, unaccepted as a victim.[29] Indeed, Jingô and other shrine guardians have the right to such cattle that do not accept their own deaths (again, an absorption on their part of pollution or "waste").

A Jingô consultant insisted that Jingô were slaves long ago and that their sacrifices were not voluntary. She asserted that in the past Jingô were slaughtered for every major ceremony on behalf of Ndramisara, that is, at the annual Great Service as well as at royal funerals.[30] She explained further, "When a monarch died or as the Great Service approached, the Jingô used to run away." For a week the victim was well fed and dressed and adorned with gold jewellery. The sacrifice took place on the Monday of the Great Service—the same day that the relics were bathed. A Jingô man added that in the past, several Jingô might have been killed on the death of the monarch. He insisted that the victims were forced and went so far as to liken it to the actions of Hitler.

The termination of human sacrifice was explained by at least one Jingô consultant with respect not to colonial injunction but, characteristically, to a story that emphasizes the relations between *karazaña:* "One day an Antavarabe man killed a cow and hung the hindquarters in his courtyard. People were astonished; the ruler threatened to kill the man and demanded an explanation. 'Why are you trying to act superior to us?'

The Antavarabe replied, 'You kill something you can't eat or use. At least when I kill a cow I use the meat and skin. Which of us is in the wrong?'

The monarch perceived the lesson and stopped killing Jingô, replacing them with cattle. The Jingô were delighted and since then Jingô and Antavarabe have been joking partners (*ziva*); when an Antavarabe dies, Jingô provide the cow to be killed at the funeral."

Interestingly, some of the sacrificial victims seem to reappear as *tromba*—minor figures who possess the same mediums who manifest the senior *ampanjaka* with whom they died. They are referred to as *marovavy,* a term that designates female attendants of the ruler. Their role as sacrificial victims is not openly stated. These spirits rarely appear in public. They have short individual names and are portrayed as children or childlike but with innuendoes of sexuality.[31] The one I met (risen in a male medium at his home) stayed shyly behind her veil and enjoyed collecting and playing with tiny, virtually worthless coins. She was not happy to gossip about her masters, as I had been led to believe, but pleaded ignorance of royal affairs.

The *marovavy*, named Soa, described herself as "a creature of the royals (*biby ny ampanjaka*)." "*Biby*" refers in everyday language to animals, especially insects; here it indicates an unmarried consort of royalty. As she went on to say, "There is no mother-in-law, no daughter-in-law." She added that her role was to serve (*manompo*), emphasizing less the sexual than the menial labor. The first task she mentioned was clapping hands, a form of musical accompaniment. She said, "We cooked and laundered; in those days there were no irons so clothes had to be folded very carefully and pressed under cushions." She said she had not yet given birth and was still young at age twenty-three and identified her father and his village. Tired of all my questions, she began to query me ("Can't I ask questions too?") and observed with a neat combination of tartness, naiveté, and accuracy that the marks I was making in my notebook "look like fly shit."

Insofar as the sacrificial victims may have been the monarch's concubines, it is interesting that a woman of Jingô background told me in another context that Jingô are made ill from sexual relations with royalty. A Jingô person who married royalty would even be at risk of death. This is an instance of "back talk" or "insubordinate discourse" (Aucoin 2000), but one that may also reinforce the sense of vulnerability and subordination to the ancestral order.

Human sacrifice illustrates the extent of stratification among "kinds" of people as well as the degree of coercion that must have existed. The *marovavy*, Soa said that had she refused to serve she would have had her throat cut. But there is more to it. Today, one of the full-time Jingô attendants at Ndramisara's shrine told me that on the death of royalty they still get very nervous wondering which of them will be taken, even though death now occurs by means of ostensibly natural causes. Another said that if any Jingô fall sick at the Great Service they become very frightened. He related that indeed every year one of them does die; and that the previous year an Antankoala resident at the shrine died just before the festival, "taking the Jingô's place."

In other words, Jingô attribute a death by illness or accident that follows the death of an *ampanjaka* or occurs around the time of the Great Service to be an act of the ancestors. It is as though they cannot escape the burden of sacrifice for royalty. Their internalization of these fears exemplifies how even today, and despite the choices that bilateral descent ostensibly provides, some people find it virtually impossible to escape the kindedness with which they are attributed and the subjection it entails. The blame is not put on direct coercion from the living. As a Jingô attendant said, "Ancestor People cannot escape/abandon Ndramisara (*tsy mahazo miala Ndramisara*)."

Albertine

Where does coercion come from, and why is it effective? Jingô with whom I have spoken attribute their burden to their *own* ancestors. Albertine, a delicate middle-aged woman, explained how she received a Catholic education and was even confirmed: "But my Jingô ancestors would not accept this. They forced me to abandon Christianity and go to work at the shrine. I would die if I didn't." When I pressed the point, asking whether the managers coerced her, she repeated, "It was my own ancestors who obliged me to do so." Her companion, a Sambarivo, said the same. "It is like an inheritance (*kara lova*)," she explained. Like Folo, they saw their grandparents threatening them in their sleep if they neglected the work.

Albertine arrived at the shrine as a young girl when her parents were sent there by a *tromba* (again, like Folo) to bring something for the Great Service [thus the Great Service served as an occasion for recruiting personnel as well as acquiring goods]. Both Albertine's parents were Jingô, though her husband was not. Her children are Jingô, and one of her daughters will eventually take her place. Albertine succeeded her father at the shrine; he selected her and reported the decision. The fact that workers are replaced one by one and their number remains fairly steady means that parentage is destiny for only one member of each sibling set.

During the mid-1990s, Albertine lived in a tin shack in a suburb near the shrine village and walked to work. She said she preferred living outside the shrine because there were fewer restrictions to observe. But in 1998 she was building a house at the shrine on a plot granted her by the *fahatelo*. Albertine complained that her deceased mother was making her sick for failing to live up to her promise to devote herself fully to the shrine.[32]

Such experiences further entrench people in the obligation to serve, yet at least they redefine the situation on their own terms. I said earlier that it is only royal ancestors who count in the system; yet here in the heart—right next to the *ampanjaka*—are servitors who explain their position in terms of the authority of their own ancestors; people who presumably held the same jobs in earlier generations and specifically bequeathed them to their offspring.

Albertine also takes the royal ancestors seriously. She is medium to a few spirits, and when she is actively possessed living royals pay their respects. When, after our first conversation, I gave her a bit of money, she said that she would offer part of it to the shrine to offset any offense that her having spoken so freely to me might have produced.

Other Shrine Residents

I asked a woman why she and her husband, who were Tsimihety and hailed from a distant town in the interior, lived permanently at the shrine. She said outright, "Because I am the spouse of a Jingô. After we came to Mahajanga, our child fell sick and we went to a diviner (*sikidy*) who explained that my husband was Jingô and should live at the shrine; he couldn't live so close without actually being part of it. He consulted his parents who confirmed his Jingô status and explained the duties associated with it. So we moved to the shrine and have had many healthy children since." The husband worked at a factory but asked for days off when he was needed at the shrine.

Thus, people living at or adjacent to the shrine who have regular duties there are free to take other work as long as they provide their services when needed, especially at the annual festival. In return, shrine workers are periodically remunerated with a portion of the offerings that supplicants bring to the shrine. There has been a good deal of concern in recent years over the size of the remuneration, which is certainly not enough to live on. While the shrine manager looks sleek and well fed, the servants have become increasingly impoverished, producing a situation that is probably quite different from that experienced by the ancestors who coerce them to remain on site.

The feline population provided an index of the impoverishment of shrine dwellers in the early 1990s. Individual spirits are partial to particular animals, and Ndramisara likes domestic cats; but the shrine dwellers no longer had the means to feed them. The unhealthy condition of the cats was the occasion of some outrage on the part of town dwellers.

Other people attributed their subjection to an alternate source. Approximately half of the shrine residents were sent from Analalava by Ampanjaka Soazara.[33] "After all, the relics are her ancestors too," explained the leader of this group, Veloma Soloteña (*ampanjaka* through a minor line). He said in 1993, as though it were yesterday, "Our party of 62 people arrived on the fifth of December, 1960. We had left Analalava in May of that year. We brought cattle: 26 cows left Analalava, and 32 arrived, some having been born along the way. We handed the cows over to the shrine manager. We have been waiting ever since for the call to go home. In the meantime, we have begun to bury our people here." They received no salary from the ruler or the shrine but earned their living through rice cultivation, fishing, and wage work in town. The old man described himself as Sakalava and expressed no resentment except to the Merina as foreign conquerors. "Where else in the world" he wanted to know, "are there still *ampanjaka?* Israel? Kuwait?"

The group was sent bearing money and other gifts and cannot leave until their mission is completed; something, they say, that requires the settlement of the conflict between the royal factions. The people caught in this situation, some of whom are by now in their third generation, belong to various *karazaña*. Some asked to go home and were refused. They left anyway—and died on arrival. Someone clarified, "You should ask leave of the rulers but can do what you like so long as you are not afraid." In the course of his conversation it became clear that the *ampanjaka* whose permission he would ask were not the living, but the *tromba*.

Critical here is not the specific "kinds" of persons, or their rank, but the fact that a group of people—from a variety of clans and offices—were specifically commissioned and must abide by royal decision. To emphasize further that the power relations in question are not simply imposed downward, expressing inequality between living rulers and their subjects, but subjecting everyone; remember that (according to sources in Mahajanga) Ampanjaka Soazara, at whose behest the group set forth from Analalava in 1960, is herself partially immobilized. If her subjects cannot leave Mahajanga on pain of death, she herself cannot visit it for the same reason.

CONCLUSION

The relationship of *ampanjaka* to *razan'olo* is evidently not simply a matter of hierarchy. There is an interplay of rank and function, and the weight placed respectively on each shifts along with the balance of power. Moreover, within the division of labor and the assignment of positions to specific individuals there is a complex dialectic of freedom and constraint derived largely from the bilateral ancestors.

To take the first point, *ampanjaka* and *razan'olo* are said to be complementary; there can be none of the former without the latter. Indeed, it is understood that *razan'olo* produce *ampanjaka* and preserve their sanctity. The patron of the most sacred shrine, Ndramisara, himself, is sometimes said to have been a *razan'olo. Ampanjaka* put themselves in the hands of the *razan'olo*. As the late Ampanjaka Voady explained to me in 1993, "Even *ampanjaka* have to request permission (*mangataka lalaña*) of the shrine officials in order to enter the temple and pay respects to the ancestors. They are our ancestors," he said, "But we put other people in charge of them."

In controlling access to the ancestors, *razan'olo* also determine the status of the living. As noted in chapter 2, the shrine is constituted through a chronotopic maze of closures and disclosures. Most people entered the sacred enclosure (*valamena*) through the west gate; the sacred south gate was, in principle, restricted to royalty, Ancestor People, and royal *tromba*. At the Great Service an Ancestor Person stood guard, selectively admitting or barring individuals. Gatekeepers publicly clarified the ambiguities of kindedness among the living in a binary, digital way (Rappaport 1999). Admission was a vivid performance of identity. Yet there was danger; occasionally someone with legitimate claims was barred through the gatekeeper's ignorance, producing humiliation and conflict.

More recently, there was general criticism that the shrine managers let in state and provincial administrators and other powerful public figures who had no rights of entry on the basis of traditional calculations. This has weakened both the sanctity of the relics and the function of the shrine as an arena for the display of privileged kindedness.

As a great diviner, Ndramisara was able both to discover and to determine people's lot. The "genius of Ndramisara," as this is put, was to come up with a system of organic solidarity that downplayed its hierarchical underpinnings and allowed for at least a semblance of a balance of power, if not of privilege. The idea was to make everyone feel a stake in contributing to the whole. Ndramisara's model engaged commitment (and may once have ruthlessly enforced it) by developing a set of complementary tasks and distributing them among different kinds of people, all of whom thereby were or became subsumed as "Sakalava."[34]

Power in such a system does not express itself outright in domination or exploitation, though it has elements of both, but in subjection (Foucault 1983: 212). Servants at Ndramisara's shrine grumbled at their poverty and low remuneration and recalled direct coercion and fear of physical punishment in the past. Jingô were revolted and frightened by certain tasks. "Preparing royal bodies in the mortuary ritual is like a horror film," said one participant, "and the smell is nauseating." Guardians of the relics were frightened that Bemazava would attempt to seize them by force. Shrine servants risked death around the time of the annual ritual, yet they did not abstain from the work.

Insofar as the division of labor is conceived organically, Ancestor People are part of a whole. It is less the case that they are directly servants to living royalty than that both

Ancestor People and royalty are players (personages) within, and subject to, a larger cosmological scheme. The overt aim of the division by kind today is not domination, expansion, or exploitation, so much as stable reproduction. The ceremonies, as Feeley-Harnik has observed (1991a), are directed largely to the service of the dead monarchs, not the living ones, and everyone is engaged in that endeavor.

Put another way, the organic division of labor is not to be reduced to its political function or to social solidarity. The most critical distinction is that between the living and the dead. Royalty are associated with the living and must be removed from all association with death. But this in turn requires a class of servant whose identity is determined precisely by his or her proximity to and absorption of death. This is the classic product of a social hierarchy defined by purity (Dumont 1970). To keep members of the royal clan pure there must be another "kind" (*karazaña*) of people to remove and absorb that pollution. However, the result is not a caste system (cf. Lavondès 1967) because the various "kinds" of people are neither endogamous nor hypogamous, and they do not inevitably pollute each other.

There is, in principle, a good deal of individual autonomy. Not everyone who could activate the status of Jingô does so, and those who do so could draw on other aspects of their ancestry to emphasize other dimensions of their identity. Through bilateral descent everyone is multiply kinded. However, the element of choice is offset in certain instances either by dispositions acquired during socialization or by the experience of direct ancestral interpellation—either from royal *tromba* or from people's own parents and grandparents.

Those who serve do so in part because their subjectivity is formed and given value through a particular social identity. In this hierarchical system Jingô identity is one in which people come to understand and accept themselves as profoundly "other" to the dominant categories of prestige and yet in such a manner that to break the hierarchy or abandon the discursive and disciplinary practices would be more threatening, if not shattering, than to accept them. Their identity places them within a larger whole and gives them the satisfaction of dependency, connection, and of doing an important and necessary job. Their work is not only meaningful but self-constituting.

More than reproducing or curating the ancestral order in an objective way, people reconstitute it in their very being. In fearing death or falling sick at the occasion of a royal funeral, Jingô continue to embody their historical position. They are not at "arm's length" from the past but within it, as it is within them.

In sum, subjection, as I describe it here is evidently the product of the subtle workings of power of the general order described by Foucault, in which subjects are constituted through practices of dividing (here into "kinds"), self-recognition (here, through ostensible interpellation that informs which "kind" is relevant *for me*), and ethical self-constitution. These practices are somewhat successive, and not everyone reaches the stage of conviction of a unity of self and value, especially under the materially difficult and exploitative conditions in which some shrine servants live.

Beginning with the organic division into distinct "kinds" of people, in the end this chapter has been about what it means to speak of Sakalava subjects. If we ask who the Sakalava are or attempt to distinguish Sakalava within the social flux of contemporary

Madagascar and a polyethnic city like Mahajanga, we can say that Sakalava are the people willing, and in some cases obliged, to bear *this* history, to carry out *this* historical labor, and to divide it up in *these* ways. It is their commitment to a certain history that distinguishes them from their neighbors. However, if history is a burden for Sakalava, it is a burden that is distributed among the population. Not everybody bears the same portion or bears their portion in the same way. Not only are portions distributed unequally but not everyone feels equally obligated to shoulder the portions cast in their direction.

One evening after a long day at the shrine when we had heard some complaints about reimbursement of the servants, two friends and I were walking back along the dusty road into town. Solange, a new medium who had spent several years in France, said that she just could not understand why people did not demand accountability for the money brought in at the Great Service. I asked whether shrine residents at least had rights to harvest and sell the mangos from the heavy trees. Mme Doso remarked that I could only see the disadvantages of living at the shrine and was still seeking the advantages. At this, and simultaneously answering her own question, Solange observed, "Strange as it might seem to foreigners, Sakalava find it an honor to serve."

 **6**

Personal Particularism, Mediumship, and Distributive Memory

I n the previous chapters I examined two forms of the division of Sakalava historical labor, concerning, respectively, the segmentary quality of Sakalava polities and the complementary tasks of various "kinds" of people within them. This chapter elaborates a third way in which the tasks of bearing Sakalava history are distributed, which I refer to as personal particularism. It follows from the fact that the ancestors are not lumped together as an amorphous mass but distinguished as discrete individuals. It is a division of knowledge and labor by historical personage or character. These personages are borne primarily by tomb custodians and spirit mediums. Respectively, they guard and host specific royal ancestors, not ancestors in general.

TOMB GUARDIANS AND SPIRIT MEDIUMS

If royal ancestors constitute, signify, and represent the past, each of them evokes only a single portion of it and each is remembered in a particular way; each carries his or her own particular weight. Ancestors are distinguished according to their individual places in the royal genealogy, the locations of their demise and their burial, and, in a few cases, of bodily relics and artifacts. They are also identified according to particular traits or habits linked to the manner of their lives or deaths that are visible in performance. One *tromba,* Ndransinint', cannot bend his legs, wears a kind of turban, and likes to drink beer; another, Ndramamahaña, wears a tailored jacket, pith helmet, and sunglasses, and so on. Their appearance, comportment, dialect, quality of house erected in the tomb village, artifacts, and consumption practices all manifest the period at which they lived. As they are distinguished by such features, so they bear them forward.

Ancestors are individuated, yet their individuality is always partly constituted through relational qualities. A given *tromba's* identity is established and verified as it makes connections with its closest kin, that is, with the *tromba* of its parents and offspring in possession of other mediums. Every important *tromba* is exactly located within the common genealogical grid. When two or more spirits are present simultaneously they greet each other according to their respective positions, with earlier generations having seniority over younger ones.

As noted in previous chapters, ancestral relations have both temporal and spatial dimensions. Ancestors buried in the same cemetery often have denser relationships with one another than do those buried at diverse locations. Ancestors buried at the same place cooperate in the rituals of upkeep and renovation of the site. Although tomb enclosures usually include ancestors who were very closely related to one another—siblings and contiguous generations of a single branch of the royal family—sometimes more distant relations are present or closer relatives are missing. There is always a story associated with anomalous burials, and it has implications for the interactions of the relevant spirits (chapter 9).

Each ancestor has ideally a guardian (*rañitry*) and one or more mediums in whose bodies the spirit rises, speaks, and acts. Conversely, tomb guardians and spirit mediums are distinguished according to which of the royal ancestors they bear. Tomb guardians are drawn from the ranks of Ancestor People. But the roles of individual guardians depend less on the particular kind of Ancestor Person they are (that is, their *karazaña*) than on the identity of the person whom they guard.[1] A new guardian is selected for each newly deceased fully recognized member of the royal clan, and he is succeeded in office by his descendants. Each cemetery tends to have a numerically dominant clan from which the majority of guardians are drawn.

Rañitry means "sharp," and in this respect guardians are likened to spears, to the sharpened posts that make up the sacred palisades (*valamena*), and to the cut ears of cattle that identify them as royal. Each guardian is responsible primarily for one royal personage and lives at the village adjacent to the respective cemetery. Guardians participate in the maintenance of the entire cemetery in which their charge is found, but each is identified with the single personage that he serves. Guardians of both male and female ancestors are men.

Guardians are custodians of the ancestor's physical remains: bones, tomb, and artifacts. They maintain and possibly live in the house (*zomba*) that is built for each ancestor outside the perimeter of the tomb enclosure. The houses are constructed and renewed through the enterprise of living monarchs on behalf of their deceased kin and of mediums on behalf of their spirits who send money or materials to the guardian. Mediums and their followers have access to the respective houses when they visit the tomb village.[2]

The position of spirit medium differs from that of guardian in a number of respects. First, accession to mediumship lies entirely outside the system of stratification by kind (*karazaña*). Persons of any kind, including royals, Ancestor People, commoners, and outsiders, can become spirit mediums. Thus, the guardian could also be a medium of the ancestor he serves, or of other ancestors. Frequently the wives of guardians are mediums of the very spirits their husbands guard.

Second, mediums are not restricted to representing any particular spirit nor to only one. Every medium hosts more than one, and some mediums host as many as a dozen.

Third, whereas each royal ancestor has only one tomb and guardian, some ancestors appear in many mediums. Certain ancestors have numerous mediums in Mahajanga, whereas others are currently unrepresented. Given the discretion associated with power, some ancestors are "reserved" (*miavong*) and appear in only a single medium (ideally, replaced as the medium ages or dies). Others have no mediums presently living in a given locality and possibly none at all.[3] Mediums are also found across and beyond the territory over which the ancestors once ruled, and overall there are many more mediums than there are either royal ancestors or tomb guardians. A tomb guardian may have relations with several mediums of a given ancestor, whereas mediums have relations both with other mediums who share the same and closely related spirits and with the guardians of the several spirits who possess them. The guardian of an ancestor whose *tromba* appears in a large number of mediums will have a correspondingly larger social network and resource base than one whose ancestor makes only very restricted appearances.

Fourth, whereas tomb guardians are relatively stationary, live at their respective tomb villages, and make only occasional visits to town, spirit mediums are mobile in multiple ways: Most can readily change their primary residence, and they frequently make visits from their homes to services held at tombs and shrines. In some instances, individuals are assigned primary responsibility as mediums for given ancestors at specific places, and then their mobility is somewhat but not entirely restricted.

Mediums and guardians divide up the labor of bearing history in that each is directly responsible for only the particular deceased personages that directly concern them. For each ancestor, the medium's and guardian's tasks are complementary and they should cooperate (chapter 9). Mediums have higher status than guardians, especially, of course, when they are in a state of active possession. Guardians occupy themselves with the material remains of bodies and are associated with death and its polluting qualities. When a medium dies, the *tromba*'s clothing and jewelry are inherited by the relevant guardian. Guardian and mediums share the same taboos of their personage. Thus, there is an identification of guardian and medium with a particular royal ancestor and, hence, with each other.

Dadilahy Kassim, a senior medium, explained the relationship between guardians and mediums this way: "The guardian is the lieutenant (*solontena*) of the *tromba*. The guardian is responsible for (literally, holds the key to) the ancestor's remains and for fixing up and maintaining his house (*zomba*). The guardian is appointed only once the ancestor first rises in someone (*mianjaka*). It is then up to the mediums to supply the guardian with what he needs to do his business. It is the mediums who raise the money for the *tromba*'s house which the guardian then builds."

Each of the ancestral personages and, hence, his or her respective mediums and guardians, is associated with specific rights and responsibilities (*anjara*) with respect to the royal service at specific places. For example, in describing the space at Bezavodoany, Kassim commented that cattle entered the outer enclosure through the gate associated with the royal ancestor Ndranavia. "Cattle," he said, "are Ndranavia's *anjara*." Ndramañaring is responsible for the key (*fangalagadra*) at Bezavo. We were puzzling over Ndramañaring's genealogical location, and someone ventured that he must belong to a relatively early generation because he held such an important job (*anjara*). *Anjara* in the ancestral polity resynchronizes office; that is, offices are distributed among specific personages who lived at different times but who have cooperated in a unified organization and scheme of practice since then. This is one of the primary vectors of the Sakalava chronotope.

The distribution of tasks is largely a matter of relative status among the ancestors. Genealogical position, gender, and relative age at time of death are each factors. *Tromba* of men who "passed on" in the prime of life are expected to take the active roles that would not be fitting for their seniors (irrespective of the gender or age of their respective mediums). Mbabilahy takes an active role in looking after the shrine of his ancestors in Mahajanga, but in his own cemetery at Betsioko where he is the senior figure, the active managerial role is left to his son.

Some *tromba* are proprietors of houses at the Doany Ndramisara. These included Mbabilahy, Ndramañonjo, and Ndramamahaña; there was also a house belonging to the Antandrano (Zafinifotsy) spirits, and another was inaugurated for Ndransinint' (figure 2.2). This seemed an odd group of ancestors—some senior and some not—but eventually the logic became clear. Each house was linked to the group of personages residing at a particular cemetery. When they visited the Doany Ndramisara, each constituency had a place to call home. Ndramañonjo stood for the entire group from Bezavodoany, for whom he was the active organizer; Ndramamahaña's house was that of the spirits from Nosy Lava (Ampanjaka Soazara's line). The Betsioko spirits are at home in Mbabilahy's house—named for him because he is both their senior spirit and most active in the affairs of the Doany Ndramisara. Ndransinint's house serves personages descended from Ndramamonjy (Ndramandisoarivo's older son; chapter 4).

Medium and Spirit[4]

Being actively possessed is a form of poiesis, "pro-duction into presence, the [passing] from nonbeing to being, from concealment into the full light of the work" (Agamben 1999: 68–9). But mediums also engage in a good many practical activities, including entering trance and working both in and out of trance on behalf of both the ancestors and their own contemporaries. In these respects spirit mediumship crosscuts the mechanical and organic divisions of labor. Not only are mediums highly mobile, often practicing far from their natal regions, but each political segment has its own "cast" of resident mediums.

Neither which spirit one is possessed by nor whether one becomes the leading spirit medium associated with a sacred place is linked directly to descent or status. As one man explained, "The position of shrine manager is inherited but mediumship is not passed on by inheritance; it is a matter of luck or set by the stars." This refers to a theory of character and relative openness to outside influence that is linked to astrological considerations (Lambek 1993), but the main point is that access to mediumship is not predictable by social status or overtly expressed intention or desire. Mediums can be of either sex, and the majority are women. Although the offspring of a given medium sometimes acquire the same spirits that possessed their parent, this is not a matter of social convention but whether the spirits themselves will it. The offspring of mediums do not necessarily become mediums themselves, and they are not necessarily possessed by the same spirits as their parents (cf. Lambek 1988, 1993).

Although in practice the spirits who possess given mediums are likely to be closely related to the other spirits present in their milieux (that is, among family members or neighbors), again this is not part of the theory. The latter rigorously excludes both social constraint and the mediums' personal motivations in the match between mediums and the spirits who possess them. There are minor gender constraints: Men are almost never possessed by female ancestors, nor women by Ndramisara. There may also be religious constraints, in which case an unwelcome spirit will be asked to leave. Because of the long historical connection between Islam and the Sakalava monarchy there are many Muslim mediums and some Muslim spirits; in the 1990s there was little tension between the two systems. The relationship with Christianity, especially Protestantism, is more complex. Ancestors from Betsioko who were practicing Christians do not prevent their hosts from doing likewise. In contrast, the Antandrano (Water-Dwelling) spirits prohibit Christian worship among their mediums because of the association between Christianity and the Merina (who were responsible for their deaths). Thus, the two sets of spirits articulate quite different historical, sociopolitical, and religious positions. Moreover, through the kinds of spirits by whom they are possessed, individual mediums can articulate their own relationships to the church, whether tacitly rejecting it or embracing a hybrid or double practice.

In addition, certain spirits encourage or forbid the presence of certain other spirits in their mediums or in other members of the mediums' households (chapter 11). These patterns are further manifestations of the relationships between the original ancestral figures. Those who fought with each other in the past prevent a stable marriage between their respective mediums; those who loved each other too well are forbidden in the same medium for fear of incest. Through their possession mediums continue to reproduce the social and emotional relations of the past and simultaneously articulate their own personal lives and social relationships by means of the spirit idiom.

The onset of spirit possession is often linked to personal suffering on the part of the medium, and establishing a regular relationship with the spirit is a form of cure; but this does not determine possession nor the identity of the particular spirit. At base, two things are critical. First, the presence of the particular spirit must be at some level acceptable and of interest to the medium. Sense can often be made of the "match" by examining the medium's past history, childhood relationships, social connections, conflicts, desires, and so on. Second, the medium must "play by the rules," and possession must accord with the social realities of the present. Legitimate possession by a given spirit entails regulating the timing and performance of trance (especially vis-à-vis other mediums) and observing the associated practices (for example, monthly bathing in the ocean) and taboos. Possession is a highly coherent cultural form of articulation. Moreover, spirits are understood as sacred and dangerous and must be treated accordingly.

Mediumship is a lifelong commitment. Spirits generally manifest themselves overtly in early adulthood, but this is often preceded by signs of prospective possession or interest on the part of particular spirits. The possession "cure" is a kind of initiation, establishing performative competence and legitimating the presence of a given named spirit in a given medium and thus the medium's subsequent activities embodying the spirit. Whether or not the medium remains active, the presence of the spirit continues as a force in her life. Sometimes spirits who arrive relatively early in the life course are overshadowed by more powerful ones arriving later; however, exceptional cases aside, all of the spirits whose presence has been legitimated remain with the mediums until the mediums' deaths.[5]

Accession to mediumship appears to differ at the present time between members of the Bemihisatra and Bemazava royal factions (chapter 10) and has undoubtedly shifted over time as well. The relationship between individual mediums (as vehicles of ancestral royalty) and contemporary live royals (who are their descendants) is also historically contingent and complex. Living royalty are neither as sacred nor as potent as their ancestors, but live monarchs are often more powerful than the manifestations of their ancestors in particular mediums. Spirits can speak jocularly and persuasively to their living kin, but most mediums lack the authority to be too bold or critical (chapter 10). *Tromba* are sometimes referred to as gods (*ndrañahary*); living royals never are. Both senior spirits and living royals are greeted with the honorific form of address "*koezy*" or "*koezy ampanjaka.*"

The relationships between mediums and the spirits who possess them are also complex. The mediums of younger, junior spirits enjoy the opportunities for entertainment and public display that are supplied with their roles. At the same time, many mediums complain about the burden of having to amuse their spirits or follow their taboos. Some mediums become respected healers; others support activities at the shrine or cemeteries. Some become artistic and ethical virtuosos. Others enter trance irregularly in support of royal service, to assist in the spirit ceremonies of others, or for the fun of it. Most importantly, in bearing royal ancestors, mediums embody qualities of royalty and ancestry themselves. Their connection is understood as metonymic but is nonetheless real, especially for mediums themselves. Mediums not only emulate royal ancestors, but in some instances they regard their own comportment as a model that living royals ought to emulate. These mediums transcend their subjectivity, transforming it into new forms of moral agency (chapter 11).

Although there is scope for mediums' creativity and each performance seems fresh and vital, little about the spirits is fortuitous. Even the particular squeals or growls uttered in the transitions into and out of states of trance are part of the structural code. I once made a rather facetious remark to the effect that mediums of Ndramarofaly always seemed to be fat. To my astonishment, Mme Doso replied seriously that Ndramaro did prefer bulky mediums (*saha botrabotra*) because he bounces up and down so much on his hard wooden bench. Mme Doso herself lavishes attention on two white cats. It is no accident that her pets are cats or that they are white; these are the preferences of one of her spirits. Not all mediums of this spirit might take the cat idiom so far, but it is there for Mme Doso to draw upon and elaborate as it suits her.

The more senior the spirits, the less random and more contained are their actions. On the other hand, senior spirits in fully authorized and respected mediums can, if they are careful, make significant original interventions in contemporary affairs. Not merely embodying history but actively interceding in it, spirits can bring their weight to bear upon the present.

There are no general rules for the selection of a chief medium. Mme Doso was appointed *saha be* of Mbabilahy by Ampanjaka Désy. When the chief medium of Ndransinint' was selected after the death of the previous office holder, there was a "vote" (*vôty*). Four names were written on pieces of paper, and a child was asked to pick them up. When the same name was picked twice, that person was elected. Candidates are selected among people considered kind (*tsara fañahy*). As one medium explained, "We don't want people who are standoffish or quick to anger, but those who are empathic, deal well with others, and can handle lots of guests."

LIFE AND THE NEGATION OF DEATH

If "kindedness" is an ideology central to the Sakalava monarchy, it is one that mediumship transects in a peculiar way—overcoming the kinds that divide living humans to mediate and complicate the division between the living and the dead. The fact that transformation into *tromba* is largely (though not exclusively) restricted to deceased royalty also reinforces their distinctiveness of kind.

As distinct kinds of people, royalty (*ampanjaka*) and Ancestor People (*razan'olo*) are associated with the living and the dead, respectively, and must keep them clearly apart. Whereas ordinary people in nonsacred contexts acknowledge that death follows life, part of sacred labor entails patrolling and maintaining the boundary between life and death. "Collapsed" royals have the decaying flesh removed from their enduring bones by the *razan'olo*. Their non-death is manifest in the *tromba* and their mediums.

Given the significance of ancestors and mortuary practice throughout Madagascar (Bloch 1971, 1982; Astuti 1995; Middleton 1999; Cole 2001), it would appear that spirits and mediums carry this "ancestor function" for Sakalava, manifesting the transcendence of ordinary life. *Tromba* are understood as not alive. Dadilahy Kassim, a senior medium, himself, was emphatic on the point. "How can someone dead return to life?" he asked rhetorically. But clearly, risen in their mediums, the spirits are not simply dead either. Kassim described them as "wind" (*rivotro*). If neither alive nor dead, spirits can be understood as a negation of death that does not imply a simple return to life. In effect, *tromba* negate the negation of life.[6] This is a kind of death "under erasure" (*sous rature*); not death, but ~~death~~.

Mediums hold an ambiguous position. They must mediate between their own life and the nondeath they bear. When a spirit takes active possession of the medium, the performance is one that often begins by illustrating the throes of death followed by emergence from the shroud. On leaving the medium's body the spirit returns to the shroud. Mediums thus regularly traverse the boundary between life and death and are pervaded by death. Death is both manifested and transcended by the mediums.

Taboos held by mediums can be seen in part as forms of purification from the taint of death that their spirits impose. These are also the taboos of royalty. All living royals must strenuously avoid association with death and the dying, stay away from funerals, and so on. *Tromba* must also avoid their own, prior deaths. Thus, although mediums, like living royals, avoid death in general, the additional taboos practiced by given mediums concern the deaths of their particular spirits. Mediums avoid anything that brings to mind the specific circumstances of the death of the royal personage whose spirit they bear. No spirit can rise on the day of the week on which the royal ancestor "collapsed." Food taboos comprise the dishes eaten by the royal personages at their last meal and that may have poisoned them. Practicing these taboos imposes and impresses the royal personages onto the bodies and into the lives of mediums and sets them apart from others. Perhaps observance of the taboos also protects the mediums from suffering the same fate as the royal personages they bear.

The tabooed substances and actions remind *tromba* too strongly of their own deaths. Spirits are angry to be so reminded; rigorously following the taboos protects mediums from the anger and offense of their spirits. But, of course, if following taboos protects spirits from remembering their fate, for the mediums the practices simultaneously call continuous attention to what must be avoided. Death is thus an open secret, continuously *sous rature* for mediums and spirits, present in its marked absence and absent in its presence.

We now have a clearer idea of one of the burdens of spirit mediums. They bear death within themselves and yet must strenuously keep death separate from life, under erasure. The same may be true of living members of the royal clan who must avoid death and whose deaths are never to be spoken of and yet who are, in a sense, continuously pregnant with death—their future death as spirits evoked by all they do, especially by the cultivation of their ancestors and the curatorship of the cemeteries with whom and where they will one day rest.[7]

Too great a contact with things sacred and ancestral—too intense an interest in spirits—is dangerous. Possession, itself, is a vivid demonstration of how excessive interest on the part of the dead could extinguish the living. However, possession is also life-giving. *Tromba* offer blessing by aspersing their supplicants or offering them drinks of water in which white clay (kaolin, *tany fotsy*) has been mixed. The source of this clay is the relics at the shrine, and it is evocative of ancestral bone substance. Mediums conjoin in their persons both the horror and pity and the power and attraction of the dead and provide a measure of the boundary.

The marked separation of death from life produced by the social and spatial classificatory and practical schemes is accompanied by the presence of the *tromba* as evidence of a kind of life in death available to some kinds of people under some kinds of conditions and approachable by anyone who wishes to offer them respect. But at the same time, the work of separating death from life, which is evident in the practices of the mediums, bears with it an implicit recognition of the fact of death in life. Such implicit recognition, always partial and conveyed by means of its negation, is in essence ironic. Irony is a quality of discursive practices that refuse their reduction to a single unambiguous referential meaning. Irony is intrinsic to the double voicing of spirit possession and the split between spirit and medium (Lambek n.d.) and it lends Sakalava ancestral practices a depth, richness, and subtlety of "knowing yet not knowing" equivalent to that of great works of literature or philosophy (cf. Nehamas 1998).

MEDIUMSHIP AS PARTICULARISTIC PRACTICE

That royal personages are not socially dead is evident in the roles that they play—not only in the everyday lives of the mediums and their clients but in the cult of the relics itself. *Tromba* too must labor on behalf of their ancestors. Everyone—living and dead— has ancestors in generations that preceded them. This ancestral practice on the part of the *tromba* has profound implications for their mediums whose proxies they are.

Mediums must participate in the services (*fanompoa*) of the shrines or tomb sites with which their own *tromba* are associated. Their activity is based on a combination of the ancestor's place in the genealogical hierarchy and gender, age, and mobility at time of death. As noted above, *tromba* of royals who passed away in active adulthood often are expected to take on executive tasks; *tromba* of younger generations but very old in terms of years of life lived are not. The age of the medium is not directly considered; the work that individual mediums pursue is largely determined by which spirits they host.

Possession varies in its physical effects; the incarnations of some spirits cause more pain than others in particular mediums. Male mediums are often as taut as coiled springs when embodying the most powerful spirits,[8] which is one reason why the most senior spirits do not stay long when they rise. As Tombo, my teacher on Mayotte, pointed out (Lambek 1993: 314), the experience of possession is like a *coup d'état*. Some mediums accept it from the start, whereas others resist. In some cases, resistance is sufficient to keep the *tromba* from ever fully rising, stating its name, or developing a regular and publicly acknowledged place in the host. In others, resistance increases the intensity of the performance. But a medium who performs regularly becomes, as one said, "used to the experience."

Possession is significant to mediums not only during the moments of trance ("active possession"). Possession pervades their lives; it means living with someone else— sharing one's body with powerful others. The omnipresence of the spirits is manifested preeminently in the practice of taboos and in talk about them. Taboos bring aspects of the spirits' prior lives to the present. Mediums can never forget that they bear spirits, both because of internal constraints, that is, punishment in the form of illness consequent on failure to observe the taboos and because of public recognition. Their presence, as vehicles of specific ancestors, may be requested at numerous events, and they are expected to serve as conduits for funneling money to the shrines.

If mediums are the most faithful exemplifications of the values of royalty, so too do they form royalty's most loyal subjects. Indeed, whereas living royals begin to liberate themselves from the constraints of tradition, mediums continue to express their subjection to it. Being a subject is not only a matter of ostensible passivity and constraint; mediums are often precisely the people exhibiting the most energy with respect to the ancestral cult and responsibility within the community. They take care of

things, organize (*mikarakara*), acquire clients, offer treatment, convince clients to contribute to royal service and contribute even more themselves, and serve as key nodes
in networks of exchange. Mediums generate wealth and direct the major portion toward the shrines. They enliven flows of people, interests, and gifts.

In contrast, guardians stay still and absorb what is used up, finished, and dead and
conserve and consolidate the rest.[9] They provide the stable background against which
the mediums move and generate movement. Feeley-Harnik's important observation
(1984) that since the onset of colonialism the Sakalava kingdoms have reoriented toward death and the care of tombs and dead rulers needs to be supplemented to take
into account the liveliness of spirit possession. Indeed, the democratization of possession (which Feeley-Harnik also observed), that is, the spread of mediumship—the
possibility for "the king or queen in every man and woman"—means also its converse;
namely, the expansion of the tribute base and of labor recruitment. Possession creates
among mediums and their clients a corps of earnest participants who are motivated to
execute, manage, and even initiate particular forms of service.

Distributive Memory and Implications for Fieldwork

The explicitly distributive and performative nature of Sakalava history distinguishes it
from both a generalized collective memory and from the Western tradition of specialized historical scholarship. It is comprised of commitment in and to specific historical
figures and engagement in crafting history (poiesis) from the particular standpoints of
these figures rather than inspecting or investigating it from a distance (*theoria*).

Each actor controls only a small portion of a larger picture. But piece by piece the
larger picture continues to be reproduced. There may be no one who can see the entire large picture or who, if they did so, could speak about Sakalava history from that
perspective (from the outside, as it were). Yet although virtually no one aspires to this
position, it remains a matter of importance to everyone to know that somewhere each
part is being carried out in its proper place.[10]

Although a given person may use the vehicle of more than one personage from
which to view the world, each personage offers its own perspective, its own standpoint
from which it speaks and acts, and its own contribution to the broader historical picture. The large cast of historical characters and their contemporary conveyors provide
a real richness in that they do presume a whole.

The plenitude is one of multiple voices but also the product of the juxtaposition
of at least three registers. Ancestral knowledge is not only embodied and voiced in
spirit mediums but inscribed in narrative and on spatially distributed material remains
and architecture and the movements between them. There is some tension but also
much flexibility and interplay between registers. Each constrains and contextualizes,
confirms, substantiates, enlivens, challenges, and invites the others to produce internal
coherence.[11]

Thus, although Sakalava understand that their system forms a whole and has "laws
or aims or charters definitely laid down" (Malinowski 1961: 83), the latter could be
found in the orchestrated movement of artifacts and persons between sacred sites as

readily as in a discursive register. Buildings materialize and objectify intentions and ideas. Bodily relics and artifacts are not merely signs of particular rulers but condense ideas about the relationship between ruling and reproduction (cf. Feeley-Harnik 1991a). The successful completion of the annual Great Service and other events that require coordination of the various divisions of labor indicate that synchronicity is still working. That is, the constitution of the Sakalava polity is realized in the effective enactment of the main ritual events.[12]

One of the most striking aspects of the embodied register is the mobility it permits. *Tromba* representing deceased personages spring up over widely dispersed communities—hundreds and even thousands of miles apart. The same spirit can appear in Antsiranana and 600 km away in Ambongo, or even Mayotte, La Réunion, and France. One and the same spirit can appear in domestic contexts and in highly public ones, in major centers and in the remote bush.

Whether and how such representation is recognized by others and understood as consistent is something of an open question (Keane 1997).[13] Some interventions provide or require more space for public recognition than others and can actually instantiate changes in the way things are done. Spirit mediums are not cognizant of all spirits. Guardians at one tomb enclosure are unaware of the details of the work of guardians at another. Yet the signifying practices within and between registers fit together, articulating the royal line and its history. There is a shared sense of a relevant whole with reference to which all are working.

The register in which one might expect to find a more comprehensive, less particularistic picture of the whole would be that of narrative. This is both because of the qualities of abstraction inherent in discourse and because, unlike material remains and incarnate ancestors, Sakalava narratives are not ostensibly the preserve of specific individuals. However, there is both a generalized sense of discretion and a link between specific narratives and personages (chapter 3). Most narratives are either not widely known or not widely spoken. Some people are reluctant to speak, whereas others are proud to recite formally (*mitantara*) episodes or fragments from a larger chronicle or repertoire. Recitations are understood as partial, and no one claims complete authority for them.

One of the most knowledgeable men in Mahajanga during the 1990s was a flamboyant medium of Comorian origin, Amady bin Ali (since deceased), renowned for his ability as a public speaker to generate enthusiastic responses at ancestral events through his wit, risqué jokes, and mannerisms. Amady Sarambavy (Queer Amady), as he was called informally, knew large chunks of royal genealogy by heart and was able to present them fluently; though he grew impatient as I stumbled over the long names. Although Amady's mode of providing information was not always quite legitimate, I found his version confirmed by written records preserved in a notebook shown to me by a prominent member of the royal family as well as by published sources and the shorter oral fragments presented by others. It was also confirmed in the relationships I saw among the *tromba* and served as an invaluable reference.

Some Sakalava intellectuals, including Amana Ibrahim, had comprehensive anthropological accounts, like Baré (1980), on their shelves. Baré's book concerns the segment located in Nosy Be, and a number of people expressed the hope that I would be able to provide the equivalent for Mahajanga. Others were interested in having me transform their own accounts into writing. But of course the medium does play a role in the message. What is placed within a book is objectified, and unless it is like the manuscripts reputedly hidden in the shrines or the homes of royalty, it is shamelessly open for inspection. Mine is no exception. Knowledge so inscribed loses its potency even as it becomes a tool in a different politics of objectification and display. What books may conceal is their own partiality (in both senses of the word). Oral, embodied, and performative histories are explicitly particular and immediate, refusing to stand still for inspection or to register finality. The different versions compete less to be fully heard and seen than, like the hidden manuscripts, to insinuate their existence. No one has to rationalize or justify contradictions. Such history refuses homogenization.

Amady, Amana Ibrahim, and some others did hold comprehensive knowledge or made the attempt to systematically compare distinct narrative versions or compile a master version in their heads. But most discursive knowledge consists of positioned, "internal" accounts focused on highly specific segments of the genealogy. For example, I conversed with the guardian in charge of the tomb of one of the most senior royal ancestors at Bezavodoany while he was in town. The guardian appeared quite open with me; but if so, it was remarkable how this significant person—who presided over the remains of Ndramboeniarivo and legitimated his mediums—knew so little about other ancestors or the activities at other cemeteries. He had his specific work to do, and that is what he knew or could speak about. In general. the historical knowledge over which guardians and mediums individually have authority is quite limited.

When describing the growth of their own knowledge or comparing its breadth with that of others, adepts often say that if one wants to learn things, one must travel from sacred place to place. The path to each shrine is marked, and it is in the act of following the paths that the concrete historical and political links are learned. Travel, itself, imparts embodied knowledge; in addition, there is much that can be learned discursively from the various gatekeepers, custodians, mediums, and ordinary people along the way, as well as from merely observing local sites and practices.

To visit Ndramboeniarivo's guardian at the tomb village I was told I would have to travel first to the residence of Ampanjaka Amina, the living monarch who presides over the district in which the shrine is located. Her residence at Mitsinjo is actually on the far side of Bezavodoany from Mahajanga. First-time visitors to Bezavo ask permission of Princess Amina, pay respects at the shrine containing the relic of another ancestor at Mitsinjo, and then visit the minor monarch residing at the base of the hill on which Bezavodoany rests. Since there are specific days of the week and month on which one can and cannot pay these various respects, as well as set out for, enter, or leave a given place, the trip can be of many days duration and requires careful scheduling. In addition, visitors to a remote shrine like Bezavo must bring their own food, as well as be ready to submit to the many taboos of the place. To visit Bezavo, people assured me, I would almost certainly have to present the inhabitants with a cow. Pilgrims described their fears both of inadvertently breaking taboos and of confronting the various strange beasts reputed to live in the surrounding forest.[14] This illustrates what I mean by the chronotopic maze. Not only is the journey simultaneously spatial and temporal, but knowledge is difficult and costly (*sarotro*); serious pilgrims must take their time and not rush things.

Thus, although one must travel in order to learn and piece things together, and although the act of traveling in this manner is itself a form of instruction (providing, one might say, both knowledge and "deutero-knowledge," or practical lessons about learning), at the same time, travel is not exactly encouraged. This difficulty is a lesson in itself. Not only are distances long, obstacles many, and patience a virtue, but as one elderly medium explained with asperity why she had not been to any of the shrines or tomb villages in the district, "People don't go wandering about (*mandehandeha*) unless their *tromba* sends them."

CONCLUSION

The Sakalava polity is characterized by a complex division of labor in remembering, or, more literally, in bearing, history. Not only are tasks distributed, albeit unequally, among members of society, but knowledge itself is distributed. There is no one who remembers it all or can speak from all perspectives at once, and there is no location from which to gaze comprehensively. Rather, the memories and perspectives of individual ancestral personages are reproduced through generations of guardians and spirit mediums alongside those of other ancestors. These memories are not isolated or exclusively localized but interrelated. With the exception of tomb custodians, Sakalava are highly mobile, not only "wandering about," but going on pilgrimages to specific shrines and often settling permanently in new places. Senior mediums generally work at shrines far from their place of origin. Moreover, individual mediums often host several spirits, and *tromba* themselves are highly mobile, appearing all over Madagascar and even in France.

What was striking during the 1990s was how the whole was maintained, emergent through the continuous reproduction of the parts, with each part recognizing itself *as* a part and in relation to other parts, rather than by means of central direction. This decentralization provides a kind of diffuse autonomy and an egalitarian flavor to the division of labor. To be sure, shrines have managers who plan and direct annual events, and some attract more pilgrims than others; but the rights and duties of each person are circumscribed. Managers must cooperate and coordinate their activities in complex ways with mediums, servitors, and royals, though the managers of one shrine have no say over what occurs at neighboring ones. Similarly with respect to the distributive nature of peoples' knowledge, some devotees have knowledge of several of the parts, but all of them are conscious of gaps in what they know, aware that there are yet other shrines and other knowledgeable people in which different elements of Sakalava history are preserved, tended, and reproduced. Each person is engaged in cultivating his or her part. Position is always explicit, and nowhere more strongly than in the voices of the *tromba* themselves.

The fact that many people say they do not know much or that they provide only fragments is not simply evidence of the decline of the traditional system. Rather, it indicates that the system was never designed for everyone to know the whole; in fact, explicitly the opposite. Historical knowledge is reproduced, and those who seek it can discover many things; but it is not simply on the surface and available for ready consumption. The feelings, dignity, and privacy of the ancestral personages need be respected. Sakalava history is person-centered but it is no *People* magazine.

Those with the most intimate knowledge on particular topics often had nothing they could tell me since they were expected to be the most circumspect or the most endangered by casual revelations. Those with the freedom to speak often know less and speak as relative outsiders to the subject. Their mistakes are not necessarily corrected. There are said to be people, such as deceased slaves, who have no taboos and are willing to divulge secrets, but I have not come across them.[15] Hence, again, what is important to individual Sakalava is less that they know their past than that they are assured that it is knowable, that there are people—the right people—who know and bear responsibility for each part, and that what ought to be discreet or hidden remains so.

Attention to the division of labor is not to evade the fact that some people are skeptical about the claims of certain others' rights, knowledge, or skills or that they may disagree with accounts or decisions when they hear them. Personal particularism is not without points of friction. However, it would be wrong to distinguish skeptics from believers as distinct categories of person. Everyone is skeptical of some elements in some contexts and committed to others. A man told me one day that the reliquaries were empty and that he never attended the shrine; on the next day he sent his son's shirt for a blessing in the hope that it would help him pass his final exams. There are many factors that contribute to determining where a person stands in relation to a given act or utterance of a given other. Position is always critical and a product of contingency no less than structure.

I see all of this not as a hindrance to getting at the "real facts," behind the fence, as it were—at the heart of the maze—but as the facts themselves. There is no clearing without the thicket through which one treks to reach it. And secrets wither on exposure. To the degree that Sakalava are concerned with value production, it is better if certain things remain undisclosed. Indeed, as noted in chapter 2, *undisclosure* is explicitly a central precept.

Finally, Sakalava positionality generates a kind of cosmopolitanism. Not only do some people move easily between worlds that anthropology has too often claimed are far apart, but many are sophisticated about the cultural and intellectual diversity existing within the Sakalava milieu itself—understanding the differences that descent, mediumship, or religious affiliation can make and being able to take them into account in their conversations with one another. People recognize that each party makes a distinctive contribution and holds a different balance of knowledge and specific perspective.

Positions are not freely available, and they entail their own constraints and demands (Tehindrazanarivelo 1997), but people have some mobility and also often hold multiple positions. Descent is bilateral, "clan" affiliations numerous, residence and religious commitment frequently optative and additive, and mediumship a matter of circumstance. Siblings often articulate distinct social profiles from one another. In shifting between different portions of their identities, people are aware of positionality, itself, as a factor in perspective. There is a structural predisposition for tolerance and an empathy, within limits, for other points of view. This is evident in the multiple voices of the *tromba*.

Particular Voices

This concludes the structural overview. In keeping with the larger theme of "bearing" history, the analysis has been conducted by means of three intersecting dimensions of the division of labor. As noted, these can also be understood as alternate modes of a social imaginary that together comprise a "heterogeneity of hierarchical belonging" (Taylor 1998: 40). Speaking of her participation in the distribution of raw honey at the Doany Ndramisara in April 1994, Mme Doso said that she ate or could have eaten from Ampanjaka Désy's portion through her association with him as medium of his ancestor, Mbabilahy; from the Tsiarana portion as a member of that *karazaña;* from the portion set aside for the monarch and people of Nosy Be since that is the region where she was raised; and from a portion given to Somotro, the medium of Ndramisara, who invited the chief mediums of other spirits to a share. In effect, she received honey according to her position within each of the three dimensions along which, I have argued, labor is divided and identity imagined. She belongs, "mechanically," to the Nosy Be segment; "organically," to the Tsiarana clan; and along the "personal particularist" dimension with Désy.

At the honey distribution each of these positions provided her with rights, but each position also brings obligations. Just as some people are quicker to exercise their rights, so some enact more fully (and feel more deeply) their obligations. I explore obligation and sense of duty in the following chapters. The primary emphasis will be on the duties of persons along the "personal particular" dimension, as loyal subjects, and especially as mediums, of specific ancestral personages.

In the course of my research I have worked with people in a range of positions both within and outside the organic division of labor, both part of the Mahajanga segment and, while resident there, connected to other segments, both Bemihisatra and Bemazava. As in previous fieldwork in Mayotte, some of my best working relationships have been with spirit mediums, and it is with their perspectives that I have developed the greatest affinity. Perhaps I have found mediums congenial and relatively easy to work with because their practice offers them a unique combination of personal modesty and authority, as well as the reflexivity that accompanies (and may be necessary for) taking on alternate positions and voices. In their roles as patrons and therapists, mediums also have experience dealing with clients with varying needs and develop empathy and didactic skills. Perhaps they found my presence less disconcerting than did others.

Whatever the case, mediums are interesting because they epitomize the creative ways in which people build their lives and conversely draw on their life experiences to make imaginative interventions in the community. Mediums combine poiesis and phronesis, aesthetic craft and virtuous practice, in ways that I have found irresistible and that I hope I can convey.

Among the many smart, dedicated, and enthusiastic mediums I have worked with, two have been key, distinguished by their sense of purpose, agility, thoughtfulness, and creative energy, and singled out in this manner not only by me, but by the public, by whom they are held in deep but not unqualified respect. Dadilahy Kassim and Mme Doso live in the same urban neighborhood and know and respect one another, but their spirits orient them to distinct cemeteries. Although she is older, Mme Doso addresses Kassim as father because Mme Doso's senior spirit, Mbabilahy, is the son of Kassim's spirit, Ndramboeni. They do not work closely together, but they have a good deal in common. Both are self-assured and bright company, and are able to move in various milieux and yet maintain respect for and comfort with traditional modes of life. Both take an interest in national, regional, and urban affairs. Both are nominal Muslims, but neither prays regularly nor spends the majority of his or her time with Muslims. Each has to be careful about money but has the means to sustain a public persona; both like to dress and eat well. Kassim will cross town on foot to purchase the freshest fish. Mme Doso stops at a grocery, where the owner, a classificatory child, supplies her with yogurt.

Mme Doso has a wealth of stories about spirits, royalty, and the whole milieu. She is a warm person with a lively sense of humor; her wit, forthrightness, and impatience sometimes make enemies. In her seventies, she was tireless, trekking out for nights of spirit possession or traveling to distant shrines. Kassim is more contained but can wax eloquent on moral issues and is a superb orator. Both have managed to observe discretion while teaching me a good deal. And both have drawn on their respective spirits to specify, diversify, and enlarge the positions from which they act within and speak about Sakalava ancestral tradition and contemporary life in urban Mahajanga. Together, they will substantiate what I mean by position, the distributive burden of history, craft, virtuous practice, historical consciousness, and agency.

The next chapter examines preparations for the Great Service by following the activities of Mme Doso. Chapter 9 highlights the practice of Dadilahy Kassim. Their voices appear throughout the book.

Part III

Serving the Ancestors

7

Popular Performances:
Paying Homage and
Gaining Respect

Every Sakalava subject is offered a stake in bearing history. Service is acknowledged; individual acts are recognized as making a difference. Subjects participate in the Great Service along all three modes of identity that the previous chapters have described. In the work of the Great Service one can see also the intrinsic relationship of history to renewal, of acknowledging the past to ensure present and future, and of acknowledging present obligations to ensure the continuity of the past.

The biggest event in the calendar and the one for which northern Sakalava are best known throughout Madagascar, the *Fanompoa Be,* or Great Service, is held annually at the shrine of Ndramisara on behalf of the "four men." The Great Service itself will be explored in chapter 8. But to begin there would be quite misleading. Like equivalent ceremonies of royal blessing in other parts of Madagascar, the Great Service is initiated by acts of obeisance. These acts are constituted by the giving of gifts, of which the most important in the 1990s was money. Gifts are generated among the public in an activity that might be called "fund-raising." The process culminates on the first two days of the Great Service when the totals are announced, but it begins well before. It is found most vividly in the sequence of performances that take place in town in the weeks preceding the Great Service. It is here that ordinary people most directly experience their own participation (their will, intentionality, and agency) and identification with the ancestral system, which can thrive only through their commitment to service.

Service is constituted through a series of prestations in which ever larger sums of money are amassed and channeled into the gifts that are finally brought to the shrine.[1] The main participants are spirit mediums and their clients and the *tromba* (spirits) themselves. Prestations take place at neighborhood events at which some of the royal ancestors rise. These provide a primary locus for popular engagement with the ancestors—occasions where commoners and royal ancestors cooperate in serving the most senior ancestors whose relics are preserved at the shrine. It is, in part, this work and

the collective achievement of relatively poor people in amassing large gifts that is celebrated during the Great Service itself.

Fund-raising is activated largely by women—both spirit mediums and the officials
known as Great Women (*bemanangy*) who represent specific segments of royalty. In this
chapter I trace the series of events that preface the Great Service by following the activities of Mme Doso, a medium heavily committed to the process and reflect on historical changes in the system of reciprocity and redistribution. The material in this
chapter is restricted to what takes place in town immediately before the Great Service. In the next chapter, I follow events at the shrine, culminating in the Great Service itself.

MME DOSO AND THE COLLECTION OF MONEY

Known popularly as Dady Bonbon (Granny Candy) for her penchant for handing out
sweets to children, Mme Doso is the Sakalava widow of a Comorian soldier. She rents
out rooms in the compound that he built and left her, and she supplements this income making mats, preserves, and cakes. She has developed dense relationships with
clients, former clients, and especially with a group of women who are also involved in
spirit activities and for whom she forms the fulcrum. She is host to a senior spirit,
Mbabilahy,[2] whose popular descendants possess many of her entourage; and she has
several of the latter herself, including Ndramandaming. Mbabilahy is the apical ancestor of the Bemihisatra branch that controlled the main shrine in Mahajanga; their burial ground is at Betsioko, near the home of the current ruler, Ampanjaka Désy, at
Ambatoboeny. As son and successor[3] of Ndramboeniarivo, the youngest spirit whose
relics reside at the Mahajanga shrine, Mbabilahy serves as the person who the "Four
Men" have appointed manager.

In the 1990s there were many hosts of Mbabilahy in Mahajanga, but Mme Doso
was recognized as his chief medium (*saha be*). This meant that it was generally she in
whom Mbabilahy was called up for consultation at the shrine. She thus had a public
role and many obligations that went with it. She was expected to convey large sums
to the annual Service, generated from her own and Mbabilahy's followers. She formed
a node in a wide network of spirit mediums and had to juggle her position in the public domain with her practice as a *tromba* healer in the neighborhood in which she
lived. In the latter respect she served in a maternal relation, and her spirit in a paternal relation, to many clients to whom she showed generosity and solicitude. Mme
Doso had a strong feminist streak and demonstrated particular concern for women
and their problems. Her position was somewhat precarious, especially as she aged.

As a leading medium of a senior ancestor and as medium to several others, Mme
Doso's life was very full. She had to monitor her activities to ensure that she maintained the taboos of each of her spirits. She kept track of the day of the week and so
knew what she could and could not eat, say, or do. She tried to get up to the shrine
on auspicious days, especially at monthly "openings," to pay her respects to the ancestors. Her clients visited in order to call up her spirits on personal or family matters.
Many were troubled by spirits themselves and were in various stages of becoming

mediums in their own right. Each of her clients worked with one of her spirits in particular. At curing ceremonies several, generally closely related, spirits rose among participants. Mme Doso had various mediums who she could call upon to support her in these events. Likewise, frequently she was asked to embody one of her spirits at other people's events.

"We received a letter . . ."

In the weeks and days leading up to the *Fanompoa Be,* Mme Doso's life became even busier. To begin with, she received a letter from the shrine, signed by Ampanjaka Désy and the *fahatelo,* requesting money for the Great Service (figure 7.1). The shrine is the place where the money ends up—after several phases of donation and reception along the way—and the source from which blessing (*hasing*) is distributed in return.

At her own expense, Mme Doso photocopied and mailed copies of the original letter to contacts in other cities. The recipients were leading mediums of Mbabilahy or closely related ancestors. As she did on May 24, 1998, Mme Doso wrote to a medium of Mbabilahy in Réunion and to a medium of Mbabilahy's son, Ndramarofaly, in Vohemar, a town on the northeast coast of Madagascar. Each recipient forwarded the letter to her clients in turn and also gathered followers at her place of residence. She then sent the money collected to Mme Doso who, in turn, on the day before the Great Service began, passed it on to Great Woman Honorine, who represented Mbabilahy and all those buried at Betsioko. At the shrine, Honorine publicly announced the money gathered by each of these people in each of these places but counted as part of the larger package offered on behalf of Mbabilahy and the Betsioko segment. In this way, the honor accruing to the gift resounded on those at each level of the chain—on the coordinating mediums in Réunion and Vohemar (and implicitly on their clients), on Mme Doso, and on Honorine.

Sometimes collectors arrived from their hometowns to announce their gifts at the shrine personally. Whether it was a Great Woman or a hired announcer who made the final prestation, it was always on behalf of the ancestors from a specific political segment. The city, or in the case of Mahajanga, the quarter from which each gift derived, was named as well as the senior mediums who sent significant sums. A given quarter or town could send money in the name of more than one ancestor, via multiple mediums, and to more than one Great Woman.

Mme Doso also received a letter from Great Woman Safia who was responsible for the Nosy Be segment. Although the spirit Ndrankehindraza was not named in the letter, she referred to it as his portion and informed mediums of Ndrankehindraza in her immediate entourage. They brought money in his name, which she then took to the Great Woman.[4] Mme Doso contributed to the Nosy Be gift not as a medium of one of their ancestors but because she was raised there.

Thus, mediums and others who wished to contribute to the Great Service could do so along multiple routes, according to various aspects of their heterogeneous identities, and implicating one or more mediating figures in the form of particular ancestors, senior mediums, and Great Women. There was friendly competition to see who

Mahajanga, faha *10 Jun 1996*

Manantany Fiarena MAHATOMBO

Fahatelo ny Doany ASINDRAZANA Edouard

Ampanjakaben' ny Boina RANDRIANIRINA Désiré Noël

sy ny mpiandraikitra ao Doany Miarinarivo

———————

Dia manasa **ANAREO** mba ho tonga hanome voninahitra ny **FOTOANDEHIBE**
FITAMPOAN' ANDRIAMISARA EFADAHY manankasina sy ny Fankalazana ny **DOANY**
Miarinarivo ao **MAHAJANGA**

Ny Alatsinainy faha 29 Juillet 1996

Maniry ny ahatongavanareo tapaka sy namana rehetra miaraka amin' ny vady
amanjanaka izahay jiaby mpiandraikitra solon-tenanareo ato amin' ny **DOANY**. Ataovy
ny fomba fanao tsy miova ny **PARE** sy **OMBY** manitra avy aminareo malala-tanàna
arotsay ao **DOANY-MIARINARIVO**.

RANDRIANIRINA Désiré Noël

7.1. Letter from Ampanjaka Désy Requesting Contributions to the Great Service

could announce the larger amounts. Those who pooled the money also attempted to
reach larger sums than were raised the year before.

Within Mahajanga, Mme Doso sent her letter to loyal contacts in two other neighborhoods and also informed people in her own quarter and the adjacent one. Mediums held small ceremonies at which they invited people who had spirits closely
related to their own and collected money on behalf of the spirit hosting the event.
The group of congregants, in turn, took this money to the medium of the senior
spirit. In effect, then, money flowed to Mme Doso on behalf of both Mbabilahy and
Mbabilahy's descendants; both from within Good Progress and from three other quar-

ters. The money was forwarded to Great Woman Honorine, as was money collected by other mediums in this and other neighborhoods.

I examine this process by following Mme Doso over a critical two-week period in 1994 just prior to the Great Service itself, as she attended and organized "fund-raisers" (*tetibato*). This is supplemented with observations from 1998.

Mampiaming's Fund-Raiser, Saturday, July 9

A small collection ceremony was held at Mme Doso's, hosted by a neighbor—a medium of a spirit known as Mampiaming—whose own house was not big enough. Mampiaming is a distant cousin of the present Bemihisatra ruler, Ampanjaka Désy, and a generation younger, who is buried at Betsioko. The medium was a mild young woman who normally dressed in jeans and worked as a clerk in the electricity service. The event lasted from 3 P.M. until it got dark (around 6 P.M.) and included a group of some fifteen women plus a male accordionist and a woman playing rattles. The scene was much as described at the beginning of chapter 3. Mme Doso, who was not in trance during the event, introduced me to the five spirits present. Ndramarofaly, her "offspring," wore a red cloth, had his hair tied in a topknot and carried a stick of light-colored wood. Ndramandaming, her "grandchild/descendant" (*zafy*), sported a felt hat. Ndramandaming is a mother's brother of Ampanjaka Désy. Mampiaming, embodied in the hostess, wore a hat with brim, a sparkling white dress shirt, and carried an ebony stick embossed with silver. Ndranaverona, a male cousin of Désy's generation, sported a jaunty straw hat with a black band around it painted with vertical white stripes. Finally, Ndrankehindraza, popular "man about town," wore a red fez and had a band of white kaolin painted around his right eye.[5] All of the mediums were from the local quarter.

The spirits discussed the relations among them and knew the appropriate kin terms. Even if they did not identify all of the intervening links, their accounts matched the genealogies I received from Amady and others. It was clear that they were also largely Mme Doso's "kin"—the spirits were descendants of Mbabilahy and the mediums and guests were people who had consulted or received treatment or initiation from Mme Doso.

The moment I arrived, the spirits requested drinks and sent me out to purchase cold beer and rum.[6] Of those present, only Ndramaro drank rum, but he hardly touched what I brought and advised Mme Doso to bring it to the next event at which he would be present (not necessarily in the same medium). Mampiaming and Ndranaverona drank lemonade. The hostess purchased three large bottles of beer and two of lemonade, plus two packs of cigarettes and matches.

The spirits, along with the women present, sang, laughed, and chatted. They briefly discussed the price and appropriate color of felt hats of the kind Ndramandaming wore and wondered whether I could bring replacements from Canada. When I passed Ndramandaming the beer, he immediately poured glasses for me and the women present. These were neighbors devoted to *tromba*. Most had spirits of their own who would rise at other gatherings. A given spirit can only appear in one host at a time, and only one host is invited to embody him at any given occasion.[7] Mme Doso exhorted people to attend a fund-raiser at another medium's house on Wednesday. One of the mediums observed that it was not so easy for working people and they should not be criticized for not attending every ceremony.

Mme Doso announced that fmg17,400 had been collected from the guests; in 1994, this was approximately US$5.[8] Given the income of most of participants and the fact that they were contributing on multiple occasions, it was not insignificant. At the time a kilogram of rice cost between fmg900 and fmg1,050, and a kilogram of fresh fish or meat, between 2,000 and fmg2,800. The accordionist was paid fmg1,000 from the total.

The spirits flourished their hats and left one by one, each giving a little performance on the way. One covered himself in his white shroud and lay on the floor with his head cradled in Mme Doso's lap. She admonished him to go gently, remembering his host's age. When the medium re-emerged, Mme Doso massaged her briefly to ease her discomfort. Mampiaming lay in the arms of another woman but prolonged his parting by sitting up and lying down again several times. The accordion stopped abruptly as the last spirit left.

The women changed and departed except for a couple who helped Mme Doso to clean up, move her living-room furniture back into position, and take the sheets off the mirrors of her two armoires. Mme Doso explained that this was the first of several events to channel money to her. She said she would hold a similar event on another day for Mbabilahy directly and would eventually receive money collected in equivalent ceremonies in two other quarters of town from mediums who were her friends. She would then take the money "in three envelopes," bearing the names of the respective quarters, all in the name of Mbabilahy, to Great Woman Honorine.

It is clear that things at Mme Doso's that afternoon transpired at multiple levels. The spirits gathered to raise money for their ancestors, but they were also entertaining themselves in the process. Neighbors contributed money, but they too received drinks. It is a characteristic and critical feature of the collecting activities that those who give always receive something in return. The return is smaller, but like the original gift, it is a sign of respect and sociality. The money and other gifts signify what is really at stake, namely, the transmission of honor (*vonahitry*), which in turn is productive of sacred potency (*hasing*).

Past, present, and future were imbricated in the event. Ndramaro stalked about as though he were hunting in the forest during his lifetime, while I overheard another spirit suggest to an unmarried woman that I would become her spouse. The double set of relations among the spirits and among the women set up a complex and ironic play of meaning and affect. Overall, the impression was one of female solidarity, warmth, and intimacy, even as the central characters were men. Alongside the doubling of gender was a doubling of wealth and status, as debonair royal spirits discussed both stylish hats and the concerns of the urban poor. As Mampiaming's medium observed the next day, having a *tromba* is expensive in large part because of the wardrobe.

In the course of these events not only is money collected, obeisance made, and hospitality enjoyed, but the entire ancestral system is rehearsed. In town after town and compound after compound, groups of royal ancestors rise to greet and acknowledge each other, amuse themselves, and affirm their existence. As they play, they also display themselves, reproducing at a popular level the voices, personages, and relations that constitute Sakalava history.

Monday, July 11

I came upon Mme Doso at the shrine where people were gathered to ask the senior spirits advice on the Great Service (chapter 8). Mme Doso had spent the entire night at a ceremony held for Ndramarofaly in town, where she had been invited to embody Ndramiantarivo, an older brother of Ndramandaming. She had Ndramianta's red fez, together with a bottle of beer and fmg1,500 that she had received. A different medium had embodied Mbabilahy.

Mme Doso reviewed the schedule for forthcoming events. She was anxious about the success of her own fund-raiser (*tetibato*), planned for the following Sunday, and emphasized how she needed to generate money. She explained that participants include people who have received medicine from the spirit in whose name the event is held. However, most of Mbabilahy's clients (that is, of his incarnation in Mme Doso) had migrated elsewhere and hence were not available. That was why Mbabilahy was urging everyone to attend. Because of his seniority, Mbabilahy's fund-raiser would be last in the sequence and people would make larger donations. Mbabilahy is the senior-most ancestor collecting for the *Fanompoa* because his father is among the very people (the *efadahy*, the "four men" who comprise the relics) for whom it is taking place.[9]

As we were speaking, someone dropped by to invite her to an *arbain*, the Muslim feast forty days after a death, which happened to fall just on the day of her fund-raiser. Though raised a Catholic, Mme Doso had been married to two Muslim husbands and had long ago converted to Islam. She told me gamely that she would attend the *arbain* in the morning and return home for her fund-raiser in the afternoon.

Wednesday, July 13

Late in the afternoon Mme Doso went to a fund-raiser across town for the *tromba* of the Nosy Be lineage. Although she was not possessed by any of these spirits herself, she attended because she was raised in Nosy Be. She was also invited to embody (*mianjaka*) Mbabilahy at an event held to accumulate liquor for Ndransinint's house-warming at the shrine. To enter the new house properly there had to be "happiness," and that meant a night-long party (*tsimandrimandry;* henceforward, "all-nighter").

Thursday, July 14

Mme Doso attended the large all-nighter for Ndransinint' in the Ambalavola quarter where, the spirits witnessed the money, liquor, and rice collected (*mijery zaka*). I stayed until 12:30 A.M. by which point the densely packed crowd—squeezed under an awning between two poor dwellings—had risen to 300 people and the number of spirits was fifteen. Mme Doso, herself, did not enter trance until around 4 A.M. Male and female mediums came prepared with the clothing of the spirits that each of them had been asked to embody but entered trance in a seemingly spontaneous and flamboyant manner. Around midnight, Ndransinint's guardian (*rañitry*) from the cemetery at Anjiamañitry announced that he had brought three cows and invited people to the housewarming at the shrine the following night.[10]

Friday, July 15

Ndransinint's goods were taken up to the shrine, and that night the mediums officially "entered the house" (*miditry an trañ:o*). Some of his key supporters must have engaged in three straight nights of trancing, partying, and very little sleep, although the celebrants were not identical on every night.

Saturday, July 16

Mme Doso returned home after spending the night at Ndransinint's housewarming with the half-kilogram of beef distributed to each of the senior mediums. Happily, they slaughtered the cow in a Muslim fashion so she could eat it. They do not always do so; conversely, some Muslims never touch meat from animals slaughtered at the shrine.

Mbabilahy's First Fund-Raiser, Sunday July 17

At 3 P.M. Mme Doso sat in front of the altar in her living room arranging her white dish containing water, kaolin, silver coin, shell, bead bracelet, and various sticks and seeds. She lit incense and placed tobacco and a pile of *tromba* clothing on the altar table. The accordionist began to play. A plump woman, dressed in a wrapper printed with Indian chiefs and tomahawks, presented Mme Doso with fmg5,000. Mme Doso brought out beer (saved from the ceremony attended by her spirit Ndramianta) and a bottle of the soft drink known as *Bonbon Anglais*. The plump woman took up the rattle as more women arrived—one offering coins, the other, fmg2,000. Mme Doso put kaolin on the palms of everyone who contributed money and said to me, "May your book bring you much honor at home." A girl brought fmg1,000, received kaolin, and left again. Even at this lowest level, the exchange is always one of money for *hasing,* the sacred potency signified by the white clay.

By 4 P.M. eleven people were present. Mme Doso prayed for a moment facing the altar, her hands raised, invoking the spirits. An elderly woman prepared herself for trance, smearing wet kaolin on her cheeks, arms, neck, and chest. Women untied their braids. The spirits arrived, one after the other. One woman held a rich purple and yellow cloth (*sobahiya*) over her body, shaking and fluttering the cloth as the spirit rose. An armchair was brought for Ndramandaming. Clothing was elaborate: one spirit wore a long-sleeved mauve dress shirt, pale blue sarong, belt, a *sobahiya* cloth tied as a bandolier, and a towel over one shoulder with another over his knees. As the spirits arrived, they dressed, drank from the dish, lit cigarettes, and doffed their hats in greeting. Ndramandaming sprinkled water from the dish into his hat before putting it on. Mme Doso spoke to each spirit as they arrived; they were almost the same set as those on July 9: Ndramarofaly, Ndramandaming, Ndramanisko, Mampiaming, Ndrankehindraza, and his comrade Dadilahy ny Hussein. The spirits radiated energy, and people greeted them avidly.

A little girl, along with her mother clapped eagerly to encourage the spirits. Ndramandaming beckoned me over to thank me for the drinks, put more kaolin in my palm, and said I would have grandchildren. Later, he remarked that Mme Doso was a good healer. A woman danced in and rapidly entered trance. A man crawled across the mats to greet the spirits. Another man, and then two women, drummed. The plump woman danced, smoked, and loudly demanded rum. She was not in trance but was referred to as a wife (that is, another medium) of Ndramarofaly. Mampiaming drank discreetly behind a white cloth.

During a pause in the music, Mme Doso formally recognized the spirits and announced the amount of money collected: "34,850 (*eny arivo sivy anzato fitompolo*)." People clapped; the *tromba* cried "Ooy, ooy ooy!" The two spirits wearing fezzes stood and danced, making a series of poses, and cooing. A man danced with them. They placed a fez on me and engaged others in similar fashion. The wide circles of clay around their eyes enhanced the comic effect. One by one, the spirits exited in time to the music, putting down or passing on their cigarettes. The congregants helped those still in trance to leave it. Mme Doso passed fmg1,000 to the medium of Ndramarofaly and fmg2,000 to the accordionist, but because the total fell short of her goal of fmg50,000 she would host another event on Thursday.

Monday, July 18

Mme Doso spent the day at the shrine participating in the Tsiarana cattle offering (next chapter).

Tuesday, July 19

Tuesday was a taboo day. Mme Doso sewed a blouse for a relative attending the *Fanompoa Be* from Antsiranana and visited a sick friend. She purchased a new outfit in a design specially ordered by Great Woman Honorine so the women representing Mbabilahy and his descendants would look smart together in procession. The cloth had yellow fish on a green background and the saying, "Do good, receive good (*Manova soa, mahazo soa*)." Mme Doso, who was always well dressed, had two new outfits for the Great Service, costing fmg13,400 each.

Wednesday, July 20

It was taboo (*fady*) for Mme Doso to sew, fix her hair, cook rice, or participate in spirit activities on Wednesdays before dusk. She prepared cucumber salad and ate it with a baguette. She did not attend a neighborhood fund-raiser for Ndransinint'.

Mbabilahy's Second Fund-Raiser, Thursday, July 21

Mme Doso explained the frequent presence of Ndrankehindraza, who was not closely associated with Mbabilahy nor buried at Betsioko: "He 'passed away' at a young age, when he was still active. The ones who pass away in old age don't do the work of fund-raising (*mamory tetibato*); it is up to the stronger, younger ones. Ndrankehindraza collects for Ndramandiso, since they are both buried at Bezavo. He is tabooed Mondays but he will break his own taboo if his ancestors hold their service on a Monday." Monday was the critical day of the Great Service. Thus, historical contingencies—here age at death—were incorporated into the ancestral system; Ndrankehindraza's ubiquity had less to do with his life than his usefulness following his demise. It helped that he had a jaunty style, good humor, and colorful costume.[11]

The atmosphere was similar to the previous occasion. Several spirits rose, and the younger ones counted the money and announced the total: fmg19,700 (10,000 was from me). Mme Doso assured everyone that at Honorine's she would say that the money was from all of them. They replied that she looked after them. Mme Doso was content; somehow she had a total of fmg60,000 to bring Honorine.

There had been much noisy discussion: Mme Doso explained the issue was how to pool money while accounting for the separate gifts. The system is complicated; money is brought from several sources and, depending how it is earmarked, it goes on to one of several Great Women. People may contribute to her on behalf of individual ancestors; they might say, "This is the portion (*anjara*) for Mbabilahy, this is the portion for Ndramandaming."[12] One woman had brought along money for Ndrankehindraza that originated in Nosy Be. In fact, the money collected at Mme Doso's over the past weeks was going to at least three different Great Women: representing Bezavodoany on behalf of Ndrankehindraza; Betsioko on behalf of Mbabilahy, Ndramaro, etc.; and Anjiamañitry on behalf of Ndransinint'. Mme Doso combined the money that came in on behalf of the various Betsioko spirits. It all counted as Betsioko money, and she distinguished the various sources for Honorine only by medium and town or quarter of origin. Honorine tabulated her own totals by town or region, and this is what she formally announced at the shrine.

At each stage the money was counted and witnessed; it became a distinct act of giving. The various announcements indexed the strength of commitment to persons at each node. Despite the fact that all of the money ended up at the shrine, Mme Doso wanted a significant sum to deliver to Honorine. Clearly the path was as significant as the goal.

People also debated which of several all-nighters each of them would attend. By contributing along various paths, through their presence and their donations, different dimensions of donors' identities and relations were affirmed.

I too felt the pull of competing invitations and obligations. I was a bit annoyed when Mme Doso committed me publicly to Ndramarofaly's all-nighter when I wanted to attend that of Ndramandikavavy in order to see some different spirits and mediums. But I realized that she had to share me with her group.

Ndramarofaly and Ndramandikavavy's "All-nighters," Friday, July 22

Because Mbabilahy was so senior, Mme Doso said she need only put in an appearance at the all-nighter that precedes the send-off of the money. This was held by a medium of Mbabilahy's son, Ndramaro', in the Mahavoky quarter. The group who gathered there was part of her "team" (*équipe*) and would forward the proceeds to be included in Mbabilahy's offering. This money, along with money from the mediums in the other quarters to whom she had sent letters, would arrive at her house around 5 A.M. on July 23 after their respective all-nighters. As she explained in anticipation, "People will be delighted the process is completed. They will call Grandfather [Mbabilahy] as well as Ndramaro and other spirits and each group will tell them how much they have collected. He will thank them (*izy mañkashitraka, maña merci*). They will move on to Honorine's where the total will be offered in Mbabilahy's name."

A number of compounds held all-nighters on this culminating eve. One for Ndramandikavavy was co-organized by one of her mediums and one each of Ndramboeni's and Mbabilahy's. These three mediums formed a team. They assisted a client of Malagasy origins who returned yearly to Mahajanga from France to consult with them on business and family affairs. In turn, he supported the mediums with gifts and sent them an annual contribution of fmg1.5 million to the Great Service.[13] Since 1988 they used it to purchase a cow for sacrifice and a new curtain (*safy day*) for the temple. This curtain, which concealed the house containing the relics, was changed every year. It was fifty meters of white cotton and cost some fmg135,000 (in 1994). The mediums and their spirits hosted the all-nighter to send off the new curtain properly.

Mme Doso needed me to attend Ndramarofaly's all-nighter to express her patronage of the event and elicit the continued support of this large group of clients led by an ambitious medium who had already expressed some antagonism toward her. She asked me to supply an expensive bottle of *pastis* (*mason' piso*). Many *tromba*, notably the Christian ones, were forbidden rum, but not *pastis*. Having satisfied them with a nice gift, she could return home to sleep until they brought the money before dawn. She did not add that she would really need the sleep. The next day she would bring the money to Honorine's and then go up to the shrine twice; first, so that Mbabilahy could be consulted by the managers; and later, to participate in the delivery of gifts that actually kicked off the Great Service.

I returned in the evening eager to get started but Mme Doso was not nearly ready. "You told me to come at *huit heures*," I complained.

"*Huit heures, malgaches!*" she responded. (This subsequently became a standing joke between us; there were times I kept her waiting when she had meant "*huit heures, Européennes.*") She had had a series of visitors, including two young people who brought fmg15,000 on behalf of Ndrankehindraza.

Eventually we walked to Mahavoky under a nearly full moon. There were some twenty-five people in the spacious compound when we arrived, and the altar table was laden with incense, glasses, and neat piles of *tromba* clothing and hats. The leading medium sat up front, with the women on mats behind her; men and sleeping children were distributed around the perimeter. The medium displayed the money. Each pile had a piece of paper recording the amount and the donor. She called her son over to read and count it. He announced: "22,000 from Maman' Kiki; 8,000 from Rahely Soa; 10,500 from Maman' Sosiane; 3,000 from Maman' Tina; 3,000 from Martine. . . ."[14] The total was already more than fmg100,000. As people clapped, the medium folded the money in a handkerchief and placed it in a basket on the table.

I will not provide a full description. Prayers were offered, and the music started. The senior medium's husband seated each spirit as it arrived while his wife embodied Ndramarofaly. New arrivals brought money to Ndramarofaly and received drinks of kaolin. Soon there were eight *tromba* present, including a young woman with her face painted completely white, known as Moana Kely (Little Girl). Drinks were passed, and Mme Doso played one of the rattles. It was much more pleasant than the densely packed all-nighter I had attended in the old part of town.

After about an hour, the husband of the medium stopped the music and announced the new total to shouts of *Bravo!* and clapping. Mme Doso offered her respects to everyone (*koezy anareo jiaby*) and told the band to resume playing. She opened a bottle with her teeth and took a swig of wine behind her cloth. The junior spirits worked the crowd, offering cigarettes and drinks. At 11:30 P.M. Mme Doso gave her excuses, and we left.

I proceeded to the all-nighter for Ndramandikavavy. The medium's house was packed with more than fifty people so that they had to move out all the furniture. Kôt' Mena, Ndrankehindraza, and Fotsy arrived, followed by Ndramarofaly, Ndransinint', Ndramanisko, Mbabilahy, and Ndranaverona in a burly male medium who wheezed uncontrollably, staggered, and suddenly fell backward across the laps of his neighbors who covered him in a striped sheet. He then stood, finished dressing, sprinkled clayey water in his hat and went briefly to sit on the knees of Mbabilahy. The crowd sang the song (*kolondoy*) appropriate to each arriving spirit and clapped to help smooth the entrance of each. I left around 1:15 A.M. to get some sleep, missing the entrance of Ndramazavalañitry, Ndramandiso, and Ndramañonjo at 4 A.M.

Pooling the Contributions, Saturday, July 23, Dawn through Mid-Morning

Around 4 A.M. a group from the all-nighter at Mahavoky, including both live persons and *tromba* proceeded to Mme Doso's and remained to celebrate with accordion music and dancing. At 7 A.M. another group arrived from one of the quarters to which Mme Doso had sent her letter. Her room was packed. The *tromba* sat in their standard formation, but for a moment, as the groups merged, there appeared to be two of each.

Mbabilahy rose in Mme Doso and patted the heads of the other spirits who crept up to pay their respects. The group from the third quarter arrived and squeezed up to the front with their money. Everyone was splashed with clayey water and drank from the dish; Mbabilahy applied a white clay moon and dot to their foreheads. A pitch of excitement was reached. People left trance vigorously. Mbabilahy stood, exclaimed "Oi, oi, oi!" and exited Mme Doso. Ndramaro knelt, pointing his spear and shouting. He then fell into the crowd, flailing wildly, and left his medium. Maromalahy, a boy spirit (risen in an adult woman) with a lisp and a carved toy ox in each hand, wandered about. Both his face and toys were covered in a milky layer of clay.[15]

Mme Doso announced the total amounts of money. This she uttered in the older numerals that I was not used to, but it included minimally fmg160,000 on behalf of Betsioko; fmg50,500 on behalf of Ndrankehindraza (thus, Bezavo), and fmg5,000 on behalf of Ndrankehindraza's father (Nosy Be). Among those present as witness was Mbabilahy's guardian (*rañitry*), who was visiting from Betsioko.

Together with some twenty women, Mme Doso proceeded to Great Woman Honorine's. At the front of the group strode a woman bearing on her head the money

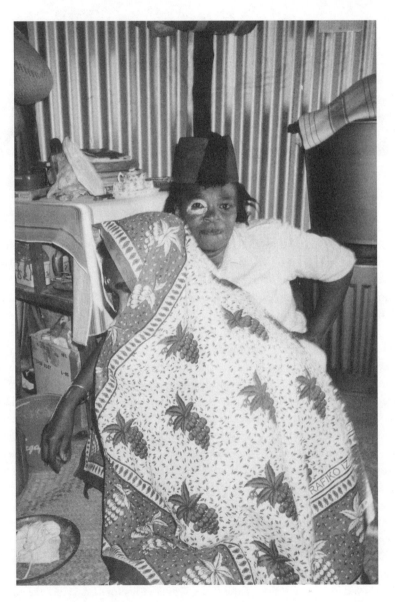

Ndrankehindraza alongside the spirit of a marovavy (female servant).

Mme Doso invoking the ancestors in her living room, Mahajanga.

wrapped in a cloth and sitting on a white saucer. We marched up the hill and then down into the main part of Mahajanga. We passed the Lutheran and FJKM (Malagasy Protestant) churches, then the Seventh-day Adventist and Orthodox churches, the city hall, and two mosques into the dense quarter of Ambovolaña. We entered Honorine's small house. Mme Doso and Honorine each made short speeches in front of her altar, as the rest sat on mats behind them. Helped by members of the group, Mme Doso reeled off the amounts, and Honorine counted the money. Another group filed in singing and sat among us, handing over money wrapped in a handkerchief. The new-comers included Ndrankehindraza and Dadilahy ny Hussein in their red fezzes, and an accordionist. No one in our group had been in trance since we left Mme Doso's. Great Woman Honorine thanked the mediums, and Mme Doso described how the money had been collected, "from women and men, mediums and neighbors, in the name of Mbabilahimanjaka." The speakers used elegant and humorous phrases, and everyone was very cheerful. The Great Woman returned a bit of money to each group and invited us back to an all-nighter that evening.

Another group passed the house on their way to a different Great Woman, and Ndrankehindraza in a male medium greeted us from the street. We continued to the house of the Great Woman for Bezavodoany. Again, people sat before an altar. Ndrankehindraza, Dadilahy ny Hussein, and Moana Kely danced in their midst, standing above them. It was very lively in the cramped space.

We moved on to the house of the Great Woman for Nosy Be. "How are you all?" she greeted us politely.

"We received a letter," Mme Doso replied. "The letter asked for money; we collected money . . ." she went on. She described the whole sequence, and all of the women with us heard it for the third time. The Great Woman thanked us formally in the name of Nosy Be, and everyone responded with clapping. She, too, invited us for an all-nighter that evening.

Mme Doso received fmg10,000 in return (*mosarafa*) from the Great Women, and as soon as we left the last house she divided it evenly among her group so that each woman had fmg500. Among them was a woman who looked Merina and whom I had observed at all of Mme Doso's fund-raisers. She had never entered trance and had seemed a bit distant. On the way, home we chatted and I learned to my surprise that she was royalty from Ambatoboeny, a (classificatory) sister of Ampanjaka Désy. She said that she attended the fund-raising at Mme Doso's every year, that Mme Doso spoke clearly and truthfully and was very strong (*matanjaka*). She contributed no money herself but expected around fmg30,000 from Désy. She explained that Désy put most of the money received at the Great Service in the bank, to be used on behalf of the shrine, but distributed some to family members. Her Merina husband had left her because she refused to abandon the services and stayed out at night. "I can't give up the practices of my ancestors!" she said. However, she was also a practicing Catholic and had married in the church. Mme Doso told me later that the woman's father already manifested as a *tromba*.

Mme Doso strode ahead, in great shape, concerned not to be late for her appointment up at the shrine (described in the next chapter). The other celebrants dispersed; the heaviest broke down and used her 500 francs to hire a *pousse-pousse*.

Great Women

Bemanangy, literally "big/great women," are representatives in town of the segmentary divisions within the Sakalava social world, that is, the political units identified with distinct regions and monarchs and also the clusters of spirits associated with particular tombs. Great Women provide local links between shrines, rulers, and the networks of spirits and spirit mediums. They stand at the top tier of the collecting process; as treasurers, they are responsible for accounting for and delivering the money brought in on behalf of their constituents. They are the spokeswomen for all those who have collected money in the name of the ancestors in their branch and rise proudly at the shrine to announce their offerings. A second and connected role is that of hostess and *animatrice*. Great Women are responsible for creating and sustaining large networks of people; they hold spirit gatherings and maintain houses at the shrine where their constituents are "at home" and where liquor is supplied to the relevant spirits and their hangers-on. Great Women are also spirit mediums.

Each political segment of northern Sakalava is represented by a Great Woman in Mahajanga. Thus, the Great Woman responsible for the Marambitsy district and Bezavodoany cemetery within it worked closely with Ampanjaka Amina, the living ruler of the district, as well as with Ndrankehindraza and Ndramañonjo, the ancestral "animateurs" of Bezavo (indeed she was possessed by the latter). Other Great Women represented Nosy Be, Analalava, and the Zafinifotsy spirits.

The Bemihisatra rulers of Ambatoboeny and their cemetery at Betsioko were represented by two powerful women. Twoiba was appointed by Ampanjaka Désy's mother during her reign; when Désy succeeded, he added Honorine. The office of *bemanangy* is not the prerogative of a specific *karazaña*. Twoiba is "pure Comorian," whereas Honorine is Tsimihety. Honorine and Twoiba both collected on behalf of Betsioko and separately delivered large sums. Honorine often accompanied Désy and his wife when they were in town. She was also a medium of Ndramarofaly and Ndramanisko.

Bemanangy Honorine received money donated on behalf of the Betsioko ancestors, not only from Mme Doso but from many other mediums. She dealt directly with representatives of large towns in the interior such as Tsaratanana and Mandritsara. She said that the largest amounts probably came from Diego and Tamatave. She also received money from Betsioko itself. She kept the amounts in her head, not in writing.

When people deliver the *fanompoa,* the verb "*maness*" is used. This means to "drive forward from behind," as one might drive a herd of cattle; but it is a sign of courtesy and, hence, of royalty, to escort or usher from behind. The gifts are carried, wrapped in cloth, on the head, and the porters are "driven forward" in this way. Heavy items are taken to the edge of the shrine grounds by taxi. Mme Doso described the Great Woman as the person who carries Mbabilahy's gifts on her head, but this is not literally true. Honorine explained that the gifts were borne by Ancestor People—it was not her *anjara*—and she herself followed behind them. "Then I tell the community (*fokonolon*) how much they have brought!" she concluded.

Goods are "driven" to the shrine on two successive days, the Saturday and Sunday before the Monday on which the relics are bathed. On the morning of Saturday, July 23, after delivering the money to Honorine, Mme Doso went up to the shrine to embody Mbabilahy and then returned on a second trip to join the contingents representing Nosy Be and the Water-Dwelling spirits (for whom Sunday is taboo) escorting their contribution around noon. She spent Saturday night at Honorine's "all-nighter" and returned to the shrine on Sunday, accompanying Honorine and Mbabilahy's contribution.

Honorine takes up all the Betsioko money that she has amassed on Sunday. In 1998, Honorine recalled her total from the previous year as fmg5,400,000. Mme Doso recollected having had some fmg900,000 from Réunion pass through her own hands in 1997, along with some fmg600,000 from Vohemar. The amount she received within Mahajanga was much less, perhaps fmg200,000.

ENGAGED WORK AND THE ARTICULATION OF SOCIAL IDENTITIES

Despite all of her activities on behalf of Mbabilahy and Betsioko, Mme Doso sent her own portion (*anjara*) with the Nosy Be group. Thus, she marked her attachment quite apart from the contingencies of spirit possession and by means of place and political loyalty, mediated by residence. At other moments of the festival she gave precedence to her status as Tsiarana (clan) and as medium of Mbabilahy and other spirits. Hence, all three modes of identity discussed in Part II were relevant.

We have seen how much of Mme Doso's time is taken up with preparations for the Great Service and how it could be argued that the Great Service begins long before it reaches "center stage" at the shrine. Mme Doso held her own fund-raisers and attended others. To attract potential donors she had to balance friendliness and generosity with strength. She did not have a reliable, magnanimous patron on whom to rely. She felt that some women to whom she sent letters were envious of her, even when she had been the one to manage their cures, alleviate their suffering, and initiate them to the status of mediums. She suspected that one of these women sent her only a portion of what she collected, passing the rest to other mediums. This she considered wrong: "You should respond primarily to the letter you receive."

Although she rarely complained, the process must have been exhausting. By 1998, she told me that she had no need to attend others' fund-raisers as she held her own. "And besides," she added, "'Grandfather' [that is, Mbabilahy] doesn't like me to attend them." She felt pressure to reach ever higher financial goals and that her reputation was continuously at stake.[16]

As senior medium of Mbabilahy, Mme Doso shared with Honorine the responsibility of feeding guests (vahiny) who arrived from out of town to serve (manompo) on behalf of Betsioko spirits. This had a considerable impact on her practice. "From April onwards, once the honey is resting at the shrine," said Mme Doso, "I put aside most of the money, and all the large bills, my spirits receive from satisfied clients in order to help feed the visitors at the Great Service. During the rest of the year I can 'eat from' the money; after all, I am a woman without a husband." Gifts to the spirits must always come in pairs. Thus, if a client gives a fmg10,000 bill plus a fmg1,000 one, she can eat from the latter. She could not use a fmg5,000 bill or anything higher. In May 1998, she put the fmg10,000 note she received from a client into the cupboard for Mbabilahy, explaining that she could no longer touch his money, and gave the remainder, an fmg5,000 bill, to the junior medium involved in the case. She said that if she had a husband, she might "split the money I made during the rest of the year . . . but probably still eat most of it."[17] She did not know how other mediums divided their income "inside their own houses."

Visitors began arriving at Honorine's as early as the Thursday before the Fanompoa Be and were fed there until Sunday when they went up to the shrine. From this point they were expected to look after themselves. In 1998, Mme Doso estimated that more than 100 people from out of town were being fed. Honorine, herself, said that she cooked a fifty kilogram sack of rice a day. In 2001, the guests I observed sleeping and eating at Honorine's were all guardians and their wives from Betsioko.

Although I have focused on Mme Doso, many mediums are active. Some are part of her entourage, some overlap with it, and many are quite independent. Earlier I mentioned a fund-raiser for Ndransinint's housewarming. The host was reasonably friendly with Mme Doso, and they invited each other to their affairs, but the medium sent the money that she collected directly to the relevant Great Women. In 1998, she held three separate fund-raisers, one for each of the three spirits who possessed her.

All of the talk about which sum is for whom, or on behalf of whom, produces a series of commitments and quite specific links in broader chains of reciprocity. These chains begin with individual people from all walks of life but mainly from the female, urban poor, and end with the comprehensive announcements at the shrine. People collect together according to ties of kinship among spirits and of neighborhood, friendship, or clientship among mediums, but the penultimate pooling and counting of the money is by region and political segment, as represented by the Great Women. Ultimately, people's loyalties to and through the political segments of the larger Sakalava world are asserted, as they quite literally stand up to be counted.

The paths of collection and the movement of gifts reproduce links among the various segments of the polity as the segments are performatively rearticulated with one another. The political system exists in these regular acts of giving, much as Evans-Pritchard (1940) understood that the political system of the Nuer was realized in moments of coming together in the feud. In addition, Sakalava affirm their loyalty to and through specific royal ancestors whose presence is embodied and affirmed in the spaces of performance. In this sense, the Sakalava have indeed composed a "segmentary state" (Southall 1956, 1999).

BETWEEN TRIBUTE AND GIFT

Contributions to the *Fanompoa Be* lie somewhere between the ideal types referred to as "tribute" and "gift." The ideology of the gift is generally quite effective, but occasionally the mask slips and the tributary quality is revealed. By briefly reviewing a few remaining aspects of the exchange process, a sharper appreciation can be gained of what is at stake.

Donors

It is clear that a heavy burden is placed on mediums both to give and to generate gifts from others. Mediums plan events, and by embodying ancestors they raise the level of public enthusiasm and enjoyment and provide people with evident reminders of what it is all about. That royal ancestors make their own contributions suggests that living royalty should as well.[18] The late Ampanjaka Tsialaña clarified, "Followers always contribute to a monarch's offering; he shouldn't 'go alone.'" At Analalava, the living ruler is always the first donor at the annual Service there because "she is 'proprietor of the ancestors' (*tômpin raza*). People drive (*maness*) the ruler's portion (*anjara*) ahead of them into the shrine."

In the 1990s, generous donors included politicians. An elected deputy who was also of Bemihisatra descent was said to give three sacks of rice and fmg150,000 one year. Another was reputed to have given fmg15 million before he was elected and sent a cow in gratitude after, as well as fmg250,000 to fix the water pump at the entrance to the shrine village.

Mediums are susceptible to extra tithes. In September 1995, Mme Doso was disconcerted to receive a letter from Ampanjaka Désy asking each medium of Ndramianta (there were few) to contribute fmg50,000 toward Ndramianta's new house at Betsioko. She had already contributed fmg100,000 toward the project. Senior mediums were asked for additional contributions at the Doany Ndramisara. As the voluntary quality of giving began to appear compromised, some mediums felt used and the burden unwelcome (chapter 10). However, Mme Doso's enthusiasm appeared relatively undiminished. She defended Ampanjaka Désy and insisted that shrine managers had to meet many expenses such as the monthly water bills of the shrine community or policing during the Great Service. At the same time, she worried about how she was going to find the money requested of her and admitted that she did not know how money received at the shrine was divided. She confessed that mediums should receive return gifts (*mosarafa*), but all they get is kaolin.

Charismatic mediums are important channels for money from external sources. Malagasy who have been economically successful abroad are often thankful to spirits manifest in specific mediums and send annual contributions to the Great Service through them. Indeed, a good portion of the *Fanompoa* is now "subsidized" from external sources. Happily, this is managed in such a fashion that it does not overwhelm the value of local contributions.

Recipients

Everyone agreed that the proceeds of the *Fanompoa* entered the hands of Ampanjaka Désy and Fahatelo Doara (the shrine manager), but opinions differed concerning which of the two men actually controlled the disbursement of the monies or had the right to do so, or on what basis they divided it. Most of the money was understood to be used for the upkeep of the shrine and, hence, for the collective good of the polity and its subjects. In addition to covering running expenses, proceeds were meant to support shrine workers, including the officials who distributed them. A former official said that the *mosarafa* "gifts" were, in effect, salary and determined by shrine officials, although the monarch might add something spontaneously, to offer "*courage.*"[19]

At base, however, a function of the Great Service is to support royalty, and the reigning monarch receives the largest amount. This could be for personal use, though in the past (see below) the monarch fed large numbers of people. Great Woman Honorine described the *Fanompoa* as the "honoring of royalty (*fañajaña ny ampanjaka*)," without distinguishing between ancestral and living members. She made the point explicit: "The monarch doesn't work (*miasa*).[20] If the people didn't support him, what would he eat?" The same held for shrine officials. Honorine acknowledged that she too received a gift (*cadeau*) at the *Fanompoa,* but that it was "very, very small (*heeely*)." Considering how hard she worked, this was more than deserved. There was jealousy among royalty as to how their portions were divided; some members of collateral branches and their supporters felt that Désy did not share sufficiently.

Conflicts over redistribution are inevitable. One medium was outraged that he had never received so much as a piece of meat from the sacrificial oxen he brought to the shrine. Another person rationalized his ill treatment by arguing, "It doesn't matter whether the Fahatelo gives me a small piece of the proceeds; ultimately it is the ancestors who will support me, give me my share. Whatever the officials do, I keep participating, in honor of the ancestors; the unpleasant politics at the shrine are immaterial in the larger scheme of things." When I pressed a former official why people, especially poor people, contribute to the *Fanompoa* even though they know the reigning monarch gets most of it for personal use, he pointed out, "The final destination is largely irrelevant to the donors. People give to the ancestors (*raza*), to the 'four men' (*efadahy)*, not to the living monarch. What the living ruler does with it, *ça fait rien*."

Engagement

Although the flow of valuables depended in large part on personal ties and the various offices and rules were not always consistently defined or applied, the system operated as a reasonably effective bureaucratic apparatus. Monarch and managers extracted money year after year and were able to carry out their projects as planned. Although the sums were obviously much smaller than the budget of the Malagasy state, they seemed to be collected more efficiently. Indeed, one intellectual in the capital remarked (humorously) that perhaps national development would be served by resorting to Sakalava methods of engaging people in collective projects.

The system is effective because it engages people in what they perceive as reciprocity. People give for multiple reasons: in hope of blessing, strength, ease, and prosperity; to honor the ancestors and, quite literally, pay their respects; out of social pressure and collective spirit; and with pride. Each act of giving links people in successively broader constituencies across space and time and both draws upon and contributes to specific dimensions of personal identity and social attachment. Collectors in town mediate "horizontally" between the wider world and the sacred shrine. They also mediate "vertically," drawing upon poor subjects and channeling the money upward to the monarchy.

Each act of giving is formally *received* with thanks, clapping, speeches, and the like, and the reception constitutes a form of recognition. Acts of giving take place in a sociable context, with music, drink, and warm fellowship. Everyone appears to have a stake in the maintenance and reproduction of the politico-religious system; they are proud to have the honor to participate.[21] If this point is not brought home in the reception of their gifts, it is certainly clarified in the mimesis of historical figures.

It is important that donations to the *Fanompoa Be* are conceptualized as acts within longer cycles of reciprocity. The raw honey with which the season begins is distributed among the Great Women and by them to the very people they expect will collect and send them money for the Great Service (as though they were being "sweetened up"). Rather than an "original" gift, then, the money is already a kind of counter-prestation; and it too entails a return. This comes in two forms. Material gifts from the ancestors back to living subjects are referred to as *mosarafa*.[22] These include the small amounts of money that are returned to participants at each step of the collecting process and the meat distributed at the shrine. The second form of return is blessing (*tsodrano*). It is transmitted by kaolin, especially by means of the clayey water that is aspersed and drunk and is a manifestation of the sacred power (*hasing*) of the ancestors. A common blessing, reiterated as the royal spirit rubs the donor's palms with white clay, is one of worldly success: "May the money you have given be returned to you many times over, may you become wealthy." And, of course, the more worldly success people achieve, the more they put it down to ancestral intervention and give again in return. Thus, the reciprocity is, as reciprocity should be, continuous. And people return to each *Fanompoa* with renewed hope and expectation. This is one reason the Great Service can be said, with some justification, to "grow."[23]

WILLING SERVICE

We have observed widespread and generous support for the royal system on the part of mostly poor people. Urban women give generously, and the final sums handed over appear tremendous to them. But they do not provide material returns sufficient to support the livelihood of the shrine servants. Here I explore why redistribution has declined and why people continue to give willingly. This entails two interconnected shifts: in the medium of exchange from cattle to money; and in the primary donors from the general populace to spirit mediums.

From Cattle to Cash

The decline in the standard of living of shrine residents is due in part to the increasing monetization of both the economy in general and the royal redistributive economy in particular. In the recent past, both relied much more heavily on cattle and rice. Though such gifts were once used by royalty for export in exchange for luxury items, it was also simple, obvious, and direct to use them to feed servants and those who gathered at royal events. Now that the bulk of what the shrine receives comes in the form of money, it is easy to bank or to use for any number of purposes other than feeding people.

Cattle retain saliency and symbolic import.[24] Each transition or phase of service requires a sacrifice and is said to "open the red blood gate." Royal burials must also be accompanied by blood sacrifice; nowadays this is always cattle.[25] For the Great Service it is important to sacrifice beasts whose donors have sincerely renounced them and dedicated them to the ancestors. During the 1990s, some six to eight were sacrificed annually at the Great Service. Most were supplied by spirit mediums who had received them from satisfied clients. Mme Doso brought five animals to the shrine over the years. Three, in successive years, were from a client of Mbabilahy grateful for a successful shop and the recovery of an ailing wife; two were from clients of Ndramandaming. Mme Doso does not offer them on her own behalf; rather, the client provides the *tromba* with a beast to offer his ancestors.

Sacrificial cattle are "fragrant" (*mañitry*) and must be accompanied by their "tether cord" (*tady*), that is, a smaller gift of money.[26] Like humans who wish to enter the sacred enclosure, they must be healthy, whole, and without sores or blemishes. Ideally, they have a pattern on their coats associated with royalty, for example, dark with a white forehead (*mazava loha*); their coat cannot be entirely without traces of white. Cattle are sacrificed inside the sacred enclosure (the "red pen," *valamena*). They are chased until they fall over and must not bellow. A beast that complains signifies that its death will not be voluntary and that the donor has not parted with it without regret. Hence, it is useless as a sacrifice and is removed live from the pen. In the past, I believe, it was given to Ancestor People; nowadays the shrine managers sell or use it for another purpose.

In practice, it is never certain beforehand whether donors of cattle will arrive or whether the beasts will accept sacrifice. Spirit mediums are proud of their ability to supply sacrificial animals, but equally they are nervous before the slaughter. Mediums gain prestige if they are able to serve as agents, but a beast who refuses brings shame to the donor whose gift is put into question. It is for this reason that the animals offered are likely to be complacent oxen.

There has always been an association of cattle with royalty. Monarchs used to have large herds and could also extract from the herds of their subjects. When the majority of Sakalava were herders and when they had large herds (thousands of head), offerings of cattle were not perceived as an onerous burden. Beasts of certain colored markings automatically belonged to royalty. They were looked after by the owners of the herds into which they were born but given on request, or simply brought to royalty or ancestral shrines on suitable occasions. In the past, an animal was sacrificed at each monthly reopening of the shrine gates. The meat was distributed and eaten by the residents, and they engaged in a mock struggle over the pieces. Slaughter is now much rarer, and officials take a large proportion of the meat.

In 1998, Kassim provided a thoughtful discussion that illustrates the radical transformation offerings to royalty have undergone in recent decades. He explained that the habit of freely giving cattle to royalty had ended. There was a great decline in the total number of cattle, in average herd size, and in the number of royal subjects who kept cattle. Factors included internal migration, which put Sakalava in a minority in their own country and decreased access to pasturage, the movement of large numbers of subjects to the city where it was impossible to keep cattle, the insecurity of the countryside ("where," suggested Kassim, "cattle thieves are armed with more sophisticated weapons than police"), the expense and difficulty of transport, and the commoditization of cattle for the meat market. Cattle became very expensive for most urbanites. Similarly, Kassim explained, "When monarchs or their representatives made a tour of the countryside, people were once enthusiastic (*mazoto*) and gladly gave rice because they produced perhaps 100–200 sacks per year. Now you are lucky if you harvest 40 sacks. People want to give, but simply do not have enough."

The decline in offerings of cattle also signifies a diversification of the economy where cows are no longer the primary measure or store of wealth. Not only has there been a decline in their actual numbers, but their convertibility into other goods has increased. Along with this, both live animals and meat have lost some of their saliency as signs of satisfaction.

Unlike cattle, money that enters the shrine is not consumed immediately, its division can be handled in secret, and there are an increasing number of projects ahead of feeding residents toward which it can be put. That is, the return of money is far less direct or "balanced" than that of cattle. The transparency epitomized by the mock fight over meat is no longer present.

In sum, living royalty were once supported by the general populace with cattle. People experienced both a relatively high level of prosperity and a return in the cooked and raw meat distributed on the occasions of royal work and festivity. The shift from a primarily rural and prosperous herding economy to a commoditized and weak urban economy has undermined both the subvention and the return.

The Emergence of Mediums as Primary Donors

In both the past and the present, the *Fanompoa* could be described as a system of re-distribution, in which goods and services are absorbed by the center and, to a greater or lesser degree, sent out again. Redistribution is sometimes treated as an ideal type distinct from reciprocity and gift giving. But in practice any system of redistribution is comprised of specific acts of reciprocity that can be understood as more or less bal-anced.[27] Where this process is exploitative and coercive, it is a form of tribute, but Sakalava attempt to ensure that it is not understood this way.

The exchange process is now defined in largely nonmaterial terms. What people give to royalty is respect or honor, of which money is the vehicle. Immediate returns define the original donations and donors as worthy and generous rather than coerced. Donors' self-respect is also enhanced by their own generosity and its acknowledg-ment. Material return is largely indirect, found in promises of wealth and fertility con-veyed by means of the sacred potency transferred in the white clay.

As the return gift moves away from immediate materiality and comes to rely ex-clusively on future potency, so the responsibility of living monarchs is eclipsed by that of ancestors. Direct, immediate, centralized power is displaced by an indirect, diffuse, and often hidden form. Once the autocratic linchpin, the living monarch now merely plays his part in the reproduction of a more dispersed power.

Whereas cattle sacrifice forms a relatively minor part of the proceedings today, it does carry a key to one of the most critical ideological elements, namely, that contri-butions be understood as voluntary—freely given by subjects out of respect for and on behalf of royalty. There is a historical (or mythical) prototype for this; the success of the kingdom of Boina was predicated on the sacrifice of what was dearest to the first king. He proved unable to perceive what this was, but his wife understood and offered herself. Today, her knife (*viarara*) continues to be used to sacrifice quiescent cat-tle, and the man who carries out the killing is one of her descendants (Tsiarana). Thus, the entire polity is based on an ideology of *voluntary* sacrifice and indeed of voluntary subjection as fulfillment of destiny.

The frequent all-nighters exaggerate the voluntary quality. As Kassim once re-marked, "If you don't hold such a party it is as though you are angry before the *Fanom-poa;* participation shows the ancestors that you are content and contributing willingly. The ancestors see this and are happy." People dance, and so do the ancestors. There is an easy camaraderie among them all. Another time Kassim elaborated, "An all-nighter is a kind of 'smoothing' that coaxes (*mitambitamby*) or softens up the ancestors. It re-solves any unacknowledged mistake (*diso*) and placates any anger (*heloko*) for a slight you did not realize having committed."

It is not without interest that the first voluntary victim was a royal wife. When peo-ple, especially women, give prestations today, they can identify with the queen, known familiarly as *dady* (grandmother) and formally as Ndramandikavavy (Noble Lady Who

Surpasses All Women). She gave her life for the benefit of her descendants, and contemporary gifts continue to ensure fertility and the success of offspring. The point is made that royals, too, are subject to the system and expected to sacrifice freely. This unity is clearly expressed in the cooperation between the living and royal ancestors in the collection endeavor. Thus, many of the figures, who elsewhere are venerated in their own right, become aligned with their subjects as donors with respect to the more senior royal ancestral recipients. This association and cooperation between royals and subjects helps realize the expectation to give willingly.

Here, then, is the second dimension of change. When I asked why senior spirits like Ndramandikavavy appeared in so many mediums, Kassim gave a thoughtful answer connected to service: "In the past spirits were limited to fewer mediums, but in the past all Sakalava provided 'service' (nanompo)." They offered cattle, and this was relatively easy to do. As the number of people with cattle and the size of herds has declined, service has become less straightforward. Nowadays one needs cash in order to give. In addition, many Sakalava have stopped following the traditions. So the burden of performing service has shifted from the mass of Sakalava subjects to those with tromba. Conversely, the tromba spread outside the Sakalava sphere to increase the base of support for service.

Mediums and tromba are happy to serve since in so doing they are also contributing to their own power. In supporting the ancestors and the ancestral system through the Great Service, they are simultaneously empowering themselves—engaged in their own self-production.

Kassim described a major transformation in the nature and circulation of power, from a system based on the obeisance of a subject population for whom tribute to powerful living monarchs was not an excessive burden (at least, as remembered now) to a community of practice in which a partially self-selected and dispersed population is linked through the diffusion of, and subjection to, ancestral power. Mediums have become the quintessential royal subjects. They are the chief transmitters of the vehicles of hasing—money to the center and kaolin back to the populace. To have access to the shrine, they must support it; the shifting field of power is one in which mediums, living monarchs, and shrine officials each carefully attempt to retain a footing.

Mediumship is thus an increased and visceral form of political subjectification. Its demands are uncompromising, and they have moved somewhat from a public sphere of external coercion to an internal sphere of unconscious motivation. Mediumship is demanding. But it should be clear that mediumship also brings its own rewards. These come in the form of expanded agency—respect and recognition and increased space and means for creative production, moral and meaningful practice, sociability, and fun. Mediumship offers a calling that links internal and external demands and opportunities into a unified whole. Mediums are exemplary Sakalava subjects and among the most effective. We have seen how hard they work, with how much enthusiasm they provide "service," and how their own performances afford them a mirroring that thickens and enriches the meaningful context in which they work and that is ultimately self-empowering.

The Great Service
(*Fanompoa Be*)

e have already seen that service (*fanompoa*) constitutes the main idiom and vehicle by which people express their connection to living and ancestral royalty, whether freely given or offered in response to demand, and whether in the form of money, goods, or labor. Service is an act at once of honoring, deferring to, and caring for the recipients and it can be compared to the Christian use of "liturgy" or church "service" in that it is understood as sacred ritual.[1] The epitome of service and the climax of every year is the Great Service (*Fanompoa Be*) at which the ancestral relics are bathed (*mampitampoko*). As elsewhere in Madagascar, this annual royal bath is a festival of renewal. It is understood as a celebration of the new year—a time both to revivify the power of the relics and to renew commitment to the ancestors, to ask for blessings, and to celebrate ancestral power.

The Great Service is a highly polyvalent event to which participants and residents of Mahajanga attribute various meanings. From the perspective of Ampanjaka Moandjy, it is the occasion at which the monarch asks his ancestors for blessing (*mangataka tsodrano*), for a good year for his subjects, for sufficient food, fertility, and the growth of children. A Muslim Comorian described it as a pilgrimage and noted the heightened activity in town beforehand as people shopped for new clothing and received visitors. A medium said matter-of-factly that it is performed to cleanse the sacred objects. A Great Woman explained the Great Service as respecting or honoring the monarchs (*fañajaña ny ampanjaka, mañamia ampanjaka vonahitry*).

The *Fanompoa Be* is simultaneously service and celebration; a recharging of the power of the relics and a public occasion for asserting and observing loyalty—the coming together of all the complementary parts to make a whole. It works in the way Durkheim described ritual more generally (1915), generating an effervescence in which members of society recognize themselves through their acts and obligations and society as a whole affirms its power and sanctity to itself. The Great Service is "religious," "political," and "social" in intent, function, and meaning—a "total social fact" in Mauss's sense (1966).

At the Service, Sakalava bear their history, in all of the senses described in chapter 1, with enthusiasm, pride, and hope. Like virtually all things Sakalava, present practice and future orientation are grounded in recognition of the past. Without proper care of the relics and without ancestral acquiescence, there would be no future. As such, service is never really completed; the celebration of each ceremony looks forward to the next, and the Great Service is described as having grown (*tombo*) rather than ended (*vita*).

Alongside this organic conception, the Great Service has developed an objectified quality. The national government reformulates it as an expression of one of the ethnic groups that comprise Madagascar and promotes it as a destination for national and international tourists; a piece of the "unity in diversity" that exhibits Malagasy and Sakalava "culture."

The Great Service takes place at the full moon that falls in the month *fanjava mitsaka* (approximately July). It lasts a week, running from Saturday through Friday. In 1994, it took place from Saturday, July 23 through Friday, July 29. On Saturday and Sunday gifts are brought to the shrine and publicly offered and received. On Monday the relics are bathed, and on Friday new mediums are tested and the successful completion of the *Fanompoa* is celebrated with a victory dance (*rebiky*). There is no formal activity on the intervening inauspicious days. The week includes moments of great seriousness and intensity, of expectation and of tedium, but also periods of play. Indeed, the area just outside the shrine village is turned into a fairground filled with food, drink, and gaming stalls.

Preparations, Consultations, Perturbations

The success of the public ritual is predicated on the performance of a number of more exclusive events leading up to it. Much as spatial access to the relics requires opening gates and permission to cross thresholds, so temporal access entails successive permissions and movements forward. Following the precept that "full containers don't rattle" (chapter 2), what transpires back stage may well be of equal or even greater importance in the Sakalava scheme of things. I briefly review the preliminary events of 1994 before describing the week of the Great Service itself.

The Great Service is part of a broader annual cycle, beginning in April, when people gather to clear the sacred enclosure of the vegetation that has grown up over the wet season. It ends in the month of Asara Be (September), with a ceremony of closure. The sequence articulates with service taking place at other shrines and for other ancestors. Most *fanompoa* take place in the dry season when roads are passable and it is slightly cooler than the rest of the year.

In 1994, the weeding of the enclosure (*valamena*) began on Monday, April 18, and finished on Friday, April 22, with no work on the intervening taboo days. The mediums of the "four men" (*efadahy*) whose relics are housed at the shrine provided the money for a cow that was sacrificed and its meat distributed to workers. A week later raw honey was distributed. The honey was eventually cooked on a Monday at the full moon (June 27), as described in chapter 5.

Around this time the Malagasy supreme court issued a judgment affirming Bemazava rights to the ancestral relics. The jubilant Bemazava celebrated at the northern shrine, whereas the Bemihisatra, at the southern shrine that contained the relics, were extremely anxious about their fate and the orderly enactment of the Great Service that year. Both groups called for ancestral advice at their respective shrines.

On Monday, July 11, 1994, two weeks before the Great Service, the Bemihisatra consulted Ndramisara. Supplicants entered the shrine as usual, giving a small coin to the doorkeeper and offering as much as fmg10,000 when they paid their respects to the ancestors inside. "The amount depends on what you feel," someone explained. Some fifteen men sat facing east to the relics housed on the other side of the curtain, with about twice as many women sitting just south of them. Led by the shrine servants, the women clapped and sang. Somotro played the drum, and someone behind the curtain blew the conch. The doorkeeper then began to acknowledge people alternately to the right and left of him to pay their personal respects. Each person expressed his or her hopes as they handed over money. The doorkeeper then intoned loudly, "So-and-so has come to pay respects today, and wants such-and-such." A woman asked that her daughter give birth safely, and the doorkeeper elaborated, "May she have twins." A man asked that his house be built. The doorkeeper gave each supplicant a small piece of dry kaolin at the end of each invocation. He pushed the baskets with money and kaolin back and forth under the curtain.

Before everyone had completed their turns one of the seated men began to enter trance. Ndramisara was lifted to a vertical position by those on either side of him as the doorkeeper reached behind the curtain to a wooden chest for his red shawl. Ndramisara stood severely bent over at the waist, draped in his shawl, and shook as he spoke. Just before he left trance, another man entered and was given a cloth striped in yellow with silver thread. This was Ndramandisoarivo. Then a third man entered trance, possessed by Ndramisara, and received the red shawl. Each spirit, in turn, conversed with the shrine managers and then quickly exited the evidently exhausted medium.

The officials asked the senior spirits for help and assurance to keep the relics and sacred objects together at the shrine. They said they had heard that Ndramisara had risen elsewhere (that is, at the northern shrine) and uttered a different opinion there, and they wanted to know the truth. "Ndramisara replied," someone explained later, "as if they were children, saying, 'Do what is right; desist from what is wrong. Do what is clean; desist from what is dirty. Follow the *fômba* (customary practices).' He rose in two people because he hadn't finished what he had to say in the first. Ndramisara never stays long in one person; he makes his medium too tired. In the second medium he told them to go ahead and plan the Service and offered them confidence (*courage*)."

For her part, Mme Doso, who was present, admitted that she had not understood the first speech of Ndramisara but said the second was clear: "As a healer-diviner (*moas*) Ndramisara listed a number of medicines (*fanafody*) they could apply for protection. Ndramisara keeps watch over the path and must give the all-clear. He said to proceed with the Great Service and advised it would go well."

Here is sacred speech that has the potential to intervene in history. However, when Ndramisara spoke in his first medium it was in generalities—generalities that were also picked up by Ndramandiso. In the second medium Ndramisara began a more decisive response, one that the assembled people wanted to hear. The fact that Ndramisara shifts rapidly among his mediums suggests that none has individual autonomy. Truly decisive and original interventions are rare, and the speech of one person can be offset by another so that it ends by reflecting as much as shaping public opinion.[2]

On Saturday, July 16, I returned to the shrine to catch the tail end of an "all-nighter" celebrating the new house built at the shrine for the spirit Ndransinint'. At 11 A.M. there were still many *tromba* (spirits) stumbling around. Despite the empty crates of beer and discarded rum bottles, I was asked from all sides to provide more drink and tobacco. In each house, hospitality was overseen by the respective Great Woman. Ndransinint' sat in the house of the Analalava spirits, flanked by Soazara's father and brother. The latter wore pith helmets (*casques*) and white shirts—with the father in sunglasses—and smoked and drank as they listened to an accordion and rattle. Everywhere humans and spirits intermingled in easy companionship. A medium of Ndransinint', originally from Mahajanga but living in Réunion, had returned for the events. She sported a large gold watch and several gold teeth and said she had provided a good deal of the money for the house construction but was unhappy with the accounting. Ndramaro and Mme Doso attempted to resolve her concerns.

Behind the scenes there was still a good deal of tension concerning performance of the Great Service and the disposition of the relics. Ampanjaka Désy was in Antananarivo attempting to explain his side of things to President Zafy Albert. The guardians slept in the temple against an expected Bemazava raid. According to one, Ndramisara rose and said people would not dare come steal the relics and that he favored a settlement rather than a fight. Ndramisara said he would not go with the Bemazava if he were not taken properly, that is, that his relic could not simply be grabbed.

On Monday, July 18, the service of the drums took place. This is held by members of the Tsiarana clan on behalf of their ancestress Ndramandikavavy (Noble Lady Who Surpasses All Women). Wife of Ndramandisoarivo, the conqueror of the North, and mother of Ndramboeniarivo, the inheritor of the kingdom, her sanctity (*hasing*) is stronger than theirs and she is treated with utmost respect. Her saber (*viarara*) is stored at the shrine and bathed and carried along with the relics.[3] She must acquiesce each year before the Great Service that honors her husband and son can commence. Tsiarana stated forcefully that they were opening the Service, not contributing to it.

Each year the Tsiarana sacrifice a cow that both signifies Ndramandikavavy's agreement to the *Fanompoa Be* and provides new skins for the sacred drums beaten during the Great Service.[4] Ndramandikavavy's power is doubly concealed. The sacrifice takes place the Monday before the Great Service, out of the limelight, and Ndramandikavavy herself never rises at the shrine. She is apprised of the forthcoming Great Service and her permission received the week before at the house of a medium in town. Mme Doso explained that her son Ndramboeny forbade her presence at the shrine because Ndramandikavavy is so irascible (*mashiaka*), but the deeper reason is that like all spirits she must avoid the instruments of her own death. In her case, this is the sacred saber. In effect, Ndramandikavavy's original self-sacrifice that established the potency of the monarchy, and without which her husband and son would be nothing, is rehearsed annually as she renews her acquiescence to the Great Service. In my interpretation, the beast is sacrificed not only in her honor, but in her place. It is "her skin" that is materially constitutive of the power and resonance of the drums. She is present at the Great Service in and through the drums.

A small group of Tsiarana—mostly women—pooled the money that they had collected from Tsiarana and from people with Tsiarana *tromba* for the sacrificial animal.[5] Because Tsiarana have long intermarried with royalty, a number of *tromba* are Tsiarana as well. Indeed, there was some discussion among the women whether all of Ndramandikavavy's descendants (and therefore the whole ruling line) were Tsiarana or whether the category was restricted to descendants from her first, commoner husband. One woman proposed that all Tsiarana could be considered royalty. Although this was not generally accepted, it is indicative of the pride that Tsiarana take in their position. Interestingly, another woman remarked that most Tsiarana were women. This perception has to do with their identification with a powerful ancestress who asserts the rights of women. For example, she does not permit her female descendants to accept polygynous unions. Among the women, one was half-Indian and another half-Mahorais.

The money was handed to the person nominated as their leader—the man who carries "Grandmother's" saber during the annual procession and makes sacrifices with it. The Tsiarana, including two junior *tromba,* then paid their respects to the relics, entering the temple by the sacred south door. The intercessor loudly invoked the ancestors, "Come, come, come!" The drum throbbed, the conch was blown, and the women stood and danced. Meanwhile, the cow was killed outside. A small procession entered through the sacred door bearing the implements used in the sacrifice—a large black saber, an ax, and a clay water pot (*sajoa manintsy*)—and replaced them behind the curtain. The women then sat outside the temple to the west and clapped and sang as the beast was skinned on the other side by male Ancestor People. Cut into four pieces and pegged at the corners, the skins were hung to dry on the inside of the south wall of the *valamena* (to be fixed to the two drums on the following Saturday.) The animal lay skinless but unbutchered. Only the *fahatelo* could divide it fairly.

Parallel activity by other groups took place. Zafinifotsy spirits sang and drank in their own house, following an all-nighter, and a party associated with Ndransinint' held another cattle sacrifice. In addition, three men from the Ministry of Culture sat in the plaza, waiting for an appointment to interview the *fahatelo.* Ampanjaka Désy arrived and held a conference with the senior officials and mediums. The animals were butchered; shrine residents received their share while mediums of Ndransinint' took a whole leg back to town to feed their guests. Tsiarana cut personal portions from a side of ribs while Zafinifotsy divided their slab behind their house.

On Saturday, July 23, I accompanied Mme Doso back to the shrine as soon as she had delivered the money to Honorine (see chapter 7). A television crew from the capital was setting up, and a few European tourists milled about. In the sacred enclosure Ancestor People were refitting the drums. Mme Doso chatted with Ampanjaka Amina and others gathered at the *fahatelo*'s house. People examined a pamphlet on the *Fanompoa Be* that the Ministry of Culture had produced. The men listened as the female mediums spoke animatedly and joked about.

A small group entered the temple to pray for a successful Great Service. Ndramisara rose in a white-haired man; as diviner, he could say whether all was clear ahead. Ndramandiso appeared in a female medium and looked around hawkishly with narrowed eyes. He was followed by Mbabilahy in Mme Doso who sat at his grandfather's feet and by Ndramboeny in a young male medium who sat just south of his father. The *fahatelo* conversed with the spirits and then took everyone outside to announce that they had the ancestors' acquiescence to proceed with the *Fanompoa*.

THE PUBLIC OFFERINGS

On the first two days of the Great Service people present their gifts at the shrine. This is a public event, held under the mangos northwest of the temple enclosure and observed by large crowds of well-dressed visitors. The event, as described in chapter 7, is the culmination of weeks of planning and a whole series of gift-giving ceremonies, ending with final night-long parties.

On Saturday and Sunday the offerings are borne to the shrine on the heads of Ancestor People who must have both parents living, followed by their principal donors. People bringing gifts gather at the north entrance to the shrine village. They lower their packages, rest, and prepare themselves. The men wear waist wraps, the women have their hair in long, loose braids, and everyone mixes with junior spirits. People dance and joke, and there is much excitement. Groups of donors then proceed south toward the plaza.

When, with much fanfare, the procession reaches the north end of the plaza, the precious bundles, each wrapped in white cloth, are passed from the heads of the devotees to those of the shrine servants who crawl forward to place them on mats before the *fahatelo,* royalty, and the multitude of observers who sit to the west, facing east. On July 23, 1994 this group included some 100 women and 20 men and grew over the course of the afternoon to many hundred. Ampanjaka Amina and her adult daughter were led out by the medium of a spirit from Bezavodoany.[6] They sat on the mango roots, together with Ampanjaka Désy, who was led out by his counselor (*manantany*) from Ambatoboeny.

A group arrived from the north pushing a cow ahead of them. They were followed by a long procession of women, singing and clapping. Those bearing goods on their heads shook as if entering trance. The donors stood on the east side facing the *fahatelo* and the large audience. Fourteen Zafinifotsy *tromba* (four in male mediums) in long white wraps were among them. The visitors announced in turn who they were, where they were from, and what and how much they had brought. The *fahatelo* responded to each offering, and the crowd listened carefully. Speeches were formal, containing mutual salutations and courtesies but also some humor. As people spoke or were spoken for, they left the standing group, which grew smaller. One or two men made most of the speeches; others prompted them with names and amounts. Some Great Women presented their own speeches, other drew on speech makers they had hired. In both 1993 and 1994, one of these was Queer Amady, renowned as a public speaker; others were dignified older men.

Several Antandrano spirits and one Zafinimena tromba outside the sacred enclosure at the Fanompoa Be.

A group of celebrants, including several Water-Dwelling spirits south of the sacred enclosure, Doany Ndramisara, Mahajanga. The "red blood" door is to the east.

Flanked by the Zafinifotsy *tromba,* Amady spoke on their behalf and the *fahatelo* accepted their gifts. They were followed by a contingent from Nosy Be. The Great Woman adjacent to the speaker stood tall and looked ahead as the sum she had brought was announced. The speaker listed all the "kinds of people" (*karazan'olo*) who had contributed to the gifts. Among the towns mentioned were Marovoay, Antsohihy, and Analalava. Amady next announced contributions channeled through Great Woman Twoiba, including women named from Mayotte, Comoros, and Réunion. The amount from Mayotte was more than fmg1 million. Consulting a list, he went on to mention Maintirano, Marovoay, Antananarivo, and Mahajanga.[7]

Onlookers clapped after each announcement. At the end, a spirit dressed in red started a *kolondoy* (song in honor of a royal ancestor) and everyone joined in singing, "*He-ee biby añala* (beast in the forest)." The officials inspected the packets, which were then placed on the heads of kneeling servant women and carried off for safekeeping as another procession with junior spirits, a cow, and packages wrapped in white cloth took their place. The packages included an even number of newly woven mats offered annually by people from Beronono.

A general air of excitement, satisfaction, and ease prevailed. Everyone was dressed in finery, including jewelry, perfume, and in many instances, hair pieces. *Tromba* strolled through the crowd, for the most part dignified, relaxed, and gracious. Some, like the former bandits (*jiriky*) who lived in the countryside, were a little less sociable. A *tromba* to whom I had earlier given money for drink offered me a tipple from his rum bottle. After the prestations, most *tromba* quickly left the plaza for their respective houses, which filled with visitors, music, and drinking. Like everyone else, I greeted acquaintances from the previous year. The Tsiarana women spoke proudly of "our drum," which was now completed. Mme Doso sought out Ampanjaka Désy to say good-bye and, on the way home, enjoyed *clarinette* (flavored ice), purchased from one of the many stands newly set up outside the shrine village.

Sunday saw even larger crowds as gifts were presented on behalf of additional components of the polity. The first procession arrived with some very fine mats for the relic cabinet. A group of some thirty donors stood together. One man leaned on an old rifle. Each speaker greeted the *fahatelo* and asked after the Ancestor People, senior mediums, and other shrine residents. The *fahatelo* returned the greetings, and the speaker thanked him. The speaker then said, "We received a letter; we acted on it, and we have brought—not one thousand, not ten thousand, but . . . [whatever the amount]." A number of large sums were announced in this manner.

During an interval, people watched the Zafinifotsy spirits amusing themselves at the south end of the plaza. Mme Doso and her fellow mediums of Betsioko spirits sat together looking very smart in their matching outfits printed with a fish pattern. I found myself perched on one of the mango roots next to a *conseil technique* of Malagasy President Zafy Albert. Cars of the bigwigs were parked near the *fahatelo's* at a spot that had been bulldozed the week before especially for the purpose. At the gate of the shrine were several buses. The butchering of a cow, south of the enclosure, was subsidized and filmed by an Italian team, not an intrinsic part of the Service.

Another round of prestations ensued, received by the *fahatelo* and monarchs, and observed by an ever-growing crowd. Honorine announced a total of over fmg4 million, whereas an offering from Diego, organized by a medium of Ndranaverona, came to fmg630,000. An envoy of Prime Minister Ravony, who was royalty from the east coast, made a prestation. Women adjacent to the speakers coached them with the correct names and amounts. There were bottles of castor oil and saucers of money, each wrapped in white cloth. The women bearing the oil staggered and shook; it is especially sacred as it is applied directly to the relics. Two bottles of oil each came, respectively, from Ampanjaka Désy and Ampanjaka Amina's domains.

Afterward people milled and chatted. Several members of the royal family greeted Mme Doso, who proceeded to instruct Désy's niece how to heal a sore eye; they joked together in an intense and possibly mutually embarrassing way. Mme Doso introduced me to royals from other branches ensconced in their own houses and to guardians from Betsioko clustered at Mbabilahy's house. Ndramañonjo, dressed in a beautifully colored waist wrap, scarf, and white shirt, strode through the crowd in an elderly female medium. He held a silver-tipped staff of blonde wood, and his face was coated with kaolin. Mme Doso greeted him politely as "brother-in-law," and the spirit bowed to her, showing equal respect to the medium of his ancestor.

Just as people began to disperse, there was a commotion as a new procession of at least forty people from Antananarivo led by a medium of Ndramandisoarivo arrived escorting four cows and bearing fmg200,000, candles, and a good deal of fresh kaolin. This prestation "from the Borzan (Merina)" was much commented upon later. In the late afternoon people sat in the plaza listening to a guitarist while Ampanjaka Désy held a quiet audience in his house for high status guests.

Writing many years later, I recall the expansiveness and sense of luxury of the occasion.[8] The Great Service is a celebration of largesse and an opportunity for *ampanjaka, razan'olo,* mediums, spirits, and commoners to renew and reaffirm their ties and positions, vertical and horizontal, mechanical and organic, personal and public. It is a great party. It is also an occasion at which people from all walks of life and from all over Madagascar, including members of the government, proudly "pay tribute" to the "four ancestors." In 1994, the total amount collected might have amounted to fmg20 million, approximately US$5,000, but worth, of course, far more relative to local incomes and budgets.

THE MAIN EVENTS

Monday is the critical day on which the gifts are brought in to the ancestors and the relics are bathed and circumambulate the temple. However, it begins rather slowly. My first two experiences of the day were each rather different. In 1993, unaware of what to expect, I spent most of my time in the plaza darting among the various activity groups. In 1994, I stayed close to Ampanjaka Désy and the dignitaries. This seemed like another event entirely. The hot, noisy, lively entertainments of the previous year were replaced by dignified conversation in the dim light of the houses occupied by Mbabilahy and Désy.

In 1993, I arrived at the shrine around 9:30 A.M. Cows wandered inside the *vala-mena*. Various officials went busily in and out of the main gate. People sat in small groups or wandered here and there, some looking a bit the worse for the wear after the night's activities; others were still asleep. In the spirit houses there was some music and drinking. From the temple came the sound of drums and singing as the shrine servants went about their preparations, including the sewing of the new curtain. People entered in small groups to pay their personal respects to the ancestors. Around 10 A.M. a cow was chased around the enclosure, pushed down, and killed. Mme Doso came up happily to say her that cow was "successful" (*nakahy fa nahazo*), relieved that it had proved suitable for sacrifice. Had it bellowed when it was pushed over, it would have been "denied entrance" (*tsy mahazo varavaraña*). In total, two cows were killed on Monday, two were "unsuccessful," and four others were taken elsewhere to graze until Friday. The carcasses were left lying in the enclosure all day.

At noon, Amady started entertaining the women seated in the plaza. He led their singing and encouraged rhythmic clapping. His antics and sexual lyrics produced much laughter. South of the enclosure the Zafinifotsy spirits made music and entertained another group. An effeminate man pranced around, inciting the ever-growing crowd to sing. The *tromba* Maromalahy, dressed entirely in white, face thickly covered with kaolin, stepped among the seated women begging for insignificant amounts of money. "*Cent francs,*" he squeaked, and "*Marzá,*" his nasal rendering of "*l'argent.*"

In 1994, I arrived at the shrine with Mme Doso, where we were promptly fined by one of the Great Women for having missed the previous "all-nighter." In Mbabi-lahy's house there were several surprises. A line of men, officials from Antananarivo, sat along the east side. To the north of them, Mme le Ministre de l'Education Supérieure was having her hair, which was cut short in European fashion, tied in tiny tufts wound in thread to stick out from her head in imitation of traditional style. Mme Doso talked with her as I sat with the men. On hearing my profession, they directed me to the Minister, who turned out to be Professor Nerina-Botokeky, an ethnologist trained in Paris who taught in the University of Tuléar before talking up her cabinet post. She was also a member of the Menabe branch of Sakalava royalty. To add to my astonishment, Mme Doso, with whom I had always spoken Sakalava, joined our conversation in reasonably fluent French.

When the Minister's hair was done, we all moved next door to Ampanjaka Désy's house, which filled with important people. Madame le Ministre sat close to the northeast corner facing Ampanjaka Amina on the north wall and Désy on the east wall. Dignitaries included another cabinet minister, royalty from the east coast (Antaisaka) and Nosy Be, Amina's two adult daughters and a small grandson who was asked to go round the room and pay his respects to everyone (*mikoezy*), and Moalimo, head of the Tsiarana, who maintains his ancestress's shrine at Bemilolo. The male minister handed cash to Désy, who passed it on to Honorine.[9] Most people sat too far back to hear the royal conversation.

People in the house were sheltered from the light, music, comedy, movement, and talk outside until a few spirits entered and shook everyone's hands. A woman entered trance and Ndranaverona too shook hands, starting with Désy, and speaking quietly to him between great wheezing breaths. Désy's wife distributed beautiful waist wraps to the men from the capital to wear during the ceremony. Most of the women left as the men changed discreetly. Désy wore a shirt and long waist wrap striped in red, black, and yellow; Amina wore a *sobahiya* as scarf and a red and white waist wrap.

The clapping outside intensified. We rose and went outdoors to receive an offering that had arrived a day late. As we sat beneath the mango and listened to the act of prestation, I was bemused to find myself under the observation of the European tourists and television crew. A *tromba* entertained a big circle of people who sang and chanted in response. We returned inside to wait, and I felt quite cut off from the festivities. Inside it was quiet and dignified, not raucous and cheerful. The separation was palpable.

A climax finally came at 1:20 P.M. As the servants stepped out of the storeroom north of the enclosure, men and women jumped up to form two lines holding hands and facing each other outside the west wall of the sacred enclosure. The gifts were carried from the storeroom along this human corridor on the heads of the women shrine servants. *Tromba,* royals, and honored guests, including the various state and provincial officials, followed the goods in through the sacred south gate, "Ndramisara's gate," and then into the temple by the southern door. Those stopped by the gatekeeper returned to enter by the west gate. This all happened very fast, and those who had not jumped quickly enough were left straining and pushing to see. During the movement, a rifle was fired four times. At the sound of the gun, everyone bowed, lifted their hands, and uttered *koezy,* the greeting offered superiors. *Ndramisara's belongings have now entered his palace.*

Despite her important role at the Great Service, Mme Doso has never participated in the procession nor even observed it. One of her spirits, Tintôño, a Zafinifotsy mortally wounded by a Hova rifle, causes her to faint if he hears gunshot. So she makes sure that she is well out of earshot before the grand entry and until the bathing is completed.

In 1993, I had no idea that the relics were bathed immediately after the rush through the gates. I watched in the plaza as the handful of clown-like *tromba* and campy men continued to entertain the significant number of people who had not entered the enclosure. At one place, a blind singer performed traditional songs, which many in the audience taped. At another, a man pretended to be the "insertee" in a coupling with a male *tromba* incarnated by a woman. At a third, a group of rum-drinking Zafinifotsy *tromba* and women danced, stamping their feet and shaking their hips Sakalava style.

In 1994, I joined the dignitaries from Désy's house and passed between the two lines of people along the western face of the *valamena*. We followed the servant women bearing the gifts southward and around the corner to the sacred, blood-red door where I and the people at the back of the line were stopped and told to enter by the public gate. This led us to the western door of the temple where the curtain had been extended the whole length of the room, effectively cutting it in half lengthwise. Those who entered by the sacred door found themselves in the eastern half where the bathing takes place. The western section was crammed with clapping people. As there was nothing to see inside the constricted space but the blank white curtain, I chose to stay outside. A couple of mediums were soon led out, overcome by the lack of air.

I joined the many people waiting outside the temple, but inside the enclosure, where we could observe and follow the procession closely when it emerged. Amady sat near me and happily answered people's questions about the past. During the bathing inside the rifle was fired and the conch blown several times.

Led by a *tromba* and by Désy, the *fahatelo*'s younger brother emerged from the southern door, holding the saber (*viarara*), his arms trembling. Other men carried two ornamental "axes" and eight spears—weapons of the ancestors and regalia of conquest. The relics came next; their bearers wore red wool caps and the names of their burdens embroidered on the cloth bundles on their backs. They were followed by the two sacred drums and cymbals, then Amina and other royals, important people, and a throng of celebrants. The drum played, the conch blew, and everyone clapped and sang. People held hands in a protective chain on either side of the procession and some swung small square fans. Within ten minutes it was over, though the amusements outside continued unabated.

"Why was the saber at the head of the procession?" I inquired later. "Queen Ndramandikavavy goes first," my consultant replied and whispered, "[because] she killed herself (*namon teña*). The saber is her weapon."[10]

After circling the inside of the enclosure counterclockwise, walking between the sacrificed cattle, the relics were taken back into the temple through the sacred southern door. I peered after them, and this time the doorkeeper beckoned me inside. The relics were put away as each bearer in turn climbed into the raised *zomba vinda* from a door on its west side just facing where people greet the ancestors from the other side of the curtain. The spears and axes were laid on a shelf under the structure. Members of the royal family lined up in order of precedence and, two by two, went in to pay their respects to the relics in their resting place. Dignitaries like the ministers were led in too. As each person exited the *zomba vinda* he or she ducked under the curtain, went through the western side of the temple, ducked back under the curtain at the southern end, and exited by the south door. Some people took sips from the big metal basin containing the remains of the mead (*tôa mainty*) and a silver chain; a Jingô woman put the liquid on the crowns of their heads. The *tromba* who had joined the procession wandered at will in a proprietary way.

In the late afternoon the animals were butchered under the *fahatelo*'s supervision and the meat distributed as return gifts (*mosarafa*). Servant men cut the meat into big pieces and handed them to boys or women to take to the households of the shrine residents. The women enacted a ritual tussle over the stomach at the threshold. It seemed that everyone who wanted meat could have some, but the best pieces were reserved for the managers. One of the guardians from Betsioko was angry that his group received so little meat from the many cattle offered in Mbabilahy's name.

The drums were beaten on and off throughout the afternoon. By evening a few *tromba* and musicians were still playing, and some people were dancing. The shrine servants cleaned up the remains of the butchered cattle. The fairground on the outskirts was thronged with young people.

In the evening the radio broadcast Ampanjaka Désy as he explained the meaning of the *Fanompoa Be*. The television news showed the goods being rushed into the temple. There were also excerpts of the entertainments. One of the songs went

Sa'mandeha, sa'mitady	One goes, the other seeks
Lalahy sasan manao sarambavy	Half the men are queer
Olo jaby samby an zakan tiany	Everyone with what pleases them
Zoky 'Mady sary ve'vavy	Older brother 'Mady is
	the picture of a woman

The Bathing Itself

I did not observe the bathing of the relics, which was conducted in secret; however, in 1999, officials gave access to a journalist. I include excerpts from his account published in *L'Express de Madagascar*, August 3, 1999, which is already a matter of record and strikes me as quite accurate.[11]

. . . On one side of the curtain are women seated on mats who sing throughout; on the other side a table covered in white cotton cloth has been placed near the *zomba vinda*. On it sit a large metal basin filled with a mixture of sacred water, sacred oil, *katrafay*, and 4 glasses half full of [cooked] honey. This layout preserves the royal intimacy that no women, other than the wives [of royalty], should encounter. Beneath the table women move raffia fans nonstop to feed the charcoal fire placed in 4 identical incense burners.

A few men in charge of the ritual are busy around the table. . . . They finally decide to take out the ancient royal weapons composed of a machete, 2 axes with pointed heads, 7 spear heads, a piece of iron shaped like a small spade. Each of these arms is carried by a man and the men align themselves in close order. Four men in "uniform" and red hats, enter the *zomba vinda* to bring out each a casket bound on their backs with the help of a brown and white silk covering with dark stripes. These four caskets, shaped in the same manner, enclose respectively the relics of Andriamisara, Andriamandisoarivo, Andriamboinarivo, and Andriandahifotsy [Merina transliteration]. Their respective names are written on the silk cloth covering each casket. The four bearers approach the table. Assisted by their fellows they untie the silk coverings and place each casket in turn on the table. A gesture symbolizing the undressing of the kings ready to take their bath.

They now bathe the "Andriamisara Efadahy." In identical and repetitive movements, each bearer takes a cloth, dips it, with his right hand, into the liquid mixture in the metal basin and in the glass before wiping the casket with care and tenderness without removing the contents. The bath is abundant. Later the caskets are put back on the backs of the 4 bathers. In this manner the idols [sic] are re-dressed. This entire scene of purification takes place under the watchful and tender gaze of prince Dezy and his notable guests. When the four caskets have been well cleansed of dust, and thus are clean, it is the turn of the former royal weapons to be washed. One after the other the bearers pass in front of the table and make use of the same cloths. They start at the end of the handle and move upward to the point. This operation proceeds in a concert of songs.

Once the bath is completed, the relics of "Andriamisara Efadahy" are paraded outside, refreshed and sanctified by popular fervor. Their weapons at the front, the royal promenade circles the temple, from south to north, passing via the east, before returning through the southern door. In the large courtyard the procession passes between four sacrificed zebu, spread on the ground. The four caskets are returned to their places in the "Zomba Vinda" and will remain there until the next edition of the "Fanompoambe," in the year 2000 (Maniry 1999).

Tuesday, July 26, 1994

Outside the shrine in the evening there was a big fete with contemporary dancing (*baly*) and wrestling (*mrengy*).

Wednesday, July 7, 1993

The shrine was quiet; the doors closed. People chatted in small groups. Among the many pilgrims, I met a man and a woman from Mandritsara, a large city in Tsimihety country. From Mandritsara it was a three day drive and cost fmg20,000 for a one-way fare, not counting food. The man brought money for the *Fanompoa Be* and announced it; he was a *fehitany,* someone responsible for "tying up" his region. He explained that during the year the *fahatelo* visits towns in the interior. He spent ten days in Mandritsara, and clearly the trip paid off; there were many mediums from Mandritsara present at the *Fanompoa.*

Celestine is one of these mediums—the daughter of an Antaimoro mother and Tsimihety father who lived in Mandritsara. Celestine identified herself as a medium of Ndramboeniarivo, but on further conversation it appeared she was host to twelve *tromba,* including several from Betsioko, one Antandrano, one from Nosy Lava, and at least one Tsimihety spirit. Celestine had been sick with *tromba* since 1968, first with Ndramanisko, and suffered fainting and blood loss during her pregnancies, from which five out of ten children survived. "Ndramboeniarivo arrived April 20, 1992, a Monday!" she exclaimed. She had been "very sick," she emphasized, and "received many shots (*picures*), but prayed to offset 'errors' (*diso*)." These lay in consuming foods that were taboo, including chicken, duck, pork, dried meat, wine, rum, and cashews, but not beer or coffee.

This was her first visit to Ndramisara's shrine, and she planned to have Ndramboeny "verified" (*fitsaraña*) on Friday, after which she could start working as a curer in earnest. She said the exam was really for all her spirits, or in a sense for her. "They are all one head (*loha raiky*)," as she put it.

Celestine's husband, who worked for a state-run plantation, did not like her being a medium, and they separated. She raised the children herself, and fed and schooled them with money earned from sewing. Two of her children had *tromba* as well, but as they were still young the *tromba* had not risen yet. Her youngest daughter, whom she had brought with her and was listening to every word she said, had Ndramarofaly. He was still helping raise her (*mbôla mitarimy*). Celestine was tabooed chicken ever since she was pregnant with her, and her daughter had never once eaten chicken! Her son was linked to Ndramisara; at age ten he was sick for four months, and she saw Ndramisara in her sleep who told her he was responsible for the illness and instructed what medicine to give him until he recovered. She anticipated that the spirits would fully enter her children when they reached young adulthood.

Although I cannot predict the outcome, this illustrates the role that suggestion can play in the arrival of spirits and how possession can be determined relationally or inter-subjectively rather than individually, and over the long term (Lambek 1993).

Friday July 9, 1993 and July 29, 1994

The last day of the Great Service provides the opportunity for mediums from outside Mahajanga bearing one of the four spirits of the shrine to have their mediumship evaluated (mitsara). Local mediums do this at other times. Friday is also referred to as the day of recompense (valing or mosarafa) of the Fanompoa. Visitors who have performed their service receive balls of kaolin at the shrine. When they get home, they distribute it among the people who contributed to the gift they brought for the Fanompoa. People from Mahajanga receive their valing on the following Monday.

On Friday, the senior spirits (tromba maventy) also come to be informed of what has been accomplished on their behalf. This, too, is a kind of closure. Finally, the success of the Great Service is celebrated in the afternoon with the victory dance (rebiky).

On the morning of July 9, 1993 some 200 people, including six candidates for evaluation, squeezed into the temple. Two candidates, Celestine and a highlander from Antananarivo, whose husband introduced himself as an anthropology student, were women. A tromba rose in each in turn, shaking and bending over from the waist until the upper part of the body was parallel with the floor. The ancestor was draped with a cloth; red for Ndramisara, striped purple, brown, and yellow for Ndramandiso or Ndramboeniarivo. Each was asked questions and then abruptly left trance as the next medium entered it. Mme Doso incarnated Mbabilahy; other important mediums surrounded those being examined. I was too far away to hear the questions and answers but learned that they were basic things like the name of the spirit and its home, and that everyone passed. Afterward the fahatelo spoke to the mediums outside, explaining the customs (fômba) of the shrine. Annual dues (cotisation) were expected, though in practice the point was to collect money from clients and send it in.

As Mme Doso explained it, once mediums have been evaluated, they are asked for money. They pay according to how many years the spirit has risen in them. Kassim emphasized that the amount was not fixed; mediums must give freely, without coercion or regret. Mediums who are being examined receive kaolin paste along both arms and across their foreheads. They sleep inside the temple from Friday through Sunday night and cannot leave the shrine grounds until Monday. They are thus initiated into a kind of guild, and their grounded liminal experience no doubt induces a loyalty to the shrine.

Many people are reluctant to come forward for evaluation. One medium who had been possessed by Ndramisara for over ten years explained that he was "waiting for the 'go ahead' (mbôla mila lalaña nazy)." The exam is valuable for mediums who establish themselves as curers or authoritative spokespersons but costly for those whose encounter with possession remains exclusively personal. Mediums may also suffer anxiety about their performance. The man with Ndramisara attended the evaluation to observe what transpired.

People say that the test used to be more difficult. In the past, spirits had to identify their individual cloth hats, which are stored at the shrine and look virtually identical. Cynics say that shrine officials have eased the test to maximize financial intake. If one considers the changes discussed in the previous chapter, this makes perfect sense. The issue is no longer to single out one strongly authoritative voice for each royal ancestor but to ensure a number of eager subjects and to increase annual revenue.

Conversely, it is notable that subjects are somewhat self-defined; many people in the region do not contribute to the *Fanompoa Be*. Of those who do, many have *tromba*. Their work as healers means that they, too, have a financial interest. Indeed, mediums are kept busy around the *Fanompoa,* holding private curing services for clients attending the event. For mediums, the *Fanompoa* is an occasion at which practitioners acknowledge their licensing body and one another's professional status.

Play

After the evaluation, I went to buy a soda and met Kôt' Mena, one of the Water-Dwelling Zafinifotsy spirits, in a male medium I later came to know well. As we spoke he casually squeezed the breast of a woman next to him at the drinks stall. She did not even look up from her conversation, and he explained that she was his "spouse," that is, also a medium of Kôt' Mena. The *tromba* took me by the arm back to the southern plaza, where the Zafinifotsy spirits install themselves. They sat facing east before white dishes of kaolin-infused water. Just behind them were musicians playing a rattle and a box-shaped stringed instrument. Some of the *tromba* were quite drunk and seemed very tired. A man in rags, possibly a leper, approached, and Kôt' Mena asked me to give him a coin. A drunk adolescent boy with Down's syndrome was requested by the *tromba,* without much success, to move from the sacred spot where he sat in dirty clothing.

The spirits soon requested more rum. When I returned, the bottle was taken from my hands by a shrine servant who crouched down and passed it to the *tromba* on her head, just like the items that had arrived for the *Fanompoa*. Kôt' Mena asked for a coin to accompany it [as gifts are always given in even numbers]. Then, he put wet kaolin in both of my hands and had me make fists and rub my pockets, suggesting implicitly that I would acquire much more. Kôt' Mena then urged on the music, and the man in rags began to dance. Similar activities went on all day in several of the spirit houses.

This brief episode encapsulates two of the most important elements of the ancestral system. First, the entire transaction of rum, coin, and clay was a minute enactment of the *Fanompoa*. *Fanompoa* is, in essence, a ceremonial exchange in which material offerings of respect are provided and ancestral beneficence is received.[12] Second, the attitude of Kôt' Mena epitomized the ancestors, combining ostensible brusqueness and proprietorship with graciousness, and pleasure with ethics. Kôt' Mena casually stroked the breast of a female medium in much the same attitude expressed in the song summarized above. The *tromba* thrive on liquor, song, sexuality, and good humor; that is the point of the "all-nighters" and the activities in the plaza outside the fence. These are, in effect, playgrounds.[13] But Kôt' Mena also channeled the resources of the relatively wealthy to the relatively poor. Seemingly without effort, he was attentive to the needy and abject people who flocked around the spirits and who were able to celebrate along with everyone else. These themes of play and ethics, here evocative of Turner's discussion of liminality (1969), will be reexamined in subsequent chapters.

Dancing Rebiky

From 1:30 P.M. onward, Ancestor People began to beat the two drums that had been hung from a stake planted in the sacred courtyard. Each drummer used a stick on the right end and his hand on the left. Near the drums lay a cow slain early that morning. The television crew arrived, and their setup became a focus of attention. The conch was blown at 2 P.M. and again at 2:30 P.M., and shortly after the *rebiky* began. The dancers had an audience of moderate size, including most of the *tromba,* inside the enclosure; the crowds outside caught glimpses and could hear the pulse of the drums. A few *tromba* continued their own dancing on the plaza during the *rebiky* while the effeminate man mimed blowing the conch.

Rebiky is performed successively by same-sex pairs. The first couple were men, and the second were women, continuing to alternate thereafter through eight sets. One pair of men were elderly; the others youthful. Dancers were led to the space between the sacred gate and sacred door at the southeastern corner. Their dress included a bandolier made of rolled red or striped cloth passed over the right shoulder and another thick band of cloth around the waist. One dancer wore a peaked red hat, rather like a hood, that hung part way down the back; the other wore a similar hat made from striped cloth. Each dancer carried a black staff in the right hand, seemingly to represent an old rifle, and a red handkerchief in the left.

The conch was blown between every round. The drum set the rhythm to which the dancers moved in very short rapid steps, stamping or shuffling. Their arms, held gracefully away from their bodies, undulated gently. Each dancer was attended by a fan bearer, and one of the Ancestor People—a woman with long braids—moved among them. They started with a prayer in the corner and ended there too; the dance itself consisted of a series of movements back and forth across the dance space in an east-west direction. Dancers watched their partners and leaned in toward each other without touching, as though chasing one another.

Rebiky has been described for Antankaraña (Walsh 1998) and for southern Bemihisatra Sakalava (Feeley-Harnik 1988; cf. Tehindrazanarivelo 1997). There, it appears to represent the ancient sibling rivalry and conflict between Zafinimena and Zafinifotsy clans, replaying, in effect, their moves from Menabe to the North. It is a form of historical mimesis that does not entail possession. People in Mahajanga did not provide specific interpretations, though one person suggested that the dancers were dressed like former monarchs. The *rebiky* was "just a Sakalava dance," people said, "performed because people are happy (*rav'rav*)," and to indicate the joyful closing of another cycle of the Great Service. However, I noticed that both the Zafinifotsy *tromba* and members of the royal clan watched with particular interest.

The individual dancers were selected by a young Ancestor Person, a Jingô who also happened to be a rising spirit medium and who danced a lead role in the first set and as one of the fan bearers in several subsequent sets. He selected dancers on the basis of their skill and probably their status as well, indicating his choice by passing on the red handkerchief. Those who wished to decline had to pay a bit of money. Although persons of any status may be selected to dance, *rebiky* belongs to royalty and can only be performed on their occasions. In his house afterward, Ampanjaka Désy acknowledged that the dance was very difficult. An elderly woman emphasized the point and laughingly went through the arm motions while sitting on the floor. She said with vigorous disdain that none of the performers knew it well. Some people pointed to northerners as the best dancers of *rebiky;* others, to Makoa.

In any case, this was the end of the Great Service. Désy reclined against a cushion, and I was too exhausted to ask any coherent questions. Two or three confident smiling women, wearing a good deal of gold, guided the conversation. A few others, both *tromba* and servants, entered to offer congratulations. One woman brought fmg5,000. A *tromba* in a female medium entered together with the *tromba's* live sibling, Ampanjaka Amina, who sat next to Désy in the sacred northeast corner. Another spirit rushed in to greet the living monarchs. Their greeting serves as a mutual recognition and legitimation.[14]

CONCLUSION

The Great Service extends, elaborates, repeats, and celebrates the process of giving. The act of the gift is staged and restaged, performed first in neighborhood get-togethers, then at larger all-nighters, then in the plaza outside the sacred enclosure, and finally in the temple. Successive acts of giving expand their recipients, from *tromba* leader to shrine manager and live monarchs, to the four senior ancestors. The spirits in the houses and courtyards are a more proximate form of the ancestral essence found at the shrine.

The repetition continuously reinscribes donors, indexing their subordination and generosity and reconstituting them from individual clients of individual spirits, to collective subjects of spirits, branch, tomb site, town, or polity, to the entire ancestral political community. With each move to greater inclusiveness the amount given grows proportionally bigger, culminating in the vast sums announced at the onset of the Great Service.

The repetition and extension of the act of giving iconically emphasizes that it is in fact the *act* of giving rather than the amount that is ultimately critical. Giving expresses political loyalty rather than financial obligation, and the rhetorical structure is one of persuasion.

The bathing of the relics is comparable to ceremonies of the royal bath of living monarchs once practiced in Imerina (Molet 1956, Bloch 1989b); to the *fitampoha* of Menabe, where the relics were bathed only once a decade but much less secretively than in Mahajanga (Nerina-Botokeky 1983; Chazan-Gillig 1991); and to the Antankaraña ceremony at which the living monarch invites all spirit mediums in the realm to a mass annual sea bathing (Walsh 1998). It may also be compared to the curing, initiation, and renewal rituals of individual mediums, which involve baths of medicated water, sometimes directly in the ocean (Lambek 1993: 367).[15] The sea is identified as potent (literally *ranomasing*), and estuaries and clean, flowing water are significant all around Madagascar. It is clear from the comparative material that bathing is not mere cleansing but also a form of strengthening; the relics are recharged through their bath.

I cannot pick up all the themes developed by other interpreters of Malagasy ritual but limit myself to a few remarks. Bloch (1989b) has noted that the obeisance of their subjects was a critical feature of the royal baths of Merina monarchs. In Mahajanga this is manifest in the initial gifts, not only from the living but from more junior generations of ancestors to more senior ones. And just as live Merina monarchs splashed their subjects with bath water, so the relics are briefly visible after their bath and people drink and asperse themselves with the liquid. Subsequently all supplicants are given water from a white ceramic dish containing concentrated ancestral substance in the form of kaolin and silver. Kaolin is particularly useful as a medium of transmission insofar as it is both hard and pliable, substantial and dissolvable, smeared on the surface of the body and ingested.

It is clear that among Sakalava it is the *efadahy* rather than the living monarchs who form the superior source or seat of sanctity (*hasing*). The decline of the power of living monarchs notwithstanding, it would appear that the relics have played this role throughout Sakalava history. What is weakened though still evident in the contemporary period is the way possession of the relics legitimates the living monarch. There is the opening of a space between the sacred and the political (see Conclusion).

Conversely, one can ask why relics receive greater attention than royal skeletons. I have rarely heard them compared, but Kassim explained the main difference as being that the "relic" (*mitahy*) began to be constructed while the monarch was still alive, as fingernail and hair clippings and possibly other bodily exuviae were collected. They were amalgamated with small bones once the monarch collapsed, but their source was the living ancestor, whereas the skeletal remains (*masenga*) are purely a matter of death. This, concluded Kassim, is what makes the relics so potent. That is, relics form unique mediators or amalgams of life and death (cf. Lambek 2001). Kassim recognized that this exegesis was no longer common and said that he had received it as "aural inheritance" (*lovan' sofing*).

As the prime public ceremony and festival of the year, the *Fanompoa Be* manifests and articulates central themes and features of the Sakalava polity. The chronotope is constituted through oppositions of inner and outer, as well as genealogically older and younger. The older and the further inside, the more sacred, concentrated, and potent. At numerous points, barriers and exclusions are made evident. Inner and outer divisions are recursive: the shrine community is "inner" with respect to the rest of the city of Mahajanga; the *valamena* (sacred enclosure and fence that surrounds it) is "inner" with respect to the plazas and houses; the temple is "inner" with respect to the enclosure. The temple itself is divided, and the inner portion contains the yet more enclosed *zomba vinda* and then the relics buried in various layers. The ritual enacts highly demarcated crossings of these boundaries, both the "inward" moves of subjects and the "outward" moves of undressing the relics (to a degree), removing layers of dirt, and taking them for a promenade "outside" (to a degree).

The relics are paraded, wrapped and covered. They are kept within walls that are within walls. Much Sakalava politico-religious action is about crossing thresholds, traversing gateways, penetrating toward the center, and conversely about all of the obstacles that prevent most people from penetrating and that distinguish relative insiders and outsiders. If the Sakalava polity rests ultimately on the "bare bones" of royal ancestors, its substance is to be found in all of the layers that lie between them and their subjects—layers that shroud the relics and enable them to remain silent, occluded, heavy, concentrated, and potent. Power is conserved in the bones and maintained by restricting access to them, honoring them by covering and concealment.

Although always salient, the division of the temple into two parts and the distinction between those who have the right to enter the eastern part and those who do not (thus those who can have access to the secrets and those who cannot) could not have been made more obvious than during the climax of the Great Service on Monday. The south gate was jealously guarded, exclusion (as I personally experienced) was blatant, and the freedom of the *tromba* to stroll through just reinforced the point. The western part of the temple was filled with people before the bath began; when the curtain was changed to narrow the space on the western side, people were forced together and some of them were squeezed out.

The amusement outside the fence is distinguished from the more serious business taking place inside. Likewise, the gravity and quiet of the house of the living monarch is contrasted to the liveliness of the plaza and the houses of the spirits. The contrast is reminiscent of discussions elsewhere in Madagascar (especially Bloch 1986, 1999) that emphasize the way ancestral order is constituted through its "conquest" or encapsulation of youthful vitality. Evanescent pleasure and nonreproductive sexuality become positive signs of that which is absorbed by and enriches ancestrality. But at the same time it is evident that this vitality encloses sanctity like another, living fence. Sanctity is surrounded and protected, spatially and temporally, by expressions of youthful vitality that the ancestral relics and the kingship both draw from and in turn bless.[16] Pleasure also signifies to the ancestors the voluntary character of submission.

The Great Service is filled with the past speaking noisily in the present and its joyous reception: spirits, rifle shots, drums, drinking, dancing. This effervescence is also valued in and of itself. The crowds, noise, and enthusiasm are part of what make a "great" Great Service. As people say when it is over, "The Service this year was really 'loud' [conspicuous, celebrated, far-famed] (*Fanompoa naresaka tóng ty*)."

The distinction of inner and outer, concealed and visible, silent and noisy is relevant with respect to Ndramandikavavy in whose honor her descendants sacrifice the first cow. Though her role is kept quiet, she is tremendously important. Indeed, she provides the prototype of how everyone, bovine and human, should give voluntarily of themselves. Although one could analyze those who offer gifts—and especially women, since they form the majority of donors and fund-raisers—as victims of mystification, it is evident that they seize upon their activity as a source of pride and potency. Ndramandikavavy is a patron of women, remembered both as a life-giver (as opposed to a male life-taker) and as someone of tremendous integrity, courage, and intransigence.

The chronotope establishes priority and rights to initiate events. Mme Doso said that Ndramandikavavy leads the *Fanompoa* and that Mbabilahy comes second. He is the one who maintains the "keys" (*fangalagadra*). This statement amplifies Mme Doso's own position, but it also has direct political relevance. Ndramandikavavy made possible the successful reproduction of the dynasty and established rules that limited legitimate succession. Mbabilahy is the significant figure from the generation immediately succeeding the last of the "four men," the successor who in turn serves as the founder of the triumphant Bemihisatra branch.

For much of the twentieth century Sakalava politics was based on a balancing act between Bemihisatra and Bemazava, whose cooperation in the regular enactment of the annual Great Service was symbolized and legitimated by their rights of proprietorship, respectively, over keys and drums. In the years when they held a joint Great Service, the keys were brought from Mbabilahy's keeping at Betsioko and the drums from his brother Ndramahatindry's keeping at Doany Mahabo, located, respectively, on the east and west banks of the Betsiboka. Keys and drums, carried in joyous procession from the respective tomb enclosures, met at the former royal capital of Marovoay and from there were brought jointly to the shrine at Mahajanga, their arrival signaling the imminence of the Great Service. They arrived the week before, in time for the drums to be refurbished with the skin from Ndramandikavavy's sacrifice. After the Great Service, each was returned to its resting place for the remainder of the year.

The movement of the sacred artifacts replicated and reversed the fissioning of the royal descent group from the two sons of Ndramboeniarivo, bringing them together annually at Mahajanga. As it rehearsed history it also provided an iconic representation of contemporary political relations, as well as indexing the balance of power or support for the respective branches and the ideal of unity. Their march through the landscape also indexed how the factions transcended locality. The conjoining of drums and keys in Marovoay and their annual procession to and entry into Ndramisara's shrine would undoubtedly have been quite dramatic.[17]

Most importantly, the movement of the artifacts performatively re-established and enlivened the royal constitution. It was said that Bemihisatra were responsible for making decisions concerning the relics, whereas Bemazava had the job of opening and closing the Great Service.[18] When the conflict between the factions escalated, Ndramahatindry was no longer permitted to bring the drums home and they were kept throughout the year in Mahajanga. The keys appear to have been kept at Doany Ndramisara as well.

The Great Service brings together all three divisions of labor discussed in Part II, demonstrating equivalence, complementarity, hierarchy, personal distinctiveness, and totality. People attend from all quarters of the Sakalava world. Ancestor People, royalty, spirits, and commoners mingle and perform their respective roles. I conclude that the

Fanompoa Be is a kind of performative enactment of the "constitution" of the politico-religious system. The relations of the constituent parts exist largely in and through the annual realization of the ritual cycle—a realization that entails the chronotopic movement of gifts, subjects, ancestors, artifacts, and relics between demarcated places and in the right sequence and that conjoins and confirms past and present.

Part IV

Practicing History

**9**

Kassim's Burden:
The Practice of an
Exemplary Spirit Medium

The Great Service at the shrine of Ndramisara is far from the only event to which northern Sakalava are called. Over the course of a year people may contribute to service at several other shrines. In this chapter I examine activities in Mahajanga oriented to Bezavodoany, the cemetery in which the monarchs who founded the kingdom of Boina lie, as well as the more general ethos of conservation and care that underlines service to the shrines. As in chapter 7, the discussion is organized around the practice of a key medium. His story illustrates the very rich and complex family relationships represented by the senior-most spirits and the polyphonic interplay between the lives of mediums and those of their spirits.

Kassim Tolondraza is an exemplary bearer of Sakalava history. He is both a highly respected medium of a senior spirit and the main agent in Mahajanga of the service at Bezavodoany. He is widely addressed and referred to as Dadilahy (Grandfather), not simply because of his age but because he is possessed by Ndramboeniarivo. People possessed by the many *tromba* who are Ndramboeny's "*zafy*" (grandchildren, descendants) started calling him "Grandfather" and it caught on. This chapter describes Dadilahy Kassim's practice with respect to the work of curating of the material ancestral remains at Bezavodoany. There is no such thing as a "typical" senior medium; the chapter shows what an engaged and authoritative spirit medium can make of his relationship to the tomb enclosure where the body of his spirit lies, and describes the interaction between senior mediums, royalty, tomb guardians, and the public more generally. My understanding of practice approximates what Aristotle called phronesis, that is, situated judgment within a tradition of values whose aim is human flourishing and dignity.

The chapter is based on many long conversations with Dadilahy Kassim between 1993 and 1998, which took place at his house in Lower Progress, often including his wife Tsetsa, known as Maman' Sera. Kassim's house was a pleasant and airy construction of tin, painted bright blue. We usually sat around the dining table in the shaded alcove between the kitchen and the shrine room. The house also included a formal living room with a framed *Ordre du Mérite de Madagascar* and a *Médaille de Travail* on the wall, and a bedroom with large double bed, wardrobe, and sacks of unhusked rice that Kassim purchased annually at harvest. The open area of the compound served as an arena for spirit gatherings and contained a shower, latrine, and duck yard. At the time Kassim was a vigorous and articulate man who looked younger than his age (mid-sixties). He dressed in a sarong of beautifully patterned material but wore trousers and a felt hat when he went outside and a suit and tie for photographs. He wore a large gold ring engraved with his initials and a gold pendant with his initials flanking an outline of Madagascar.

Bezavodoany and the Politics of Ancestral Inclusion

Located on a remote hilltop west of the Betsiboka, Bezavodoany rivals Ndramisara's shrine as a sacred place. Indeed, as the burial ground of both Ndramandisoarivo and Ndramboeniarivo, it provided the source of the material in the reliquaries of these two personages in Mahajanga. Bezavo was the site of the first capital of Boina (or very close to it), and a number of other early monarchs are buried there as well, including Ndramandikavavy. The tomb enclosure officially no longer admits newly "collapsed" bodies; but exceptions have been made. Among the most recent and, hence, most junior, inhabitants is Ndrankehindraza, the fun-loving, quasi-Muslim spirit with the red fez and circle of white clay around his eye, who is a frequent participant in *tromba* activities around Mahajanga.[1]

Kassim explained that most recent royal descendants do not lie in Bezavo because they are too distant in generation from the original inhabitants. However, he elaborated on more interesting reasons for the inclusion and exclusion of specific figures: "Mother and son, Ndramandikavavy and Ndramboeniarivo, are simply too mean-spirited and aggressive (*mashiaka loatra*) and scare off prospective residents. Ndramandikavavy wanted to prohibit the entry of all her descendants beyond her grandchildren. Ndramboeny came into dispute with his mother when he wanted to admit Ndramañonjo as his adjutant (*militera*). Ndramandikavavy refused and warned her son that Ndramañonjo would seize power (*fanjakana*) from him. They argued until finally, Ndramboeny said, 'If you won't accept Ndramañonjo, then I will leave too!' At that Ndramandikavavy relented, but she said, 'Enough (*bas*). We don't want any more people to join us.' Then Ndrankehindraza came along and it was the same story. Ndramañonjo wanted to include him. There was a big argument and Ndramañonjo threatened to leave in turn. Finally they reached a compromise; they would bury Ndrankehindraza at Bezavo yet outside the sacred enclosure (*valamena*). So his tomb lies between the outer enclosure (*fiaromby*) and the *valamena*. Those people are really mean (*reo mashiaka reo*)," concluded Kassim ruefully.

On another occasion Maman' Sera explained that Ndramañonjo lay at Bezavodoany because he had rejected the practices (*fômba*) of his own father, Ndramhotsy, for those of his great-grandfathers. That is, having refused to become a Muslim, he could not be buried in the Muslim cemetery of his father. His rejection of Islam, according to Maman' Sera, stemmed from the fact that Muslims do not drink liquor.

Dadilahy Kassim seated in his patio, Mahajanga.

These arguments about the disposal of royal corpses condense much of what I think is critical about Sakalava history-making. Muslim and non-Muslim ancestors are enabled to comment on one another and on the larger forces that their individual choices exemplify. Even more, former monarchs intervene actively in the fate of their descendants. Although Kassim and Maman' Sera do not say so explicitly, the arguments of the monarchs must have been enunciated through their spirit mediums.

Their descendants are also powerful royal figures, but less potent than their ancestors, and their fate is partially subject to decisions of the *tromba*.

In their acts, the ancestors repeat with some consistency positions that they reputedly held during their lives. Ndramandikavavy reserves the power of government (*fanjakana*) for her son. Her position is always one of narrowing royal succession, excising potential rivals and conserving a condensed, exclusive power. It is, in effect, a kind of paranoid view (rendered explicit in her excessive jealousy of her husband's possible dalliance with other women, including his female mediums). Ndramandikavavy resists the dissipation or diffusion of power.

History thus perdures as a kind of continuous political argument. Positions and controversies are played out by means of specific historical voices. The spirits do not simply represent these positions in some kind of static theater but, rather, they are engaged in arguments that have real consequences. The voices of the spirits (as manifest in their mediums) actively intervene in history. They really can decide who is buried where and thus how sociopolitical identities unfold.

The spirits also demonstrate the selfishness and ferocity of royal power; they are so mean that they would cut off their own descendants. This is something that intrudes intimately on the mediums themselves.

Kassim's account illustrates aspects of the Sakalava chronotope and ideas about power. Ndramandikavavy takes the ostensibly ludicrous position that the inclusion of Ndramañonjo within the tomb enclosure threatens Ndramboeny's power. Yet Ndramañonjo lived well over a century after Ndramboeny, in a kingdom overrun by foreigners that was only a shadow of the size and power of Ndramboeny's. At one level, Ndramandikavavy's point can only be the fear that the memories of ancestors get successively displaced by those of later generations. This is a real fear, as new generations of *tromba* speak to contemporary issues and become more popular than those who precede them. Yet her very challenge shows that the voices of the most senior ancestors are far from obsolete. It is their presence in Bezavo that renders it such a significant place.

As the senior-most ancestors withdraw from public action, and possibly from public appeal, they become all the more potent. Ndramboeny's point to his mother is that if the elders are to retain the potency that adheres in silence and immobility, they need the younger generations as their adjutants, to interact with their subjects and carry out their intentions. History is generationally and hierarchically layered with respect to potency, sanctity, and activity. Ideally, successive generations serve preceding ones. Moreover, they cannot fully displace and supersede them insofar as younger generations owe their authority to their descent from older ones. It is senior generations who bless and sanctify junior ones.[2]

Bezavodoany lies within the domain of Ampanjaka Amina, but as a reigning monarch of an "official" line descended from Ndramboeniarivo, she cannot live permanently at the cemetery. She resides in the small town of Mitsinjo at some distance southwest of Bezavo and Mahajanga. She also cares for the tomb enclosures of more recently deceased ancestors. As a result, the organization of the Bezavo service in Mahajanga is left largely in the hands of Kassim. Kassim and Amina initiate projects together, but Kassim takes on the bulk of responsibility. He is, as he says, the queen's lieutenant or representative (*solonteña*) in Mahajanga. Ampanjaka Amina gave him a signed statement to that effect, which states that in her absence he can speak and act for her with respect to collecting money for the shrines. Kassim's relationship to Amina is partly equivalent to that of Ndramañonjo to Ndramboeniarivo. But things are more complicated than that because Kassim is also the medium of Ndramboeniarivo.

Kassim's Life . . .

Kassim was born into a family of Ancestor People living at a Zafinifotsy shrine on the banks of the Loza River some fifteen kilometers from Analalava to the northeast of Mahajanga in the domain of Ampanjaka Soazara. On his father's side, he is Bevony. Kassim explained, "This is a clan (*karazaña*) whose members fled the Borzan (Merina) along with Zafinifotsy royalty and many committed suicide alongside them by drowning in the Loza. The Zafinifotsy said that those who had stuck with them—to the end—would never part from them and so made the surviving Bevony the chief caretakers of their shrines. The services of the Water-Dwellers (*Antandrano*), that is, those spirits who drowned in the Loza, must be officiated by Bevony."[3]

Kassim remembers his father's father who controlled the entrance (*nitana varavara*) to the shrine. He had long braids but no beard. Kassim explains his own reluctance to dictate traditional narrative (*mitantara*) to me to his grandfather's admonition that "tradition should never be recorded in writing." However, the grandfather also became Muslim, "like a Grande Comorian (*Ajojo*)." His children did not follow him in this respect, but at his grandfather's request, Kassim did. He went to Koranic classes (*kiôny*) until his grandfather died and he received a Muslim name. Kassim admits he has forgotten most of the recitation (*jôro*) and when I asked whether he was a Muslim, he replied, in very un-Islamic fashion, "*Métis,*" thus taking the category Silamo, in Sakalava manner, to refer more to a descent-based category (*karazaña*) than an exclusive religious faith. Kassim prays (*mikosoaly*) at Ramadan and asks his *tromba* not to appear during that month, but he only performed a Muslim marriage (*kofongia*) when a diviner reported that his grandfather was upset. Kassim also attended school in Analalava and is literate in Malagasy.

Kassim recalls a childhood that was, in some respects, idyllic: "The women would go out on the evening of a receding tide and collect lots of small fish. You could put the rice on to cook, then go down to the beach to fetch yourself fresh crab." Kassim's father made his living catching shrimp and fish and taking them by canoe to Antsohihy for sale. He used his earnings to buy cattle, and Kassim has rights in the family herd. According to Kassim, "Life was cheap. Nowadays fish are much scarcer and rice more expensive."

Kassim was raised by his father's younger brother who was possessed by Ndramañavakarivo. Kassim remembers seeing from early childhood this "vigorous (*matanjaka*) royal spirit dressed in elaborate Islamic garb. When Ndramañavaka was around, no one was permitted to sleep!"

Kassim was sick and his father sought a diviner (*moas*) who said that to be healthy Kassim would have to move south of Analalava. He had been hired in 1954 by the *Service d'élévage,* and in 1960, the year of independence, he was transferred south to Mahajanga where he continued to work in the provincial veterinary service as a carpenter. He separated from his first wife and children in Analalava and had several other partners in Mahajanga.

Kassim retired in 1994 after forty years of service and has since enjoyed what he considers a reasonable and steady pension.[4] The construction of his house was subsidized by a wealthy French client so he paid no rent. He has been able to invest a little in Maman' Sera's home village where the couple has built a simple house and purchased an ox cart, plow, and a bit of land for bananas. But Kassim has not used his retirement for relaxation or private benefit. Instead, he has embraced civic and religious responsibility. Kassim's energy, initiative, and imagination have enabled a series of successful annual campaigns for the restoration of Bezavodoany. Kassim has generated a lot of enthusiasm in Mahajanga for the remote shrine that most city dwellers have never seen and has also garnished support from the government. In the culmination of the first phase of his efforts, a large expedition to Bezavo was organized in 1993 that included many national and provincial political figures and that put the once dormant shrine on the national cultural map.

. . . WITH NDRAMBOENIARIVO

Kassim's considerable efforts on behalf of Bezavo are not unconnected to the fact that he is a medium of Ndramboeniarivo, the son of Ndramandisoarivo and Ndramandikavavy, whose reign affirmed the consolidation and success of the Zafinimena conquest of Boina. All subsequent rulers of the main line are descended from him. Kassim explained that he has been possessed by Ndramboeniarivo since childhood. He inherited Ndramboeny from his father's father who was also possessed by him and who helped raise Kassim.[5] "Ever since I was a child of five or six," Kassim said, "if I did things that were taboo to the *tromba* there were consequences. I could not close up the hen house; if I so much as held the door, the chickens expired. Conversely, if a chicken got into my mother's rice store I would get very sick, and almost die (*marary saiky maty*). My father beat me with a stick when the chickens died. But my grandmother stopped the beating, shouting to my father, 'There is a reason (*misy dikany*) all the chickens die and you don't know it!'" So Ndramboeny was already present, but it was not until 1974, as an adult in Mahajanga, that Kassim first entered trance (*mianjaka*).[6]

When Kassim arrived in Mahajanga he did not attend the service rituals and he was not active in spirit affairs. "In 1973 I fell very ill, said Kassim. "I was sick a long time and people who saw me thought I would die. The diviners could not see the problem until finally they pointed out that I had not been home in a long time and people there missed me; my father and grandfather were thinking about me (*mahacharo*) and this was making me ill. By this time my condition was grave so it was decided to send me home. I wanted to go by car but no driver would accept me. Next I determined to fly, but when I got to the airport, the pilot took one look and refused to take me. I signed a paper removing the pilot of any responsibility should I die during the flight and so I flew to Analalava. From there I was carried to the village on a litter, 'already like a corpse.'

"Now, my clan followed the customs (*fômba*) of the Water-Dwellers. Our grandfathers attended (*miambing*) them and they look after us. So my father said, 'Let's call a *tromba* from the water.' The 'god' (*ndrañahary*) came. People explained: 'Your grandchild has arrived sick from Mahajanga and we're calling you first before doing any medicine.' The god said that it was a *tromba* from where I had come [that is, Mahajanga] that was making me sick. And yet in Mahajanga for a year the diviners had not been able to ascertain this. The Water-Dweller asked my father: 'Do you promise that if Kassim gets well you will raise his *tromba*? Do you really agree?' They joined hands and then released them in commitment.

"I had been sick each time I took medicine. But now the *tromba* explained that external illness was not the source of my condition. The *tromba* gave me water to drink from a white dish. After fifteen days of drinking this liquid my symptoms were entirely gone. And everyone had thought I was dying! When I became stronger they called the Water-Dweller to say I was returning to work in Mahajanga. I was told to arrange my *tromba* within the month. After three weeks in Mahajanga I fell sick again and so I raised the *tromba*. Ndramboeniarivo emerged.

"That was in 1974. Since then I have suffered little things, like everyone, but no major illness. And since that time I began to attend the shrine, to dance with the *tromba,* to learn the practices. Before my illness I knew nothing. In fact, in Analalava during my youth there were very few senior *tromba*. My grandfather had a hard time finding mediums of Ndramboeny's family when he held his own ritual there."

I asked about Lakolako, his second spirit, who is Zafinifotsy but not a Water-Dweller. Kassim said, "Lakolako came along with Ndramboeny and is his diviner (*moas*). [Like many senior Zafinifotsy, Lakolako is from the northern tip of Madagascar (Bobaomby)]. Ndramboeny met him on a military campaign in the region and they became close." At least, so Lakolako informed Kassim.

Kassim explained that all senior spirits bring a follower (*mañaraka*) with them to a medium: "The more junior spirit is the one that generally attends to the needs of clients. For example, you might call up Ndramboeny first, but Lakolako would actually carry out the work. Ndramboeny might tell you the medicine to seek but say, 'When you return, call Lakolako to apply it.' A reason for this is that senior *tromba* stay for no more than fifteen minutes at a time; their presence exhausts the medium. With more junior spirits, the medium is not so tired." When Kassim is actively possessed by Ndramboeny, his body is extremely tense and perspiring, whereas Lakolako, dressed all in white, has a calmer presence.[7]

In September 1994, Kassim acquired a third *tromba,* a "spirit of the above" (*tromba añabo*), named Bibilahiandañitry (Male Beast in the Sky). Spirits of the above are said to be ancestral to the royal *tromba*. They leave no human remains and, hence, have no tomb sites but do maintain a shrine at Ankáboka near Marovoay. Active possession by a spirit of the above is extremely intense and brief. The spirits clasp metal rods with bells that clang continuously as they shake.[8]

Kassim first experienced the spirit of the above during the Great Service: "At the time I didn't know what was affecting me. As I sat in the temple I felt very strange; I began to pour sweat, shake, and tingle. I left quickly and only once I was outside the sacred enclosure and cooled off under the mango trees did I calm down. Now I realize it was probably the spirit of the above trying to rise. This happened two or three years; in the end the spirit made me very sick until I went to Ndramañonjo in my colleague Maolida Kely [see below] to have it emerge properly."

Kassim once observed that *tromba* know whom they like and whom to enter. His own possession by Ndramboeniarivo might be understood as an identification with, and interpellation by, his grandfather. None of Kassim's spirits are young or light-hearted, but Ndramboeny in particular has a darker side that both attracts Kassim and that he can withstand. Ndramboeny is remembered as an immensely powerful person capable of tremendous violence. In the past, that violence was turned against his wife; today, the main subject is likely to be the medium himself. Ndramboeny appears with greater intensity and potential danger in Kassim than in his few other mediums. Kassim says: "Nramboeniarivo is terrible (*mashiaka*); one small fault and he strikes you!" Kassim means that if he breaks a taboo or inadvertently offends Ndramboeny, the spirit will take it out on him.

This perception is shared by others. In both collective representations and individual commentaries Ndramboeny is portrayed as intransigent (*mashiaka*) and ready to lash out when crossed or confronted with human error. As a female colleague of Kassim's put it, "If there is something Ndramboeny doesn't like, there is no way out (*tsisy hevitry*)." The quality *mashiaka,* attributed to virtually all senior spirits, is especially applied to Ndramboeny. It refers to a character who criticizes others loudly, who is uncompromising, punitive, aggressive, caustic, violent, or just plain mean (Lambek and Solway 2001). The term evokes grudging admiration as well as fear.

A medium said, "When Ndramboeny is angry he will not even listen to his father or grandfather. He is often too proud and standoffish (*miavong*) to come when he is called, even in the presence of two or three of his mediums. The reason he has far fewer mediums than either of his parents is that he is simply too caustic." Ndramandisoarivo, Ndramboeny's father and the military conqueror and founder of Boina, is considered flexible and complacent (*malemilemy*) by comparison: "Ndramandiso recognizes that living humans support the spirits (*olon'belo mañamboatra izahay*), and he is happy to instruct them in turn. Ndramandiso comes when he is called and listens to what is asked him." People explain the difference in psychological terms: "Whereas Ndramandiso struggled to acquire the kingdom (*fanjakana*), Ndramboeny received it without effort on his part and feels entitled to do what he wants."

The story is not without its oedipal side. Ndramandikavavy, Ndramandiso's wife and Ndramboeny's mother, is quite aggressive (*mashiaka*) herself. I have been told that when the two are present together, Ndramandikavavy impatiently orders her husband away: "Get out of here! What are you still doing here? This is none of your business any more, go away!" She never insults Ndramboeny in this manner, and it is clear that she loves her son more than her husband. Ndramandikavavy says of Ndramboeny, "He is my son, my father, my brother [that is, he is everything to me]." She can get angry at the medium of Ndramboeny for not respecting him enough, and likewise Ndramboeny can get angry at mediums of Ndramandikavavy for neglecting his mother. Interestingly, one and the same woman is often the medium of both husband and wife but never of mother and son. Likewise, a married couple may be the mediums, respectively, of husband and wife but not of mother and son. The latter alternative in each instance is explicitly too incestuous.

Maman' Sera knew when she married Kassim that he had Ndramboeny. Unlike many people, she was not afraid. She comes from a minor branch of royalty in the countryside, and her father's father had several spirits so, as she said, she was "used to them." They first met at a spirit event she was holding for an adult daughter. Kassim was married to someone else at the time, but they eventually separated; a year later, Kassim and Maman' Sera got together. It was with Kassim's previous wife that he went to Bezavodoany for his accreditation.

According to Maman' Sera, "Ndramboeny cannot stand the dirt of small children. Kassim liked his previous wife and raising children, but Ndramboeny did not. Kassim built his wife a separate house, but even so, after a few months, he suggested they divorce. Ndramboeny is caustic (*mashiaka*); when he enters people they soon split up. Ndramandiso and Ndramandikavavy, by contrast, can raise children."

Maman' Sera herself is a lighthearted, sonsy woman, with spirits of an entirely different nature from those who inhabit her husband. Her *tromba* are more junior, and they like to smoke and drink. Sometimes she spends an enjoyable afternoon in trance with a few like-minded spirits. Kassim and Maman' Sera have quite an egalitarian relationship. Unlike many couples, they eat together at the same table. They often take turns attending *tromba* events, especially the all-nighters when it is wise for someone to stay home and guard the house. At other occasions they watch over each other's transitions into and out of active possession.

None of Kassim's spirits touch alcohol, but as he noted, each of his wife's three spirits is a big drinker. Maman' Sera added, "Because Ndramboeny is particularly opposed to rum, it cannot be in the house, so my spirits drink whiskey instead. Even to serve whiskey or beer at the house they have to ask Ndramboeny's permission." Ndramboeny himself is very fond of Pompeiia cologne, but unlike Ndramañavaka, he only asperses, rather than drinks, it (cf. Lambek 1981).

In part, because of his excessive exigence, Ndramboeniarivo possesses very few mediums today in Mahajanga. Of this small group of women and men, Kassim's authority is undoubtedly the greatest. It has grown as he has suffered and then withstood Ndramboeny's violence, and he is known to direct his energies toward containing it. Kassim is uncompromising in observing the taboos to keep his environment pure and his spirit contented. He keeps two old silver coins instead of one for each of his spirits. He cannot visit the dying or the very sick, though he knows other mediums of Ndramboeny who do. Kassim says he is unique to insist that anyone who wishes to call up Ndramboeny come a day ahead to inform him. Kassim then speaks to the spirit over incense and waits to see what he dreams that night. The first time I met Kassim was after someone asked him to look after me during the Great Service. He arrived at my house the next day following a night of troubled dreams in order to excuse himself from the task; Ndramboeny was angry that he had not been petitioned properly.

Another salient difference between Kassim and other mediums, he avers, is that the latter live in the rooms where they perform as *tromba*. Trousers, shoes, menstruating women, hats, and other taboo substances all enter indiscriminately. But Kassim has built a special shrine room on the east side of his house in which his spirits are called up. No one wearing shoes or trousers is ever allowed in this space, which is filled with various objects of power and kept entirely free of polluting substances from the street. The door to the shrine room is closed whenever the moon is "dead," just like the temple itself. "From his first entry, Ndramboeny announced, 'Your guests are different from my visitors; everyone in his own house!' And that," concluded Kassim, "is why I say Ndramboeny is particularly demanding (*mashiaka*) in me."

Of course, Kassim is the person most subject to these constraints. Although a nominal Muslim, he rigorously avoids visiting houses where the dead are laid out (until the premises are purified by a Jingô person after the funeral) and avoids any mortuary procession he might encounter on the street.[9] Kassim takes his trusteeship of Ndramboeny extremely seriously; it frightens him and forces him to curtail certain activities, but at the same time it permits him a broadened agency. Causticity is not simply a negative quality but indicative of autonomy and power. Kassim is at once completely subject to royalty and able to act virtually with the assurance of royalty and even from a position of greater moral authority than living royals. If he is Ampanjaka Amina's representative in Mahajanga, he is also host to her most powerful ancestor.

Kassim's authority is boosted by the fact that in 1980 he spent two months at Bezavo and passed a test arranged for him by the guardians there that indisputably established the validity of Ndramboeny's presence. As Kassim explained, "You don't attempt the test until your *tromba* tells you to. I was very sick. The diviner said the *tromba* wanted to perform the exam. They called up the spirit and he confirmed it." Bezavo is considered a dangerous place—Kassim, himself, was frightened during his initiation there—and the test, a difficult one. Few other mediums have gone to Bezavo to be tested and none bearing Ndramboeniarivo.

The exam (*fitsaraña;* "passing judgment") at Bezavo is much more challenging than the accreditation that takes place at the Great Service in Mahajanga (chapter 8). Said Kassim, "In Mahajanga you just give your name and those of your parents. At Bezavo the main trial is carried out in the innermost courtyard. But before you reach it there are tests for admission through the gates to each successive concentric enclosure." Kassim recalled, "There were lots of *signales* indicating whether I could proceed further. When I arrived at Bezavo the guardians told me I would have to wait a month. At the new moon they told me the day of the test. In the intervening days they watched for signs." Although Kassim knew nothing about it at the time, it was absolutely necessary that Ndramboeny be escorted inward by Ndrankehindraza. "So the guardians waited to see whether an appropriate medium would appear. If he hadn't come I could not have advanced. Three days before the exam was due, a medium of Ndrankehindraza showed up. He was from far off and someone I had never met." Kassim still had no idea what was at stake but the guardians told him that the first test was over.

Kassim continued, "On Friday night of the full moon the guardians held an all-nighter at which Ndramboeny rose and was asked whether he agreed to take his medium in for the test the following morning." Again, unbeknown to Kassim, the guardians watched for the second sign. "This was the appearance of a large snake (*menapitsko*) who raised its head to observe the revelers.[10] The snakes are the proprietors (*tompin tanana*) of the place where the all-nighter occurs and if none had risen I could not have proceeded."

Kassim was then called to the outermost gate on the south side (see Appendix): "People sang '*Senganay, ohh talenay.*'[11] The 'proprietor' (*tômpin*) rose and knew what to do next. Ndrankehindraza escorted Ndramboeny into the innermost courtyard with its stone wall (*lakor'vato*) where the 'really sacred objects' (*teña raha sarotro*) are kept. There he was faced with a line of some ten to twenty identical-looking white plates, each containing a silver *tsangan'olo* coin. The guardians mixed them up and said to Ndramboeny, 'Take your plate, take what is yours.' Ndramboeny could discern his possessions from afar; he went right over without hesitation and selected the right plate."

"Bezavo is *dangereux!*" emphasized Kassim. "If you don't succeed in the test, you have to kill a cow. In the past, before the French, the guardians told me, if you failed the test they killed you! Now you offer a cow instead.[12] They tell you this the day you arrive." Kassim was unsure whether mediums were given a second chance.

"And so," Kassim explained, "it was '*automatique*' that I became de facto Senior Medium (*saha be*) of Ndramboeny for Bezavo, although I was never officially appointed by either the monarch or the public (*fokon'olon*). The guardians said, 'You are the only medium of Ndramboeny from Mahajanga whom we recognize, the only one who has correctly completed the custom.'" On the Saturday following his return to Mahajanga, Kassim went to pay his respects at the Doany Ndramisara. As if to announce and confirm his status, Ndramboeny rose inside the temple. "Prior to this he had only risen in me in town," related Kassim. "Sometime later, Ndramandiso and Ndramandikavavy rose in their respective mediums [both close colleagues of Kassim's] and announced that their son was responsible for the 'government' (*fanjakana*); they passed it to him. That," said Kassim, "was how the job of organizing the services (*mikarakara fanompoa*) at Bezavo came to me."

Kassim explained that although some people liked positions of responsibility and sought them out, he did not: "The role just came from *ndrañahary* (God, or the *tromba*). The work is very difficult. It is easy to make mistakes and one constantly has to think and plan so as not to do something forbidden. Sooner or later the *tromba* express their displeasure. You receive a 'beating' (*kibay*). You don't see the beating with your eyes, but something happens to you and you realize, 'Oh, I've been punished.'"

MEDIUM AND GUARDIAN

As a result of his stay at Bezavo, Kassim established good relations with the resident guardians, especially with the Ancestor Person responsible for Ndramboeny who tested him. After the judgment, the guardian proposed kinship (*fihavañana*) between them, and they established a link of blood brotherhood (*fatidrâ*), in which Kassim was the junior sibling (*zandry*). The relationship was continued with the guardian's successor who sometimes visited Kassim in town to bring news of the shrine, discuss the next stages of the service, and consult the senior spirits.

In July 1994, the guardian stayed for several days at Kassim's, sleeping on a mattress in the shrine room. He was an old man in a battered hat and frayed sarong with glazed eyes, a shrewd, striking face and demeanor, and a lean sinewy body that contrasted with the sleek urban mediums. Like other men I have met from the countryside he seemed dignified and yet detached, possibly ill at ease, from those around him in the city. Kassim invited over a medium of Ndramandiso who could not have looked more different—dressed in Comorian style with a Muslim hat, white shirt, pressed trousers, and large gold ring and watch. Kassim explained two reasons for the guardian's visit. The first was to resolve a problem through consultation with the senior spirits concerning a fellow guardian who had eaten rice designated for the service (*fanompoa*) and become blind as a result. The second was to acquire rice for the upcoming service to renovate the shrine buildings. As his colleagues and their families were currently dispersed in their fields, the guardian needed to feed everyone who came for the service. He was planning to buy rice with his own money, but Kassim said this was inappropriate; it was up to the mediums in Mahajanga to purchase the rice. The Comorian medium could only agree and promised to contribute. Kassim told the guardian to use his money to buy his wife a new wrapper.

Throughout the conversation Kassim referred to "our father," thereby identifying both himself and the guardian with the position of Ndramboeny and the other medium with the position of Ndramandiso, the father. This rhetorical device enabled Kassim to direct the conversation while maintaining an egalitarian or deferential position as son and brother to his interlocutors.

Although mediums and tomb guardians may be defined as siblings insofar as they have equivalent relations to the same ancestral personage, their work is complementary. Kassim is the only person the Bezavo guardians know well in Mahajanga, and they are somewhat dependent on him. He provides hospitality, and in 1995 I observed him give three visiting guardians each a fmg10,000 bill "for expenses" when they left. But in Bezavo, things are somewhat different. There Kassim is no longer an active director; his only job, he once said, "is to enter trance when called upon to do so." Where Kassim takes responsibility for planning the renovations at the shrine, the guardians contribute the cattle that are sacrificed at the commencement and completion of the work. They seek animals from herd owners in the area; if a beast were refused, the entire herd could fall sick.

The next day demonstrated another side to the relations between guardian and mediums. Early Friday morning a few mediums gathered at the home of a medium of Ndramandikavavy. Wearing a tailored jacket and looking both formal and ill at ease, the guardian repeatedly called respectfully to Ndramboeny and his parents, "from a single sacred enclosure (*valamena 'raiky*)," to rise for consultation. The medium of Ndramandikavavy, her face and neck smeared in wet kaolin, donned a shiny yellow blouse and entered trance. Ndramandikavavy asked for perfume, poured some on her handkerchief, and continuously dabbed her face with it. She was attended by two other women. As Kassim explained the issue to them, Ndramboeny (in a female medium) and Ndramandiso rose. Ndramandiso was placed on a chair while Ndramboeny sat by his mother's knees and then on a low stool at her right side. Each *tromba* sat at a different level but close together. Ndramandiso shared his cloak with his wife and occasionally gave her a discreet kiss on the head. It was an idealized family tableau.

Ndramboeny spoke first, briefly and abruptly. Ndramandikavavy spoke at length, in a high-pitched voice and angrily to the guardian. Leaning far forward in his chair, Ndramandiso nodded continuously while Ndramboeny shook and Ndramandikavavy dabbed herself with her perfumed handkerchief.

When the spirits left, the guardian sat back and lit a cigarette. As Maman' Sera folded the spirits' clothing and distributed the money used in the invocation among the active mediums and the guardian, Kassim eloquently summarized the proceedings. The *tromba* had refused to withdraw the blind guardian's affliction and expressed anger that he no longer lived full-time at the shrine. They also said that he should ask for pardon in person. The present guardian was wrong to have come in his stead, and if he were not careful the blindness could spread to him. They suggested that the blind man be taken by ox cart to the medium of Ndranantanarivo, Ndramboeny's daughter, to seek pardon. The *tromba* accused the afflicted guardian of neglecting his duties and even consuming rice meant for service. In addition, and drawing on the second reason for the guardian's visit, they remarked that rice for the service should not come from one person's pocket alone.

I was rather taken aback at the senior spirits' blunt lack of sympathy, but clearly they illustrated the ruthless quality to which everyone referred. Although Ndramandiso sat in the highest position and had, as always, to be raised and lowered, Ndramandikavavy was really the center of attention. There was a coiled and rather frightening tension in her obsessive movements, and she provided both the most overtly angry and the most elaborated response.[13] Kassim proved once again a very effective speaker who took over the role of mediator from his wife and the guardian to his own evident pleasure and their relief.

INTERLUDE: Oedipe Sakalave

Kassim did not serve as medium because he was feeling ill. But Ndramboeny limits his appearances in Kassim to important matters and comes usually only just before dawn. His presence is extremely impressive. He is beholden to no one except his mother, at whose knees he sits; and he, in turn, is the one man whom she will treat gently or listen to with any patience. Kassim, himself, says of Ndramboeny, "He was loved too well. Having been spoiled rendered him intolerant and threatening (*mashiaka*); he is a good deal more violent (*mashiaka*) than his father." Kassim was afraid of Ndramboeny's reprisals and refused to tell me about his life. It was from others that I learned how Ndramboeny, hearing about her adultery with his half-brother, had murdered his wife.

Ndramandikavavy is often angry when she arrives. On one occasion the subject of her wrath was her own medium who had neglected to try hard enough to find a Jingô person to purify her after a funeral that she had inadvertently encountered. She was nice to Maman' Sera and asked after her children. "But," Maman' Sera said, "everyone is afraid of Ndramandikavavy. Even her husband is afraid of her. If she tells him to go home, he leaves. At one event, as soon as the first spirits had left, she said to him, 'What are you waiting for? Things are over, go home!' And that is why he said so little and left so quickly. He is beaten by her and never speaks back to her. Once she had dismissed Ndramandiso, Ndranihaniña wanted to leave too, but she turned to him and said, 'What's your big hurry?' She insults and chews people out (mañashiaka olon), but she never speaks this way to her son."

As Kassim says, mother and son "really like/love each other (reo fañkatia reo)." It is possible that her reason for wishing to exclude Ndramañonjo from Bezavo was one of jealousy. Conversely, Ndramboeny's (over-)identification with his mother is so strong, "He gets angry if his medium eats foods that are taboo to his mother. If I eat a certain kind of fish (karapapakia) I vomit," said Kassim, "even though the taboo is not Ndramboeny's but his mother's. For similar reasons I cannot eat beef tongue or raise duck (dokitry).[14] Ndramboeny likes his father as well and whenever father and son rise they stay next to each other."

The political message is that succession from mother to son is clear-cut, and Ndramandiso's children through other women must be prevented from competing for power. Ndramandikavavy earned the right to exclusive descent through her sacrifice. If Ndramboeny killed his wife it was because her alleged infidelity threatened his mother's rights to succession through him alone. These events stand at the foundation of the monarchy. But the political is personalized; the relationships are evoked each time the senior spirits appear. Ndramandisoarivo is never allowed to forget what his wife did for him; Ndramboeniarivo can never be confronted with what he did to his wife. The political mythos and family drama continue to be lived out in the bodies and practices of the mediums. They substantiate a local psychology of power and present an unromanticized view of the powerful. The stories are the equivalent of Greek or Biblical myths except that they are not told outright as coherent wholes but emerge in bits and pieces, signified through clever associations and original actions like the one Kassim makes in keeping his mother's taboos or the one made by the medium in whom Ndramandikavavy refused Ndramañonjo access to the burial ground, and in the close identification—performed, embodied, and presumably internalized— between mother and son. The "primal events" are rarely spoken, but hinted at, alluded to—and "somatized" by the mediums.

KASSIM'S RELATIONS IN THE COMMUNITY

On June 12, 1998, a small group called up Ndramboeny in Kassim's shrine room. Ndramboeny entered snorting and kept up the deep menacing sound throughout his stay. He wore a sarong and draped his upper cloth over the left shoulder, leaving the right shoulder bare. He sat above us, on a chair, and held an ebony stick with metal ornaments that clanged lightly as he shuddered. When the spirit left, Kassim, exhausted and drenched in sweat, sat by himself in the doorway to cool off.

Kassim is a man of considerable presence, admired not only for the concentrated force of his active possession but for his integrity and strength of character. He is known for his largesse, his honesty and uprightness, his organizing and motivational skills, and for the clarity and judiciousness of his speech. He holds an annual fete (*lamalama*) each September for his own spirits at which his compound fills with other mediums and well-wishers. Numerous spirits arrive to greet and acknowledge Kassim's three spirits. I found the fete in 1995 to be a tremendously exciting event. The entrances and positioning of the spirits appeared informal and spontaneous yet they were highly orchestrated and built to a crescendo just before dawn at which the founding ancestors of Boina were present at once, surrounded by the Zafinifotsy as well as their descendants from Betsioko, Analalava, and elsewhere. Over the course of the night more than twenty-five distinct spirits attended. Kassim spent a good deal on the event, including fmg130,000 in crisp new bills handed to the various mediums as "honor" (*vonahitry*) for their respective spirits, fmg120,000 on beer and rum, and fmg80,000 on soft drinks and cigarettes. He also gave small amounts of money to elderly people who attended, including a very drunk lady who found her way to every *tromba* event serving alcohol. Many of the guests brought contributions.

If the fete reaffirms the respect that people have for Kassim and his spirits, Kassim is not arrogant or seeking to be treated as someone outside or above the community of mediums. Indeed, he cultivates his relations with other mediums very carefully, and he consults widely on matters pertaining to Bezavo. He addresses mediums who bear Ndramandikavavy as "mother" whenever he sees them, even when neither person is in active possession. Kassim, who himself is referred to as "Grandfather," says he has many mothers in Mahajanga; he even refers to the female spirit mediums of Ndramandisoarivo (Ndramboeny's father) as his mothers. Many of these mediums are much younger in age than he. Kassim could not marry women possessed by his "mother."[15] He works closely with the senior mediums of spirits buried at Bezavo and with other officials, including the Great Woman representing Amina's domain (Marambitsy) in Mahajanga.

Among Kassim's closest associates was Maolida Kely, a woman of Grande Comorian background who was a medium of both Ndramandikavavy and Ndramandiso.[16] Maolida Kely served to ensure the proper entry of Kassim's spirit of the above. More critically, she was the senior medium of Ndramañonjo. As noted above, Ndramañonjo is a royal personage, several generations down from the founding ancestors, who was given special permission to be buried at Bezavo on the request of Ndramboeny.[17] Ndramañonjo "collapsed" in his prime and thus was of an age ideally suited to take active responsibility for the tomb enclosure and its inhabitants who were too old or senior to do the work. When Kassim conceives of a project he often turns to Ndramañonjo to help execute it. It was helpful that Ndramañonjo's senior medium in this case was a wealthy woman with a large private practice, including many clients (*zanakasing*) living in France. Not coincidentally, Amina's Great Woman in Mahajanga was also a medium of Ndramañonjo, and she too worked to collect money and ensure lively social events.

When a new phase of Service is initiated, Amina writes letters of request to her Great Women. They pass on the letter to others and collect the money. People in

Mahajanga associated with Bezavo or the Bezavo spirits make contributions to the local Great Woman who, in turn, delivers it to Kassim. Other Great Women also send him money. Thus, Honorine sends money collected as the contribution (*anjara*) of Betsioko. Kassim keeps a list of all the addressees of Amina's letters, and he ticks them off when the money or goods (the *fanompoa*) arrive. This way he knows who has complied with and who has disregarded the request.

People who deliver *fanompoa* often come in a group; the Great Woman announces the amount and Kassim records it. He has one notebook for what he receives and another for expenses and disbursements. The delivery involves speeches in Kassim's courtyard but otherwise is not as public as the announcements during the Great Service at the Doany Ndramisara.

Two men arrived at the house one late afternoon in 1995. One came from the bush, bringing fmg13,000 collected from his family for Bezavo, whereas the other served as his spokesman. They entered the shrine where Kassim recorded the sum and then gave the donor a piece of kaolin and fmg1,000. "When people contribute money to royal service they are always given something back (*mosarafa*)," Kassim explained. This is redistributed among all those who gave, serving as a receipt, but also reciprocating the original gift.

Kassim counted on large contributions from Maolida Kely, who collected from her many clients in Ndramañonjo's name. One year she gathered fmg1,000,000 for one of Amina's newer burial grounds; another, she supplied fmg250,000 for a shrine door. Ndramañonjo once rose and pledged responsibility for rebuilding Ndramandiso's house at Bezavo. This money was not collected by canvassing all of the mediums of Ndramañonjo but entirely through Maolida Kely's network.

Kassim was not only the repository for the money collected for Bezavo but the person who organized the building projects. His job was far-reaching, and he had an assistant, whom he referred to as a *manantany,* the same word that is used for a ruler's chief advisor or the director of a shrine.[18] Kassim explained that Ndramañonjo and Ndrankehindraza appointed the *manantany,* but that he [Kassim] directed him. It is quite consistent that it was these particular spirits, whose responsibility the maintenance of Bezavo is, who did so.

The *manantany* was a commoner (*olo sotra*) and, unlike Kassim, originated from Marambitsy. He lived in Mahajanga and sold fish in the bazaar. When he was away on assignment Kassim paid his travel expenses and provided money to support his wife, drawing from the *fanompoa* fund, a portion of which he kept aside for current expenses. Kassim and his wife also sent the *manantany* on local errands when they were working on shrine matters and asked him to invoke Kassim's spirit in the shrine room.

The flow of money through Kassim's hands was extensive and he was extremely conscious of the position of trust in which this placed him. After the 1994 building campaign he still had fmg4 million left over, he explained, but he "did not tell anyone and people didn't know. In 1996 when Amina prepared the next phase of renovation, she sent her letter requesting contributions first to Ampanjaka Désy to add his signature to it since he is the senior monarch in Mahajanga. He refused to allow her to canvass (*mamory paria*) because he was in the midst of collecting for the Doany

Ndramisara and didn't want the competition. When Désy told her she would have to wait until they were finished, Amina was angry." It was at this point that Kassim mentioned the money in reserve. "Thus," he said, "without requesting new money from the populace, we were able to complete our project for the year. People were astonished: 'How had they managed the renovations without collecting?'"

Kassim was understandably very proud of this achievement and the personal honesty that it represented. He suggested that most people would have skimmed off money for personal use. To demonstrate his own integrity he kept careful accounts and well organized receipts—sixty kilograms of cement for fmg1,140,000; twenty-eight pieces of metal roofing for fmg1,136,000, and so on. But Kassim also thought those who cheated would be punished by the "proprietors" (tômpin, that is, the ancestors). He himself was afraid of his "grandfather" (Ndramboeny). In Kassim, the Weberian polar ideal types—charismatic medium and rationalistic accountant—blended in complete harmony. But given the amount of time he spent working on behalf of the shrine I think it would not have been illegitimate to grant himself an occasional honorarium.

THE CAMPAIGNS FOR RENOVATING BEZAVODOANY

Whereas the Great Service at the shrine of Ndramisara is organized around the bathing of the relics but collects money to no specific end, service for Bezavo is rather different.[19] On October 10, 1990, much of the shrine burned down. The sacred objects (zaka sarotro) were unharmed, but people have been replacing (misolo) the buildings and furnishings in gradual stages ever since. Each campaign is directed to specific ends and can be expected to draw most upon the mediums directly concerned. The money is used to purchase supplies and pay the builders.

The destruction was accidental; a grass fire set some distance away to encourage new shoots for cattle swept up to Bezavo. Bush fires are common, and it was apparent that the sacred objects could not remain in houses that were so vulnerable; thus, began a project to rebuild the structures in more durable materials of brick and metal. An additional reason for the selection of solid materials was that it had become too difficult to manage annual repairs. According to Kassim, "Sakalava have fewer cattle to subsidize work parties and are less eager to participate in them than in the past."

"In order to begin," said Kassim, "people had to get the 'masters' (tômpin), that is, the royal ancestors themselves, keen (mazoto) on the project." A medium explained that Kassim and Ampanjaka Amina really had to beg them. The ancestors were only convinced when they realized their vulnerability to fire. They acquiesced in the end, on condition that four head of cattle were slaughtered, to "withdraw the taboo" against using modern construction materials with which they themselves were unfamiliar.

Kassim added a perspective that anticipated further social change: "When the tromba first rejected the proposed renovations, the planners returned with two cows. They explained that in the future it might be difficult to find subjects eager to serve the ancestors or willing to see the projects through. They said to the ancestors, 'Let's take advantage of the interest expressed now and make something that will last a long time. If not, once the flimsier constructions collapse there might be no one left sufficiently interested to repair them.'"

As it was too late to do anything in 1990, the first campaign was held in 1991. A member of the royal clan who was also a government administrator hired an engineer to draw up plans. President Ratsiraka contributed fmg2.5 million. In July 1993, Kassim held an all-nighter at his house to see off twenty bags of cement for Bezavo. By October 27 of that year, they were able to celebrate a "re-entry" to the newly completed temple (*zomba*), built with bricks and a metal roof. Kassim and Maman' Sera stayed two weeks and called upon their network to help supply food and drink for a large fete. Government dignitaries visited for the day and thirty-two head of cattle reputedly were killed in celebration.

In the following years they reconstructed the dwellings outside the *valamena* belonging to individual ancestors. These house the guardians, the ancestor's goods, and visiting mediums and are the places where *tromba* are called up. In 1994, they completed the houses (*zomba*) of Ndramandikavavy, Ndramboeny, and Ndramañonjo; and in 1995, worked on those of Ndramandisoarivo and the two offspring of Ndramboeny buried at Bezavo. In the coming years they planned to replace the enclosure fences that had largely collapsed.

Each year money and various construction materials were amassed at Kassim's house and then taken up to Bezavo. Once the reconstruction was completed, a ceremonial visit (*fanompoa*) to Bezavo took place, usually in October.[20] This involved large numbers of people and lasted several days. In 1994, the pilgrims from Mahajanga brought thirty-five sacks of rice with which to feed themselves.[21] The country dwellers were simply not able to subsidize the food, though they provided cattle. Each year, Kassim was able to arrange free transport of people and materials on the barge that traverses the mouth of the Betsiboka and in trucks owned by the sugar refinery at Namakia. The refinery also cleared the road to the shrine. "The managers are eager to help," Kassim explained, "because their factory sits on the land of Ndramandiso and they reap a big profit from it."[22] The assistance is, in effect, their "lot" (*anjara*) in the division of labor—their voluntary tax. The *fanompoa* also receives contributions from government or aspiring candidates of various political parties.

By September 1995, Kassim had collected some fmg5 million in response to letters sent by Amina (figure 9.1). Kassim had a list of thirty-five addressees, including leading mediums and Great Women who could in turn draw on their networks and neighborhoods, or what Kassim referred to as their "teams" (*équipes*). In 1994, each team was asked to supply a certain number of bags of cement or other specified materials, whereas in 1995 each was asked to provide an unspecified amount of money. Kassim estimated that he would need fmg2,400,000 simply to pay the construction workers. Kassim also sent an audio cassette to a colleague working in Mayotte requesting metal roof pieces. Amina sent additional letters by other routes to other destinations, but what was returned directly to her could be divided up among the various burial sites for which she was responsible. "This is a good thing," said Kassim, "without it the newer sites could not grow." But it meant both that Kassim could not

Mitsinjo, faha-23 Mars 1998

Ho an'Atoa sy Rtoa.

Voninahitra ho ahy Rtoa AMINA Said, Ampanjakan'ny Sakalava
ny Marambitsy, tompon'andraikitra voalohany aty Doany BEZAVO, no ano-
ratako aminareo Saha Bemanangy, olona maventy ary mpanompo rehetra tsy
ankanavaka, miangavy anareo samy andray anjara amin'ny fanompoa isay
tena mbola tsy tombo any Doany BEZAVO, isany hoe : Voalohany
.Sobahia (14) - Gora balle roa (2) ary fanarenana ilay Manda
na koravato izay efa namindra ka izao avy zavatra ilaina
amin'azy ireo :
-Ciment - Fer rond
-Omby manitra
-Vola pare
-Vary

Daty filàna sy fanaterana ireo zavatra ireo any Doany BEZAVO
dia mienga eto Mahajanga tsimandrimandry arivan'ny ZOMA 25, mienga
SABOTSY 26 manompo any Doany BEZAVO, Latsinainy 28 septembre 1998

*9.1. Letter from Ampanjaka Amina Requesting Contributions to Bezavo, photocopied from a carbon.
The letter is stamped with a large five pointed star filling a circle indicating Bezavo and the senior spirits
there. The second stamp, with a smaller star above an open new moon in Islamic style, is a sign of
Mitsinjo.*

count on Amina's money for Bezavo and that the contributors would not know exactly where their offerings would end up.

The timing was often complex. In 1995, Kassim had been waiting to hear from the
guardians resident at Bezavo that they had accumulated their share of the service,
namely six sacrificial cattle. But when he finally received their letter, he was anxious
(*tsy mandry saigny nakahy*) that he had not yet heard back from Mayotte concerning
the roofing. It was unthinkable to use rusty, second-hand material to construct housing for senior ancestors, and each building needed some twenty-six metal sheets. He
fired off another letter to Mayotte.

Kassim's own movements were constrained by his commitments. Once the three
tromba for whom the new houses were being built had been told that he was in charge,

Kassim was afraid to allow himself to be distracted by anything else until the task was completed and the *tromba* released him from his pledge. Kassim felt that most people were afraid to take up ancestral work precisely because they knew that should they re-nege on their commitments they would be subject to punishment from the spirits, likely in the form of sickness. I suggested that he was, in effect, their slave for the du-ration, and he agreed. He was not willing to take even one day off.

Kassim could offer a good deal of evidence to confirm his view of ancestral con-straint. A *manantany* who had not reported his full proceeds to Kassim fell sick. A stronger case was the contingent from Ampanjaka Soazara (chapter 5) who had sworn to fulfill their mission before setting out and had been stuck in Mahajanga ever since because the factional quarrel rendered completion impossible. They had grown old there, hostage to circumstances and their oath; those who tried to return home were struck dead.

THE INVESTITURE OF AMINA

Kassim told another story to illustrate how difficult it was to be the senior medium. This account also demonstrates his importance and sheds light on another aspect of the relationship between living royalty and mediums, and on that between Amina and Kassim, in particular. Kassim related, "When Amina was selected to succeed as ruler of Marambitsy she was taken up to Bezavo to be presented to her ancestors and receive their blessing. But the guardians explained that her intentions could not be fulfilled as there were no mediums present nor had she brought any with her. She didn't know what to do (*lany fañahy*). That was a Wednesday night.

"In Mahajanga I couldn't sleep and felt the urgent need to go to Bezavo. Next morn-ing early I set off. It is forbidden to ascend to Bezavo on Thursdays so I slept Thursday night at the village at the base. The guardian of Ndramboeny heard his 'younger sibling' was at hand and told Amina, 'Your Grandfather Kassim has arrived.' Amina was de-lighted. Early on Friday morning they came down from the shrine to fetch me.

"The same evening in Ndramboeny's house they called up the 'god' (*ndrañahary*) and said, 'Your descendant Amina has come. Tell your mother and father that she is the royal person who will be in charge of your shrine here. We have called you first. We appointed her in Mitsinjo. Now we wait for your word. Are you happy to have her lead the shrine? On Saturday we plan to take her inside the sacred enclosure. Tell us whether we should proceed. If she is not pleasing to you, we will select someone else.'

"'Did they choose her in Mitsinjo? Was she the peoples' choice?' asked Ndram-boeny. 'Yes? Then I accept her. Bring her in.'

"On Saturday, they escorted her in [that is, consecrated her]. She passed first through the *fiaromby*, then the *valamena*, and then *lakor'vato* (*la cour ciment be*—the stone-walled court). Each enclosure has a gate to be passed through. Inside *lakor'vato* is the temple (*zomba vola*). This was the first time that Amina had penetrated as far as the temple."

Because no mediums of Ndramboeny's parents were present, everything depended on Kassim. Kassim was frightened to have such an important decision rest on his shoulders. "But you weren't conscious," I said. He replied that that was exactly what had frightened him; he wondered what the *tromba* would say. "Being the senior medium is difficult (*sarotro*) [dangerous, precious, the same word used to describe the relics]," he emphasized." When they ask your *tromba* things you don't know, you are afraid!"

He contrasted the situation with the Doany Ndramisara where the officials make the decisions first and merely call senior mediums to confirm them. Kassim continued, "The medium and spirits pretty well have to go along with things, since the decision has already been made and everything is going ahead. Here it is easy and unchallenging to be senior medium. But Bezavo is different. There the guardians really know all the rules and prohibitions which have been passed down from generation to generation. But they leave the decisions ostensibly up to the *tromba*. They say, 'You're the boss (*tômpin*), we just execute your decisions (*mikarakara*). If you say go ahead, we will; if not, we won't.'" That is, Kassim anticipated his *tromba* being put on the spot by people who actually knew more than he did.

"For example," continued Kassim, "the *tromba* is asked to state the appropriate pattern of the coat of each head of cattle to be sacrificed at each doorway and to name the appropriate clan to carry it out. The guardians know, but they don't say; they just ask the *tromba*. Now, the right color and pattern may change from year to year; the guardians know exactly what it is, but they leave it for you to demonstrate you know or to fall on your face."

Yet Kassim's uncertainty was matched by the authority of his success. "Once you have been certified at Bezavo," he said, "then you really know."

"Bezavo," Kassim laughed, "it is really dangerous (*dangereux*) there!"

PRESENT AND FUTURE

Being a senior medium is perceived ambivalently as a burden and an honor, and Kassim does indeed bear responsibility to the ancestors and for the renovation of their dwellings. He bears the ancestors, in the senses described in chapter 1, both of carrying things forward and of containing within himself the presence of Ndramboeny. All of this was confirmed in my subsequent visits.

By the end of 1995, Kassim had been successful in raising the cement platforms and metal roofs of the three houses. The 1996 campaign entailed acquiring raffia wall panels (*ketikety*) and wood frames for doors and windows. Kassim had sent his *manantany* into the countryside to supervise the collection and the splitting of raffia, and to take the pieces by ox cart to the paved road where they would be picked up by a truck along with the carpentered parts from Mahajanga and then driven close to Bezavo where the final stage of the journey would again be by ox cart. The careful scheduling preoccupied Kassim. As well as the speed of the workers and availability of donated transport, he had to take into account calling the *tromba* at the local shrine adjacent to the place of work on auspicious days of the week and month before the materials could be moved on toward Bezavo. At each stage of the journey, on appropriate days, *tromba* are informed and cattle sacrifice is performed. As it turned out, he had to wait for the new moon.

The people producing raffia were not paid for their work but did it as service (*fanompoa*). Their return (*mosarafa*) would come from the ancestors, in the form of prosperity and well-being (*riziky*). The *manantany* was not paid either, but the mediums covered his expenses. This is the flip side of the burden: Spirits and mediums act like living royalty in commanding assistants but then must support them.

Once Kassim knew for certain that the wall panels had been moved to the road, he announced over the radio an invitation to an all-nighter at his compound to send off the carpentered pieces. They would call the spirits to acknowledge completion of the phase. The spirits appear at "bull-bellow" (*mitrena omby lahy*), a phrase that replaces "cockcrow" since chickens are taboo to the *tromba*. The *tromba* spray sacred water over the collected materials, thereby rendering them truly "*fanompoa.*"

One day Kassim took me to meet Ampanjaka Amina at her daughter's house. Kassim had paid for Amina's trip to Mahajanga and was delivering her return fare. He did this because she had come on his behest to collect the wall panels. He explained, "When Amina wants to go to Bezavo, the sugar refinery lends her a car. In town, if she wants a taxi she is driven for free; after all, she is the *ampanjaka.*" When I gave her some money, Kassim predicted that she would put it to personal use, as is her right. "If Amina were to collect the service money directly," he explained, "it could all get 'eaten.' So it goes to me. Unlike Amina, I don't have the right to consume it. Moreover, I am afraid (*mavozong*) of the ancestor (*dadilahy*) in my head, afraid of punishment. So I work hard and don't take the money. This accounts for the successful completion of my projects." A medium like Kassim stands as a kind of exemplary *historical conscience*.

Kassim had had a small victory in diverting funds from the Doany Ndramisara to Bezavo. Mbabilahy had risen in a colleague (not Mme Doso) at the home of Great Woman Twoiba and in great agitation had criticized her for favoring the shrine in Mahajanga, demanding that she divide her proceeds between the two sites. "After all," said Mbabilahy, "Bezavodoany is my father's place and I succeeded to power there." Twoiba is Mbabilahy's Great Woman and collects a good deal of money from all over—France, Réunion, and so forth—in his name. Twoiba was in a rather difficult position; she was afraid of Désy and could no longer announce quite as large a figure at the Great Service, but Kassim thought she was now equally afraid of Mbabilahy and would conform to his wishes.

After successfully completing the renovation of the buildings in 1998 the goal was to replace the rich burial cloths (*sobahiya;* of Middle Eastern origin) for the inhabitants of the main chamber. Now that so many dignitaries visited the shrine, Amina was embarrassed at the state of the contents. Her letter requested fourteen *sobahiya*—two for each ancestor—and two rolls of simple cotton (*gôra*). Kassim explained that they would rewrap each skeleton in plain cloth and then in a *sobahiya*. The second *sobahiya* would be laid over the top.[23]

They also replaced Ndramandikavavy's saber (*viarara*), which was used to slaughter cattle within the enclosure and said to be worn down. Although there were large spirit gatherings at Kassim's compound in order to pool money for the purchase of the cloth, the *viarara* was paid for directly by the Tsiarana clan, who wished, as usual, to take care of any purchase that concerned their ancestress.

When I arrived, the new saber was still in the shop and would remain there until they were ready to install it at Bezavo. Kassim had just sent his *manantany* to Bezavo to locate two head of cattle for the installation. The saber is consecrated as it moves toward the center, washed in the blood of a beast slaughtered on the threshold of the outer enclosure (*fiaromby*), and then in the blood of the second beast at the gate of the *valamena*. Only then can the new implement itself be put to use. Within the enclosures, cattle are always slaughtered with the *viarara* and, hence, in a non-Muslim mode.

The "kindedness" of the ritual officiants and their links to specific historical figures and acts is critical. Within the sacred enclosure, cattle can only be sacrificed by the appointed person. It is a male job, inherited from father to son. My question whether the line of sacrificers must descend from Ndramandikavavy was met by silence, since it evoked a potency that is less secret than does not bear speaking of. In any case, the holder of the *viarara* is simultaneously the guardian responsible for Ndramandikavavy.[24]

This silence has a double edge. Kassim followed it by remarking that when they entered Bezavo at the *Fanomopoa* of the previous year (on November 13, 1997, as he noted precisely) one of the invited national dignitaries brought pictures of royal ancestors and showed them to Ampanjaka Amina. Kassim reflected on the irony that outsiders often knew more than Sakalava themselves about their traditions. He attributed this to the past emphasis on secrecy: "When people queried the elders they were always told they were too young to know. This secrecy has to go, so that people can learn their own traditions." Another time, he said that the problem of transmission stemmed from the extended absences from their grandparents of village children at school in Mahajanga. His generation of active spirit mediums came from the countryside, but the next one will have to come from town.

Kassim spoke as a local intellectual on the pivot of cultural objectification or decline, acutely sensible to both past and future. At one moment, as a medium, he feared to discuss a point that I had heard repeatedly from people less embedded in the system of power. At the next, he was enthused about the book he hoped that I would write, with each page divided into identical texts in French and Malagasy.

This chapter has explored multiple levels of meaning in the historical practice of a leading spirit medium and the subtle balance of power in his relations with living royalty, shrine guardians, and others. There is a chain of legitimation that is the obverse of day-to-day relations of power: the guardians legitimated the medium of Ndramboeny, who in turn legitimated the accession of Amina; Amina, in turn, authorized Kassim's activities in collecting for the shrine and enjoyed greater privileges than he, whereas Kassim and other spirit mediums directed the guardians. But the emphasis has been not only on authority but on Kassim's moral judgment and self-constitution through dedicated practice.

Curating, in the registers of embodied possession and architectural renovation, is a form of both political subjection and historical agency. It is within that dialectical nexus of agency and subjection that Kassim lives his life and that the practice of mediums (and other Sakalava subjects) must be described. As one spirit convinced his ancestors to let others be buried at Bezavo, so do the living convince the ancestors to permit material changes. As the following chapter will show, the ancestors eventually acquiesce—as long as they are asked first and as long as it is made evident that they retain the loyalty of their subjects.

APPENDIX: THE PLAN OF BEZAVO [ACCORDING TO KASSIM TOLONDRAZA]

In a number of conversations over the years Kassim described Bezavodoany. Bezavo is deep in the woods, amid tall trees. It is far from water because it is high up. There were snakes in the past, and there are many lemurs. All creatures at the shrine are sacred (*masing*) and cannot be killed or eaten.

Bezavo is composed of several concentric rectilinear enclosures ordered according to the cardinal directions. (figure 9.2).[25] The outermost enclosure is the *fiaromby*, which is made of thin vertical posts. Cows are let into the *fiaromby*, but no farther, to trample the annual growth of vegetation. The *fiaromby* has several gates, each dedicated to specific ancestors. On the north side, toward the western end, is the "big gate" of Ndramandisoarivo and Ndramandikavavy. On the west side, from north to south, are the doors, respectively, of Ndranavianarivo, Ndramañaring, and Ndranantanarivo. This is the side of the Marambitsy (inhabitants of the region). On the south side, directly opposite his parents' gate, is the gate of Ndramboeniarivo that he shares with Ndramañonjo and Ndrankehindraza. There is no gate on the eastern sacred side. But toward the eastern end of both the north and south sides are red blood gates (*varavara menalio*).

Each person enters through the gate that specifies a portion of their identity. Ampanjaka Amina enters by the door of Ndramañariñarivo since she is from Mitsinjo and the spirit is master and has a relic there. Kassim and his wife enter through Ndramboeny's door; if I were visiting, I would probably do the same. Government notables enter through Ndramañaring's door since they are Amina's guests. Mediums of Ndramandiso and Ndramandikavavy enter through their gate. Mediums of Mbabilahy enter through Ndranavianarivo's gate as the two are brothers. Ndranavia's gate is also the one through which all cattle enter. No one enters through the red blood gates; they are extremely sacred (*sarotro*) and always closed. I suggested that perhaps they were used for a royal burial, but Kassim did not know; since he started going to Bezavo, no burial has taken place.

The second enclosure is the *valamena*. It has only one gate, on the west (*andrefa*), and, as in the Doany Ndramisara, the posts are sharpened (those of the *fiaromby* are not). The stone court (*lakor'vato*) also has only one gate, on the west. It is made of very thick stone walls, taller than a person by nearly a meter. Inside the stone court is the temple (*zomba vola*), rebuilt in 1993, containing six tombs. Although there are no names on the tombs, people know who is interred in each. The temple forms a rough square with doors midway along the east and west walls and a path between them, bisecting the room. The eastern door is a red blood gate, the entrance of royalty. In the north half, from east to west are the "married couple (*fivadiaña*)," that is, the joint tomb of Ndramandiso and Ndramandikavavy; then Ndranaviarivo; then Ndramañariñarivo. In the south half, from east to west, are Ndramboeniarivo; Ndramañonjo, and

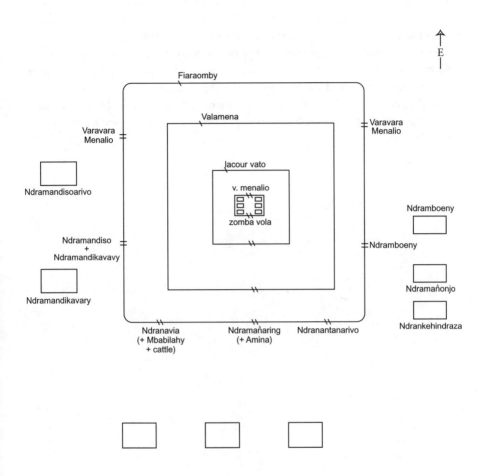

9.2. Plan of Bezavodoany According to Description Provided by Kassim Tolondraza

Ndranantanarivo. The *zomba vola* is only entered when there is a Service to fix something there. Neither Kassim nor any of the mediums was able to observe the tombs being opened. The work is done entirely by select Ancestor People, and only they see the remains.

Several more junior ancestors lie elsewhere. Ndramañaradraza and Ndramañaradreny, children of Ndramandikavavy before she married Ndramandiso, are buried within the stone court but outside the temple.[26] Ndrankehindraza, along with his close kin Tsiteraniarivo (known as Niny Moana) and Ndramteliarivo (Dady ny Moana), are buried inside the *fiaromby* but outside the *valamena*.

A visitor to Bezavo does not go to the tombs but must pay respects to each of the ancestors in turn in their respective houses that lie outside the *fiaromby*.[27] Each guardian lives outside and just opposite the respective gate of his spirit. Ndramandiso's house is north of the enclosures and just to the east of Ndramandikavavy's. Ndramboeny's guardian lives on the south side, a bit east of his gate. Ndramañonjo is slightly west of the gate and Ndrankehindraza is just west of Ndramañonjo.

Note two points in this picture: First, how completely the social hierarchy is spatially inscribed, the placement of each house equivalent to the genealogical location of the respective spirit (and also to the spatial arrangements when *tromba* rise together); and second, how closely the depiction of the layout is linked to describing movements to and through it. It is evidently chronotopic rather than merely spatial, and in the form of a maze.

 **10**

Answering to History: Conflict, Conscience, and Change

revious chapters have provided an essentially structural account of cultural reproduction. But actors experience social process as punctuated moments or unique and compelling dramas that engage specific sets of competing interests (Turner 1957; Bourdieu 1977). This chapter draws upon a series of interrelated events in an analysis of conflict over authority, reproduction, and change in historical practice.

A larger question, which I cannot really answer, concerns what constitutes significant change ("transformation") as opposed to routine reproduction. This surely depends in part on the distance from which one views the phenomena: the birth of a child that is mere "social reproduction" for a social analyst is experienced as significant transformation by the persons concerned. Political debates over what appear, in the heat of the moment, to be matters of life and death may in the long run turn out to be insignificant, or they may be symptomatic of larger forces of change that are eroding a structure rather than transforming it. With the perspective of hindsight, changes may be seen as gradual, instituted with regularity, or monumental.

My own analysis shifts between two perspectives: One is the close-up view of the ethnographer (rather than the cooler gaze of the historian), in which observations on the significance of particular events, struggles, or processes draw frequently upon the opinions of the actors themselves and, hence, tend to share their bias. These observations may need to be taken with a grain of salt, as they emphasize the disputatious, agonistic quality of life. The second perspective is one that remarks on the orderliness of certain changes and the knowing way in which Sakalava engage them. This concerns how initiatives are explicitly raised with ancestors and must be accepted by them before being put into place. In this way, Sakalava become moral, conscious agents in change, not simply passive or unwary victims of it. A yet more distant perspective might argue that participation is deceptive and that structural forces impinge in any case or at least that changes are arranged hierarchically such that the most transformative are the least voluntary. Nevertheless, the effect of cultural context—of what I have been calling Sakalava historicity—on the experience, understanding, and very means of change is not inconsequential.

The chapters in Part II that were phrased in terms of divisions of labor could be redescribed with respect to power. What a division of labor model sees implicitly as a functional balance can also be viewed as a map of tension lines and sites of potential contestation, negotiation, and transformation. Contestation emerges between branches of living royals (chapter 4); between monarchs and managers and their subjects and servants (chapter 5); and between live monarchs and individual ancestral personages (chapter 6). Royals quarrel with each other over property rights and priorities of conservation; struggle with their subjects over the changing basis of livelihood, material privilege, and political rights; and negotiate between the demands of ancestral spirits and the present state. Mediums are caught between competing demands to channel money to the center and to assist poor clients; and between serving ancestral moral order and the interests of living monarchs and officials. All this could be summarized as tension between the call of the present and that of the past; but equally it is mediated by the very multiplicity and differentiation of present and past voices. Neither past nor present speaks as a single voice.

I examine two points of tension around the shrine of Ndramisara at Mahajanga. One concerns the factional division between Bemazava and Bemihisatra branches of royalty. I do not attempt to explore their conflict in detail and certainly not to adjudicate it. My argument goes beneath their explicit positions and actions to their respective forms of cultural reproduction—specifically to the ways they handle the production of new ancestral spirits and mediums. The second concerns the position of mediums with respect to the actions of living royalty and shrine managers. This extends and complements the discussion of architectural politics in chapter 9. Whereas a senior medium took the lead in reconstructing Bezavodoany, mediums were forced into positions of reaction and resistance with respect to decisions taken over the restoration of the Doany Ndramisara by its officials.

Both struggles concern access to the relatively large sums of money raised in the Great Service and the power and influence that go with wealth. However, it would be misleading to reduce politics to purely selfish or instrumental concerns. In both the reproduction of spirits and the renovation of architecture, what are invoked and at issue are ethical matters: What are the right modes and means of reproduction, conservation, and change? How sound are the judgments of those one is disputing? Who is responsible? Whose dignity is at stake? What are the ends for which one is working and what guarantees their worth?

In sum, three interrelated issues (and questions) permeated the atmosphere around the shrine during the mid-1990s: (1) contestation between Bemihisatra and Bemazava over the relics (Who cares?); (2) concern and ignorance over what happened with sums acquired at the Great Service (Who profits?); and (3) the right way of doing things (Who is responsible? Who is respected?).

Ancestral history is inscribed in three registers—narratives, bodily performance, and buildings and material artifacts. Each entails its own politics of reproduction. This chapter addresses the latter two and, in particular, explores how spirit possession registered complaints and anxieties with respect to changes proposed in the architectural domain. The politics of Sakalava architecture centered on the question of whether building *en dure* (in solid materials) was the best means to assure that ancestral political values and practices *endure*. The issue was whether to privilege the production of lasting structures that require originally heavy, but presumably terminal, outlays of expenditure over the regular, continuous activities of repair and replacement as warranted by construction in cheaper and more traditional materials. What is more critical: finished products or acts of production? Renovation policies raise questions concerning the relative value of materials versus labor, commoditized versus noncommoditized exchange, and the strains between ends and means—technical and ethical modes of action. Ultimately, the debate is about cultural and social objectification; about whether the polity is to be constituted by means of relatively static "essences" that appear to freeze status, privilege, and the past or by means of ongoing "existence"—acts and practices by means of which people of all statuses mutually constitute each other as Sakalava subjects of a continuous past. In Mahajanga the positions are represented, respectively, by senior shrine managers and the majority of mediums.

This polarization of the alternatives (which from another angle might be tagged "modernity" and "tradition") is somewhat exaggerated. The mediums themselves took the lead in rebuilding Bezavodoany in durable materials. And despite new materials, the need for upkeep and renovation shows no signs of dwindling. As new generations of royalty "collapse," emergent ancestors need to be housed; the climate remains a harsh one even to materials like cement bricks and metal sheets; and, much as in the heartland of capitalism, the consumer demands of the ancestors show no signs of finitude.

The chapter invokes the voices of a number of mediums who have asked to remain anonymous yet among whom, on the issues presented here, there was a good deal of consensus. I repeat critical remarks and speculations made about the monarch and the shrine officials to show the issues that concerned people, not to take sides or make allegations.

SHRINE POLITICS: MEDIUMS VS. MONARCH AND MANAGERS

Asara Be is the festival at the shrine of Ndramisara that closes the *fanompoa* season. It was a low-key event in 1995 until late afternoon (Monday, September 11) when officials chose to make an announcement concerning new renovations. To the large number of celebrants present in the plaza outside the sacred enclosure, including Ampanjaka Désy and his wife, many mediums, and a few junior spirits, the shrine manager (*fahatelo*) explained that a construction company had been hired to rebuild the temple and that people should begin collecting to cover the cost. Désy also spoke to the matter and requested four head of cattle, to be sacrificed at the initiation of the project. There was much talk in the crowd until they were silenced by Great Woman Honorine, who explained energetically that the cost of the materials was included in the sum. "Let's have a house made of cement bricks," she concluded, "and once it is finished we will be handed the key!" People clapped at this.

The speeches were interrupted briefly by an elderly man shouting loudly. A companion explained that although the man was drunk, what he was saying was true, namely, that it was all very well to build a temple "of stone," but that the guardians were still afraid. He was referring to Bemazava disruption. Ndramandisoarivo had risen in a senior woman to inform people about medicine to protect the gate from intruders, but ten days after the previous Great Service the shrine had been attacked by a group of Bemazava. Women fought back with stones and men with knives; four people ended up in the hospital. Although royals, including Désy's wife, were among the defenders, the shrine servants who took nightly guard duty felt most vulnerable. However, as they pointed out, they were not the ones who had taken the relics from Bemazava in the first place.

The drunk man's intervention pointed to a widely held suspicion that people were being exploited. To begin with, although the reconstruction in durable materials was initiated to protect the relics from appropriation by Bemazava, strictly speaking this was not the public's responsibility. Most people believed that the conflict over the relics belonged entirely to the royal factions and was up to them to resolve. The public's primary loyalty lay with the relics themselves, that is, with the collective ancestors rather than specific descendants. As one medium put it, "Both sides are equally descendants of Ndramboeny."

The speeches were designed to generate public enthusiasm and ensure consent. While meant to appear as though the decision were collective, people felt that they were being confronted with a fait accompli, decided some time earlier by a select group. People grumbled about having been excluded from the decision-making process and were surprised that a new financial drive was being launched so soon after the completion of the Great Service. "How had the money collected then been used?" they wondered. "Why had the Monarch not held a single drive and then divided the proceeds among the various projects?"

The suspicion that the public was being denied its rights was heightened by the fact that the announcement came as a surprise. Indeed, officials had not advised people to stay for a meeting, and many of the celebrants had left for home before it began. More than one medium wondered whether and how the ancestral masters of the place had accepted the decision. They said that back in 1984 when officials started rebuilding the temple in brick, the ancestors had denied them permission. Now, apparently, they had changed their minds. But no one knew where they had been consulted or who the mediums had been.

The weakness of the general public (*fokon'olon*) was highlighted by another matter of concern. The monarch had unilaterally removed the doorkeeper from his office. He had been appointed by the public, but the public had not been consulted over his dismissal. He was a well-known citizen, a good speaker, and especially popular among the shrine servants for whom he had always reserved part of the money brought by supplicants. No one was happy to see him go. Even the "gods" were concerned. Speaking through Somotro, Ndramisara mentioned his displeasure to the *fahatelo.* I heard repeatedly that it was the right of the "people" to select and remove officials (a decision in which the ruler could participate).[1]

No one believed the explanation put forth by the shrine manager that the intercessor had been removed because he was homosexual. Not only had his homosexuality been common knowledge when he was appointed to the office several years earlier, but as one (male, heterosexual) medium pointed out, if the ancestors had something against homosexuals they would not choose them so frequently as their mediums. Indeed, that the monarch and *fahatelo* had nothing against homosexuals was obvious in the recruitment to their management committee. Rather, it seemed clear, especially given the dismissal of another official some years earlier, that the move was evidence of a centralization of power at the shrine and the dismissal of anyone likely to offer a dissenting opinion (or even more cynically, one less person with whom to share the income). Yet in the past the various offices balanced one another. As one medium forcefully put it, "It is taboo (*fady*) for one person to control the shrine."

I was surprised that none of the senior spirits addressed the general disaffection. People explained that mediums, themselves, cannot get involved in disputes at the shrine; their spirits could speak, but only if they were specifically called up and asked. Spirits speak spontaneously only about matters directly relevant to their persons. At least one person was adamant that financial affairs were not the ancestors' department. Indeed, when I asked whether a *tromba* would dare speak out against an action of the living ruler, I was told possibly a junior spirit but certainly not a senior one. If a senior *tromba* were too critical, the medium could be forbidden access to the shrine. Moreover, shrine officials might suggest that it was really the medium, and not the spirit, speaking. They would bring in another medium in whom the spirit would say what they wanted.[2] So, in the interests of their mediums, *tromba* were afraid to speak out. Most people objected to what was going on, but no one wanted to risk the displeasure evoked by speaking frankly.

The questions of money and personnel were not unrelated of course. According to one medium, "The bulk of the money collected at the Great Service should be used by the manager to feed shrine workers. At the shrine it is forbidden to hire workers on a wage basis (*mikarama olo*). Instead you just feed people who work. What the servants' receive is gifts from the ancestors (*mosarafa bôkan' razaña*)."

In fact, there was a good deal of confusion as to where the money from the Great Service ended up, in large part because the sizes of the divisions were never made public. One man wondered how the *fahatelo* could afford the construction of a personal house outside the shrine grounds. Another said simply that, unlike mediums, members of the royal clan were not afraid to put part of the proceeds in their own pockets. When Désy was reputed one year to have requested contributions from the funds accumulated by several mediums and a Great Woman just prior to the Great Service, he used it to meet his travel expenses to Antananarivo to lobby with government officials against Bemazava claims to the relics. Depending on one's point of view, this could be seen as a legitimate expense.

Mediums also compared the way they split their proceeds among the various shrines that called upon them for assistance, with the monarchs' tendency to devote attention exclusively to the shrines closest to their individual interests within the royal clan. Factional interests on the part of royalty contrast with mediums who may be possessed by spirits from more than one faction or by those who lived prior to the segmentation. Mediums thereby appeared as relatively disinterested.

The division of labor at the shrine was also unclear. The *manantany* should serve as the senior responsible manager and "maintain the keys" (*mitana lacle*). The *fahatelo* should appoint people to tasks, direct the work force, maintain and activate the list of mediums, and keep everyone informed. In fact, the extremely capable *fahatelo* seemed to have control over everything. But as one medium stated forcefully, "The order of Ndramisara is exactly contrary to this. That is precisely why there are so many kinds of people (*karazaña*). Responsibility should be divided so that each task has its agent. It is not good for one person to do it all."

The State and the Erosion of Sanctity

Whereas mediums saw things in local terms, the *fahatelo* and monarch had determined that the success of the shrine depended on maintaining regional and national ties. Hence, they devoted both time and money to extensive travel and lobbying. There was a segment of the population who thought such activities subverted local control of the shrine and manifested a kind of "selling out" of Sakalava tradition to wider political interests.

This difference in perspective was particularly evident during the Great Service. Just as ordinary people judged that they were being excluded from the process of decision-making, so they felt that inappropriate people were being given access to the secrets of the shrine, specifically that government officials whom the monarch wished to impress were being let through the sacred door and shown the bathing of the sacred artifacts.

That is, ancestors and sacred artifacts were not being properly honored (*mañaja*). A medium described how disturbed and angry he became during the Great Service of 1995 when the bathing of the relics was postponed several hours: "And why was customary practice not honored? Because the monarch was waiting for the arrival of the provincial governor! This gentleman [presumably a highlander] went to work as usual in the morning, then went home and had his lunch, followed by a siesta, and only then did he show up at the shrine. By waiting for the governor, they put him above Ndramandisoarivo. But Ndramandiso should not have to wait for anyone!"

In effect, it was not only Ndramandisoarivo who was kept waiting, but "the crowds of people who should have been celebrating by then," the medium continued. "And to top it off, when the politician's car finally arrived at the shrine, the monarch ran out to greet and usher him in. That is forbidden (*Tsy mety*)! Hosts should wait in their houses to receive guests. And especially on the day of the Great Service no one should be above the monarch who is running it—not even the President of the Republic." The issue for the medium was less the discourtesy of the governor (though that was bad enough) than the lack of sense in not proceeding without him.[3]

Thus, whether Sakalava royalty are higher than government officials is negotiated and enacted at the Great Service. Officials show respect by attending, yet they alter this—subtly or not so subtly—by their comportment. The monarch, in turn, must decide how to receive them, and considerations of realpolitik do not always please the upholders of tradition.

That monarchs must know how to handle the state is accepted as a fact of life, but their indebtedness and subordination to the state does not sit well with everyone. People were shocked the year that traditional monarchs were offered cars by President Ratsiraka. As one medium said, "It is taboo for royalty to mix in secular politics (*miditry an'politiky*). You can be turned any which way. No outsider should command the ancestral polity. Royalty should just pray (*mijôro*) to their ancestors and not follow anyone else." People were impressed with Ampanjaka Soazara, who refused Ratsiraka's invitation, saying if he wanted anything from her he should come and ask her. Important people do not go visiting but get visited. Another medium summarized the more general situation: "Money destroys customary practice (*fômba*); with money nothing is left taboo. Custom [for example, entrance through the sacred gateway] is not for sale."

These concerns were very real. The lack of control over time and space exhibited the breakdown of social categories and furthered the replacement or encompassment of Sakalava hierarchy by divisions of class and power determined externally with other centers of gravity. As mediums observed, the deliberate transgression of taboos at the shrine could only contribute to a loss of sanctity and power of the relics themselves: "You can't tell something sacred a falsehood and get away with it. Sacred things (*raha masing*) can always distinguish the truth. In the past the shrine was like a poison oracle (*tangena*) used to discern witches. If you drank liquid kaolin from the shrine while swearing falsely, you would die. If you stepped through the forbidden gateway, the red blood gate, without the right to do so, you would fall. But nowadays so many things once forbidden are done, so many rules of the shrine infringed, that the sacred things can no longer always distinguish truth from falsehood. They are still sacred, but not as strongly as before." Another medium declared: "People are suffering because the sacred power (*hasing*) is used up. All the taboos are broken now; everything is changing."

That is, acts of pollution erode sanctity and truth. It is human exigency in maintaining the chronotopic taboos—the order of the shrine—that guarantees the sanctity of the relics and enables them, in turn, to maintain the standard of truth and value. To the degree that they are protected—fenced off—from daily life, the relics remain a kind of "ultimate sacred substance," a meta-truth and meta-purity.[4] Yet sanctity is in part a product of people's knowing actions; the entire community plays a role in "bearing" it. The ability of the shrine to discriminate truth from falsity is in turn dependent on people themselves making such discriminations.

FACTIONAL QUARRELS AND THE PRODUCTION OF NEW SPIRITS

There were moments when certain mediums suggested in frustration that if the present managers could not manage the relics properly, they should be removed from office. They looked to the model of national government that entertained elections, transfers of power, motions of no confidence, and so on. Although they disapproved of moving the relics around and doubted that either royal faction would prove better than the other at honoring ancestral custom rather than simply using the relics for their own ends, they jested that perhaps it was time for the Bemazava to have a turn. Why was this not a serious suggestion? And why are Bemazava not more popular in

Mahajanga? Answers to these questions lead us to the mechanisms of reproduction of the Sakalava order.

The complex history of Bemazava and Bemihisatra relations was introduced in chapter 4.[5] At some point, the factions reached a working agreement in which Bemihisatra had responsibility for maintaining the relics, and Bemazava for running the Great Service. This "constitution" was represented by the movements of the drums and keys at the beginning of each Great Service (chapter 8) and the cooperation of shrine officials, including a Bemazava *manantany* and a Bemihisatra *fahatelo*.

The compact broke down just prior to Malagasy independence as the emerging political parties and candidates attempted to elicit support from the royal factions who, in turn, played them off. The relics changed hands several times (chapter 4), and a long series of litigations eventuated in a Supreme Court ruling in favor of the Bemazava.[6] Bemazava impatience at Bemihisatra determination to ignore the rulings heightened tensions during 1994 and 1995 and influenced Désy and his officials to rebuild the shrine in more secure materials. Because the quarrels were exacerbated by competition between various national political parties and both royal factions took their cases to government courts and depended on the authority of government rulings and intervention, it is clear just how fully the Sakalava kingdom has been embedded in and subordinated to the order of the state.

Mediums and supplicants often said they did not care which side won; their loyalty was to the relics themselves, and they advocated a unified shrine as in the past.[7] The relics sanctify the kingship, not the king, but as long as they remained in the hands of Ampanjaka Désy he was more powerful and wealthier than his rivals. As Bemihisatra controlled the relics, so they were successful at attracting mediums, supplicants, and donations. Related to this, as both cause and effect, Bemihisatra probably had more ties to people in government than did Bemazava. Through their own history of intermarriage and subsequent practices, Bemihisatra represented a coming to terms with the dominant Merina, whereas Bemazava retained a more autonomous, guarded, and resistant—some would say isolationist—view. It was understandable, then, that the national government supported Bemihisatra to offset "ethnic" tension between Merina and Sakalava. Bemihisatra laid claim to a royal Merina ancestor and emphasized the alliance rather than the conflict between the two polities. Many embraced a life style closer to that of Merina, including Christianity.

In addition, the sheer public appeal of Bemihisatra made government representatives, ever conscious of votes, supportive of their claims. Their popularity was expressed in the number of mediums bearing Bemihisatra spirits (that is, ancestors buried at Betsioko) compared to mediums of Bemazava spirits. Indeed, while living Bemazava were steadfast in their claims to rights to the royal relics, some of them, too, were possessed by Bemihisatra spirits.

It is not simply control of the shrine that accounts for the relative popularity of Bemihisatra *tromba*. An additional factor is that mediums must adhere to the religious preferences and prejudices of their spirits. Bemihisatra ancestors who were Christian do not forbid their mediums from practicing Christianity, whereas most other spirits do. Water-Dwellers forbid their mediums Christian practice precisely because it implies association with Merina, who were responsible for their demise. Bemihisatra spirits thus make possession accessible to practicing Christians (including large numbers of Tsimihety, Merina, and people from the east coast). In addition, Christianity accommodates the production of new spirits more readily than do the strict mortuary rules of Islam. Those Muslim monarchs who succeed in receiving a proper Muslim rather than a Sakalava burial cannot reappear as *tromba*.[8]

There is a stronger and even more interesting reason for the relative paucity of Bemazava mediums. From interviews I have conducted with members of both factions, it appears that Bemazava actively discourage the proliferation of mediums of their ancestors by exerting tighter control over the production of new *tromba* from their deceased kin than do Bemihisatra.[9]

At the collapse of royal personages, the names by which they were referred to during life become taboo. Sometime after burial, necronyms (death names) are bestowed. These generally have three components. They begin with the prefix "*Ndra-*" (indicative of royal status, or "noble") and end with the honorific suffix "*-arivo*" ("thousand/s," "a thousandfold"). The middle portion, describing some feature of their life or death, is selected in an assembly. Never having observed it, I cannot describe the process with authority, but the assembly appears to include members of the royal clan, officials, and servants. However, where Bemihisatra say the choice of name lies in the hands of the people, Bemazava attempt to limit participation.

In conversations in 1995, Bemazava royals emphasized that the new name is chosen "inside the sacred enclosure (*valamena*)." Only close family and their advisors are present. Their point was that the new name is not widely known—neither spread among the general populace, nor chosen by them. The name is kept relatively secret until the spirit rises in a medium and announces it. Relative to Bemihisatra, Bemazava take longer to select the name,[10] the assembly that does so is more exclusive, and the name is released to the public more slowly. Bemazava thereby slow down the first appearance of the new *tromba* in a medium and limit the social field from which mediums are likely to emerge since a spirit who cannot give its name accurately is illegitimate. Thus, a medium who might otherwise be ready to incarnate a recently deceased personage is hampered by not having a name to announce.[11] Unlike Bemihisatra, who may start to rise in mediums before the mortuary rites are completed, Bemazava *tromba* generally do not appear for several years after a death. Moreover, indicative of the way power was once quite centralized, mediums of Bemazava ancestors are frequently close to or even members of the royal families themselves. One person added that Bemazava mediums are more likely to be considered genuine incarnations by family members.

Bemazava continue to adhere to an older tradition in which the number of spirit mediums for any given royal ancestor was severely limited. Among Bemihisatra, the number of mediums for any given spirit is often very large and extends well beyond Sakalava themselves. Bemazava say that they intentionally keep the numbers down; their *tromba* tend to have single mediums who generally live at or near the royal burial ground. It is said that the first medium of a new *tromba* always dies quickly (thereby also perhaps inhibiting possession). Thereafter, the spirit can move on without killing subsequent mediums.

For Bemazava, Bemihisatra practice shows a lack of respect. They are also disdainful of mediums associated with Bemihisatra who incarnate large numbers of spirits rather than just one or two. A Bemazava said, "Many of these spirits don't appear to know the customs of the ancestors; in one case," he laughed, "the spirit could not even say the name the monarch had held when he was alive."

In sum, Bemihisatra spirits are more popular because they provide more avenues for participation. More people can join in naming the deceased, and the transformation from living person to *tromba* is much more rapid, taking place while sentiment remains fresh. Mediumship is more accessible and less dangerous, and the spirits rise more frequently. The facts are easier to ascertain in order to become a medium; it is acceptable for there to be many mediums of an individual *tromba;* and a given medium may host many spirits. Mediums and spirits are more sociable, and there are far more occasions to engage in possession activities with other mediums, especially those that include entertainment and drinking. All of this has significant consequences for the circulation of *tromba* among mediums and, hence, ultimately, for the political well-being of the royal branch. This is because, as argued earlier, at present, mediums and their clients are the main source of committed royal subjects and, hence, of material, financial, and sentimental support for royalty.[12]

However, to say that Bemihisatra *tromba* are simply more popular than Bemazava places too much emphasis on human agency for the local model. In explaining why many of their ancestors possess only one medium—often a person living in the countryside—Bemazava repeat the fact that they are reserved (*miavong*). *Miavong* has the negative connotation of "standoffish" or "haughty," but used here it means "proud" or "discriminating." That is, they have too much dignity to appear as *tromba* shortly after their collapse or to spread themselves over many mediums.

Finally, Bemazava discretion is manifest in their reluctance to see the emergence of spirits of those who collapsed in childhood. Bemazava say that they are appalled to observe very young *tromba* among Bemihisatra. Such *tromba*, seen clutching their dolls or toy oxen, evoke pity and sentimentality and render their live parents distraught with grief. Deceased Bemazava children would not appear, or at least not so quickly, after their deaths.

As one Bemazava said, "In the past spirits were *miavong* (proud, reserved), but now even little children rise in people. This shouldn't be. Bemazava and Bemihisatra are one people but their practices have begun to grow apart."

Asked whether it was good or bad that *tromba* were *miavong*, another person responded, "Good, because if they return and rise (*mianjaka*) everyone is sad and cries. People remember with longing (*mahacharo*). There is no more conversation. But among Bemihisatra, even the little ones, the children, rise. They play with their dolls; it makes the family very sad to see them." I asked about her own mother. The woman responded, "She appears in only one medium, at Boeny [the heart of Bemazava country]. When she rises, we [her children] cry. She was fairly old when she collapsed and a Muslim, and began to appear as a *tromba* only several years later. Her medium is a kinswoman. We are Bemazava."

Poignancy is enhanced by the fact that *tromba* generally appear in the form in which they lost their lives. One person died when he was badly burnt. I was told, "The remains were buried according to Sakalava custom. The *tromba* appeared as though every touch was painful, and he thrashed in agony. The medium was his own widow." Such an expression of grief also suggests a strong identification between the first medium and the deceased.[13]

Possession here is more intimate than among Bemihisatra but also much more political insofar as but a single authoritative medium close to the immediate family serves as a mouthpiece. This is probably closer to the way mediumship worked in the precolonial period, and it is something even those who attend the active shrine of Ndramisara remain positive about in principle. As a Water-Dwelling spirit commented, "Ndramboeniarivo remains quite reserved, rising in only a handful of mediums, while Ndramandisoarivo now appears 'in every little person.'" The authority of each Bemazava medium was presumably more secure and, once in place, unpredictable mediums were less easily replaced by more complaisant ones.

THE AFFRONT TO MBABILAHY

In 1996, events heated up. In May, shrine servants held what they called a *grève* to protest that they had received no money all year. The previous doorkeeper had given them fmg1,000 at monthly openings. Embarrassed, the *fahatelo* asked them to wait until the Great Service. The walls of the old temple were torn down, and by June new foundations put in place. The Great Service was postponed by a month so that the new walls could go up.[14] The cost of the renovation was reputed to be fmg28 million (about US$7,000). Spirit mediums of each of the "four men" were levied fmg10,000. This was separate from the Great Service and collected not by the Great Women but by an enterprising male medium of Ndramandiso who had become secretary-general of the newly constituted Association of the Doany Miarinarivo.[15]

Formal organizations of royalty had been constituted in the past as bodies that could be recognized by colonial and postcolonial government bureaucracies, and the medium explained the existence of this one in similar terms: "Everything nowadays needs papers." But it appeared to be a real planning and decision-making body and possibly a way to evade inconvenient ancestral custom, unwelcome voices, and when necessary, popular opinion.

Many mediums remained skeptical about the wisdom of the reconstruction. The situation was quite different from that of Bezavo, where the ancestors had come to realize their vulnerability to fire. Ndramisara was never really keen about the idea. A medium said, "He had to be coaxed very gently. In the end he accepted because he realized his descendants wanted the change, but people knew he wasn't happy about it." Mediums were also concerned about the extra financial burden, about hiring a construction firm, about the number and range of people admitted to viewing the secret artifacts, and about the legitimacy of the changes.[16] "People are tired of shelling out money," concluded a medium, and the *fahatelo* himself could only agree, sighing, "people nowadays have 'tight fists (*tañana mahery*).'"

The *fahatelo* assured me that the senior ancestors (*tômpin*) were not angry. He made decisions only after consulting the relevant *tromba,* and he took their advice. Mme

Doso agreed that this was how things ought to work. "As the 'key' to the shrine comes from Betsioko, and as Mbabilahy is the senior ancestor from Betsioko," so, Mme Doso explained, as the senior medium of Mbabilahy she [Mme Doso] was a significant figure in this process.[17] But in the heat of the moment she or Mbabilahy might be overlooked, especially if what they said were not in keeping with the practical intentions of the officials. Shortly before my return in June 1996, this is what happened.

The first thing Mme Doso told me on my arrival was how sick she had been. She had a pain in her side that "pulsed like *pulmonaire,* a sore right ear, and [as I could see for myself] a swollen right cheek. Mbabilahy was very angry and he caused it." She explained, "Ampanjaka Désy together with the custodians of the shrine had asked permission and advice of the most senior spirits, the *efadahy,* before proceeding with the renovations. Because the construction materials departed from tradition, the spirits at first refused the request. Mbabilahy is responsible for the shrine on his ancestors' behalf and so he worked to establish a compromise. He instructed the managers to carry out procedures including the slaughter of a cow at each of the four corners of the foundations [the four beasts Désy requested at Asara Be]. When the shrine managers disregarded his instructions and began construction without even informing the spirits, Mbabilahy was furious but said nothing."

This was confirmed by one of the servants at the shrine who described how the "four men" had been called up inside the temple. Speaking through Somotro, Ndramisara had insisted that he wanted his temple made of palm thatch (*satraña*).[18] Ndramboeny and Ndranihaniña, speaking through female mediums, Ndramandisoarivo, speaking through the secretary-general, and Mbabilahy, speaking through Mme Doso, all refused the changes. A guardian explained, "Désy and the *fahatelo* begged them. They explained they needed to change the construction materials because the night guardians were afraid of attack." The guardians were not able to speak for themselves at this point, but they resented being invoked in this fashion. They were much more afraid of the wrath of the ancestors than of invasion by the Bemazava and they did not want to be embroiled in the quarrel between the royal factions. "However," the guardian continued, "as the youngest senior ancestor, Mbabilahy mediated and convinced the others to agree, but the shrine managers subsequently violated his conditions and slaughtered only two head of cattle and not at the right occasion."

The following day I heard Mme Doso tell a friend, "Some time later, when Mbabilahy was called upon at the shrine to initiate plans for the Great Service he didn't rise. Both the shrine manager and the Great Woman tried to coax him." Mme Doso said, "I felt my skin burning as though I were on fire. Finally, after a long time, Mbabilahy arrived, exclaimed, 'I'm no longer involved!' and abruptly left again. After that I fell sick and my cheek swelled. The Great Woman also fell ill and nearly died. Then the spirit's clothing disappeared from his locked box in his house at the shrine."

Mme Doso told me that her physical condition was a result of the spirit's fury. But what of the missing clothes? This was a mystery. People speculated that whoever had taken them would be eventually struck dead by the wrath of the ancestors. Various suspects were discussed. The community of spirit mediums collected funds to purchase a new outfit and called up Mbabilahy to inform him that his clothes had been stolen and that they had purchased replacements. Kassim explained: "Mbabilahy rose very angrily until his 'father' [that is, Mbabilahy's father, Ndramboeny, rising in Kassim], calmed him down and said things were like this now and not to be upset. Ndramboeny also reaffirmed publicly, 'I gave my son the rule (*fanjakana*) and you must show him the respect he is due.'" So Mme Doso began to get well.

Kassim added to the picture what he had seen and heard from Great Woman Honorine (who had also been struck) after he left trance: "The room was densely packed with mediums and Ancestor People from the shrine, all of whom witnessed what transpired next. To everyone's astonishment, Mbabilahy announced, 'Since you never do what I ask, I've gone home to my mother. I took the clothes myself! You were baffled by the missing clothes, but it was I who threw them away.'"

Mme Doso, herself, never mentioned this and consistently maintained that she had no idea what had happened to the spirit's clothing. She shared in the expense of purchasing new material and went to a good deal of trouble in preparing the garments in the old style worn by Mbabilahy. In the end, a night-long ceremony was held at Mbabilahy's house at the shrine to formally hand over his new outfit. Some thirty to thirty-five distinct spirits, comprising virtually the entire royal descent line stretching back to A.D. 1700—with the exception of the living ruler—as well as many other people, witnessed Mbabilahy's formal acceptance of the new clothing at this extraordinary event.[19]

The anger that spirits exact on their mediums is a sign of both their transhistorical power and the limitations of that power in the present-day political domain. The affair demonstrates the weakness of the ancestral spirits vis-à-vis the living ruler and shrine managers at the time. In effect, the only outlet for Mbabilahy's anger at the insult he had received was against the medium herself, who appears here rather as the battleground of the whole affair. Yet, while exacted on the body of the medium, his violence was an expression of righteous indignation that had political and collective significance. This was made evident, and indeed recouped, in the clothing ceremony. In addition to being a vivid sign of his withdrawal from the shrine and the eclipse of his agency, Mbabilahy's destruction of his clothing was a ploy successfully calculated to demonstrate the community's support. The gift of clothing enabled Mbabilahy to regain face. Although the resulting ceremony had no direct effect on Ampanjaka Désy or the shrine managers and did not change their plans, it asserted and indexed the level of support behind Mme Doso's incarnation of Mbabilahy. It was certainly a display of solidarity among both ancestors and mediums that celebrated their existence, continuity, and outlook.

In witnessing the receipt of Mbabilahy's clothing, people were affirming that Mme Doso still held authority as his medium. Indeed, her position had been at some risk. People said that if she were not already established as senior medium she would not have been invited back by the managers to speak at the shrine. The political statements were made more safely and authoritatively by means of embodied symptoms and acts than by speech. Her suffering stood as a sign of the truth and justice of her position. Despite their recognition of the cynical motives and actions of others, mediums in this situation cannot be cynical about their own practice. Nor was it coincidence that the

additional means of action was clothing; it is through the public receipt of new clothing that all mediums are "initiated;" each new incarnation of a spirit in a medium is
recognized by a few other mediums of the same and closely related spirits in a ceremony at which the clothes are displayed and worn for the first time. Clothing is critical to the constitution of the spirit's person. Mbabilahy's receipt of a new outfit thus
drew on a well-established idiom.

The feeling that the shrine managers were at fault was widespread in the community. One person despaired that officials no longer understood how to follow ancestral practice. Another elderly medium told me, "The current changes are wrong, but
what can one do? The 'masters' are angry but the shrine managers avoid consulting
mediums who tell the truth." Indeed, Ndramisara had spoken through Somotro a second time to repeat his opinion. It was widely thought that people "in the bush
(labrousse)" were also opposed to the changes at the shrine, and one person went so far
as to speculate that Désy and the fahatelo might be impeached. "Désy had been the
people's choice," he said, "and the people could also remove him."[20]

Mme Doso's stigmata demonstrate the subjective quality of mediumship. Despite
the public significance of the episode, it is clear that the wrath of the spirit is intro-
jected, working on the medium in the way that guilt does. Is this the anger of the
powerless, or is it anger that has been in some sense sublimated—channeled to higher
ends? Mme Doso's subjectivity is also evident in the creative, dramatic, and apt way
the spirit handled the situation: destroying the clothing, leaving the situation a mystery, and then solving it with a flourish as he revealed himself as the agent. The
medium is personally subjected, yet she never forgets that she is acting with respect to,
and as an agent of, a wide regime of power.

It is worth mentioning that individuated incarnations of historical figures like
Mbabilahy also have personal meaning as parent-like figures to their mediums (cf.
Obeyesekere 1981). Personages may have first been experienced in possession of parents or grandparents, and in many instances a medium succeeds an ascendant in bearing a particular spirit (Lambek 1993, 2000a). Although demanding, irascible,
uncompromising, and quick to anger, spirits are also understood as caring for their
hosts and their families, protecting and empowering them and looking out for their
welfare, as long as the hosts respect them and observe their taboos. Often they inter-
cede in their mediums' lives long before the first signs of overt possession. To the degree that hosts come to identify with spirits who already have parental identification,
we may speak of introjection of the parent.

In bearing the brunt of Mbabilahy's wrath, Mme Doso was also identifying with
him. By carrying the burden of his indignity, she was acting as ancestral conscience.
And the injury she suffered—a "slap in the face"—was, in effect, as public a sign of
what had happened—of what she bore witness to—as the missing clothing.

THE ETHICS OF CONSERVATION AND THE POLITICS OF CHANGE

Irrespective of the personal affront to Mbabilahy, the central issue concerned the
right way to proceed with the renovations. One medium complained that the man-

agers forgot to routinely consult with the spirits at each step of the way, as though they knew better than the "proprietors." Another medium summarized Mbabilahy's request to the shrine managers. He had told them to plant a new white ceramic dish containing a silver *tsangan'olo* coin, a piece of crystal, kaolin, and the blood from a sacrificed cow directly beneath each of the four corner posts of the new temple.[21] He also specified the patterns of the four sacrificial cattle:[22] "What the managers did was simply to kill two head of cattle as soon as they received them rather than coordinating the sacrifice with the planting of the foundational plates and corner posts. This is what infuriated Mbabilahy. How can the two remaining beasts be sufficient for the four corners?"

A rumor circulated that shrine servants were frightened; because the cattle had not been properly sacrificed the ancestors might take a human life in their stead. Some of the guardians began to stay away, and even the construction crew worried about "accidents." No one was injured, but it was interesting to note how the Ancestor People saw themselves as vulnerable to ancestral displeasure to a degree even greater than the mediums (see chapter 5). The state of *being subject* to the past, and the power and salience of ancestral royalty was evident.

In 1998, the new temple was up and in active use. No major disasters had struck. The shrine managers were planning on rebuilding the sacred enclosure (*valamena*), replacing the sharply pointed hardwood posts last erected in 1985 with similarly shaped cement columns and bricks pillars. Mediums had been asked to contribute an iron rod and a sack of cement each and were looking for the means to do so when they heard that they were next expected to deliver bricks. "When," they wondered, "would we have the time to make them? On our taboo days?" Mediums felt caught between the poverty of their clients and the excessive demands of the shrine. One predicted. "The new campaign will only be successful if the 'proprietors' themselves [that is, the "four men"] are really eager for it."

I heard a spirit in a town east of Mahajanga speak his mind. The Water-Dweller, a talkative and ordinarily genial fellow, complained to a supplicant that clients all wanted things from him, but where were they when he needed to accumulate money to send to the shrine? The spirit turned to me and said, with some passion, "An *ampanjaka* has no strength (*hery*) without followers. And to have followers, you have to give people something back." He was angry that they wanted to replace the *valamena*. "They are demanding too much of the mediums," he said, "and besides, there is value in tradition. Look at the Antankaraña mast-raising or the bathing of the relics in Menabe; they stick to tradition and are popular as a result."

The spirit was particularly upset with the way Zafinifotsy (white *tromba,* of which he is one) were treated: "They think we can be mollified and kept happy with a bit of liquor, that we don't see what is going on. But we do see! We know perfectly well we are being cheated." He had noted that the Betsioko spirits were not appearing much at the shrine. "Only we Zafinifotsy attend regularly and we come to be taken for granted. We are like egrets [*kilandy,* white birds always found in the vicinity of cattle]. I got angry and didn't go to the shrine for awhile, and everyone said, 'Ohh, where is Kôt' Fanjava?'"

The spirit could perhaps speak more freely than most since, as a Water-Dweller, he was outside the main line of power and more accessible to ordinary people. He thought that the money brought in for the reconstruction served in part to line the pockets of the managers. Unfortunately, the spirit's tactic of resistance—withdrawal from the shrine—was precisely the sanction that the managers could use against him. "If any of the spirits spoke out they could be forbidden access," he said. "The managers could always find a medium who was afraid of them and in whom the spirit would say what they wanted. As soon as the managers get the answer they want from Ndramisara, they spread the word outside—yes, yes, we can go ahead!

"When they ask for four head of cattle for a ceremony, they only slaughter two of them, and sell the other two for their own profit. They say the cows have disappeared, wandered off, but how could this be so?"

I asked whether he would attend the Great Service. "We go to honor (*mañaja*) our ancestors, not to serve (*manompo*) the living," he answered.

The spirit left, and the medium addressed himself to the client saying, "Don't come only when you have problems; we need your support all the time." But the woman provided only fmg2,000 as she made her retreat. The medium continued in the same vein as his spirit, "Why do they need the new fence? The points are nice and sharp! The spirits don't change their style of dress or the food they eat. What makes a place special is the way it keeps to tradition. . . . And what are they giving back to people?"

In sum, spirit and medium were feeling squeezed to channel more money and goods to the shrine—things that they in turn could not easily get from their followers. Ordinary people were being pushed too hard to meet the ends of the shrine management, and they were not receiving sufficiently in return. The argument drew on a perception of increasing inequality and perhaps even class differentiation, but it also hinged on two major (and connected) issues that have been themes throughout this chapter and some of the previous ones. The first concerns service, via cash or labor, as a form of political subjectivity and the way such participation is changing. The second concerns legitimate ways to make or recognize changes.

Being Subject and Having the Right to Serve

The Sakalava system was predicated on giving. Gifts of both labor and cattle were once ideally unmediated—direct, spontaneous, and offered eagerly. But the present reality is that demands are high and reluctant donors must be coaxed. Often the meat and labor are commoditized—purchased on the market by ancestral funds rather than given freely to the ancestors. In the past it was forbidden to purchase beef for an ancestral ceremony. Cattle with the four patterns requested by Mbabilahy belonged to the gods/ancestors (*ndrañahary*). As noted by Kassim (chapter 7), "The owners of the herds into which they were born recognized this and the animals simply awaited being brought to *fanompoa*. As a result, Service never suffered for cattle. But now they appear to belong to the herd owners."[23] Ancestor People were marked for royal labor in much the same way that certain kinds of cattle were marked as offerings. But other people could also manage a month or two away from home in royal service—time that is no longer so freely available. "This, concluded Kassim, "was the real Sakalava (*teña Sakalava*) way."

I suggested above that the renovations were initiated to protect the relics from Bemazava. But shrine managers argued on the basis of labor as well. An occasion was described by Mme Doso at which "Ndramisara continued to voice his disagreement with the renovations, while Ndramandiso was agreeable 'as usual.'[24] The managers responded, 'In the past we invited 100 people to a *fanompoa* and 1,000 came; now we invite 10 and 1 comes. Building out of durable materials will avoid future problems with labor. Thatch falls quickly but brick and metal last.'" And irrespective of the circumstances pertaining at the Doany Ndramisara, reconstruction in durable materials was happening at shrines throughout the region and explained in the same terms.[25] It had become easier to collect money for commoditized building supplies and, in some instances, wages than to secure voluntary labor.

Service was an expression of direct political subjection and religious subjectivity. Kassim said that only the tomb guardians were still fully subject to the polity, fearing that if they left their posts they would suffer or die. But despite a greater degree of freedom, mediums too remained subject; and many others said that you need to serve if you wish to thrive. Whatever their fears and their complaints about excessive demands, people did not want to lose the opportunity to express their subjectivity through service. Service not only anchored and directed social identity, but it provided access to respect and to ancestral blessing.[26] Service was thus understood as a kind of basic right.

Commoners rejected the argument about the scarcity of labor and resented and suspected the politics of architectural durability; withdrawal of the opportunity for service was perceived as a threat. Repairing was a key form of service and an avenue of both popular devotion and expression of political will. Permanent structures threatened to curtail participation and implied a hierarchy in which the elite no longer depended so heavily on popular will for their authority. It is quite instructive to follow the Water-Dweller's lead and compare the neighboring Antankaraña, among whom public will and the role of the people in producing their monarch is expressed in the raising of a wooden mast every few years. President Tsiranana's offer to replace the "fallible" wooden mast with a permanent metal pole was rejected with alacrity. It would have meant the end of what is a regular symbolic affirmation of the monarchy but also an instance of essentially democratic process. Indeed, Antankaraña deliberately select a species of wood that collapses after a few years (Walsh 1998).

Although Sakalava may be mystified by their role in the production of ancestral sacred value, they are quite well aware and protective of their role in ancestral politics. The source of their power lies in the energy displayed in giving and caring for the ancestors and royals. This is public and pleasurable, evident especially in the all-night parties and the entertainment on the plaza. The primary contrast is not between living rulers and subjects but between the silent, hidden potency of the relics and the vitality displayed by the living. The public does not wish to leave the relics entirely in the hands and control of the living monarch and his officials; and it resents displacement by outsiders, which the public sees as evidence of its own disempowerment.

Those who serve expect that the dignity of their position will be affirmed and confirmed; that they will be regarded and addressed. They expect a response that values and acknowledges their share of the burden; to be paid respect (*vonahitry*) and to receive ancestral gifts (*mosarafa*). Yet acknowledgment from royal authorities no longer comes as easily as in the past. It was said, "Royals sometimes seem to forget that it is *their* ancestors whom mediums bear in their heads and on whose behalf they work."

The idea of *mutual* respect and recognition, expressed through the idiom of the gift, is critical and even implicates Mahajanga's massive social problems. "The 'four men' used to protect the city," one person said. "In the past no one went hungry here, but now there is a lot of suffering. If the proprietors (*tômpin*) themselves have no wealth, if they are not given any money or if it is taken from them, so in turn they do not give to the people." It was even suggested that the managers' disrespectful behavior and deliberate ignoring of tradition might leave the entire city subject to the wrath (*voa heloko*) of Ndramisara. An event on the order of drought or famine was mooted.

Recognizing Change

Everyone acknowledges that political and labor relations have changed, but the ideal of mutual respect remains critical. Although the Water-Dweller talked about the importance of retaining tradition, his example of sticking to a wooden fence was a bit misleading, especially considering how many rural shrines were in the process of renovating their enclosures in brick. The issue was not the material change itself, but how it was undertaken. The ancestors provide guidance for accomplishing collective goals. As long as they are properly consulted and informed and their instructions for carrying out changes are followed, then change is acceptable. Managers should offer a sacrificial beast for each compromise that they expect the ancestors to make. This is precisely what did not happen with Mbabilahy and why he took offense.

Ideally, when Sakalava stop carrying out certain practices because the world they live in can no longer match up with the world in which the ancestors want them to live, they do not do so by outright rejection nor with forgetful abandonment. Rather, changes are worked out in conversation and debate with ancestors, with respect for and recognition of the past. The conflict or discrepancy between ancestral tradition and present contingency is made explicit. The present generally wins, but not until the ancestors understand the need for change and accept it. When, through their mediums, they acquiesce and grant their blessing, they are compensated with blood sacrifice; there is an exchange. Thus, although a specific change may not be initially desired by everyone, it is enacted on the basis of negotiation, compromise, reciprocity, mutual understanding, and respect. Ancestral voices are heard; they are not left behind but are present as witnesses. While always speaking from their original locations in the past, they are also "kept up to date." Thus, change, even when it is not welcome, may be perceived as orderly and, hence, neither destructive nor alienating.

Change is acceptable as long as it is engaged upon consciously and is properly authorized. When changes are deliberately introduced to the ancestors and the reasons carefully explained, everyone is made conscious of the reasoning. When ancestors from different generations respond differently, each according to their own historical or personal experience, then the complexity, density, and richness of the past is confirmed. When the ancestors reach consensus, a historical trajectory is affirmed. In this polyphonous and existential process, the perduring, yet fluid, cast of Sakalava identity and historicity is reproduced.

It is also important that the order of change correspond to the hierarchy of sanctity as expressed in the master chronotope. This means that changes to the outside are more easily accomplished than those to the inside. Although the temple was rebuilt in newer materials, people assured me that the inner structure, the *zomba vinda,* containing the relics, never would be. Likewise, change should proceed from the outside in. As noted in chapter 2, at least one critic argued that the enclosure fence ought to have been replaced before the temple.

Similarly, it is easier to make changes to things initiated in the more recent than the more distant past. The mediums of the spirit Ndransinint' once went to his shrine to petition that the spirit, who had very weak legs and only sat, might be able to stand. As a result, he can now move about a little. The mediums were able to intervene in the very way their spirit presented himself—in his mode of being in the world. This change in the register of performance is analogous to the production of architectural change, but it would have been much more difficult with respect to more senior spirits.

Ancestral spirits are not expected to initiate change, spontaneously supply information, or issue directives. "In the past," explained a medium, "these were responsibilities of the live monarchs. Nowadays, when living monarchs and shrine managers frequently no longer know the right way to do things, mistakes are made. The *tromba* know the right way, but it is not their responsibility (*anjara*) to point it out. Mistakes are signaled indirectly and after the fact, for example, by a sacrificial animal who refuses to lie quietly. Ancestors thus wait for their opinion to be asked. Living royalty can call them up for advice but ancestors do not rise on their own to criticize the living. Thus, although they were sad, they never told the king he was at fault for deposing officials without replacing them."

When I asked why the ancestors remain silent, the medium replied, "Children learn from their fathers. Once they are grown and their fathers have died, they should know what is right by themselves. Ancestors don't come to instruct on their own. They will always will respond if you ask them. But you have to ask."

When ancestors acquiesce, change is orderly and meaningful. Ancestors do not merely sanction change; they sanctify it and demand the rituals that performatively establish it.[27] Changes established in this way become new precedents. Conversely, changes that take place in the face of ancestral rejection are doubly problematic. Not only might they be troubling in and of themselves, but when the changes have not been sanctioned by the ancestors the very means by which change and order are regulated are threatened. So are the grounds of sanctity. Sometimes such disorderly changes are registered and addressed through the symptoms and acts of individual spirit mediums like Mme Doso.[28]

This is why thoughtful spirit mediums worry that the sacred power (*hasing*) of the shrine may be under threat, and why the shrine cannot discriminate between truth and falsity as effectively as it once did.[29]

This chapter has explored political tensions among various players in the ancestral system in Mahajanga during the 1990s, especially from the perspectives of the mediums. Where reigning monarchs had once received a broadly based tribute and had operated from a sure sense of power and privilege, their power now depended on controlling the wealth generated at the Great Service. So whereas in the past the Great Service may actually have served the shrine, supporting its well-being and the work of its servants, now the shrine's ability to attract wealth supports the reigning monarch and officials. Mediums were concerned about changes at the shrine and in the conduct of service but did not have the authority or power to address them directly. Despite their evident limitations within the political space of the shrine, spirits and mediums remained critical in accepting and sanctifying changes. They also made imaginative work of their performances and practical interventions across a broad field of interests. This is the subject of the following chapter.

**11**

The Play of the Past:
Historicity in Daily life

Just as an Israeli cannot turn into a Palestinian, so we should not try to give up our traditions.

—Sengy, spirit of a Ste. Marien sailor, speaking through Lidy, Mahajanga, 1998.

To the present I have addressed the role of spirit mediums in formal public activities concerned directly with the shrines, emphasizing the earnest and imaginative ways they devote themselves to their work. Mediumship also permeates domestic life and less formal social encounters. In this chapter I present vignettes that show how the lives of mediums are infused with historicity, resonant with the multiple personages they carry. Although Mauss, himself, did not address the relationship between the personage and the psychological individual or self, he recognized "that there has never existed a human being who has not been aware, not only of his body, but also at the same time of his individuality, both spiritual and physical" (1985: 3). Hence, self-aware individuals must live in some form of lively tension as or with their personages rather than being entirely subsumed by them.[1] However, rather than seeing the self and personage as radically different in origin or in some sense opposed, it is more fruitful to examine how the distinction is produced, worked out, and eventually transcended through creative practice.

Spirits are at once a burden and a richly creative force, even a kind of artistic destiny. Yet, despite first appearances, this is not a matter of Platonic mimesis in which artists or audiences are simply captive to their art. "Art" here is an integral dimension of social experience, not disembedded as a discrete "institution." Living with spirits entails the continuous exercise of judgment, what Aristotle (1976) termed phronesis. In this chapter I show the interplay of poiesis and phronesis, of making and doing, of subjection in the service of bearing truth and playful agency. This brings us full circle to themes raised earlier in the book: Poiesis (chapters 1 and 3) is linked to practice (chapters 7, 9, and 10), forming a kind of poietic practice or practical poiesis in the everyday lives of spirit mediums. The presence and voices of the spirits enable a richly historicized mode of existence. Historical consciousness becomes part of the habitus of mediums and those whom they regularly address.

My advocation of practical poiesis draws on classical Greek concepts without adhering strictly to their original usage. Agamben (1999) has complemented the story of the disembedding of art in the West by deftly showing how the once distinctive concept of poiesis became incorporated in Western thought within praxis and subsumed to the will. He reveals the (pathological) implications of lending excessive weight to the agency or will of the artist or the praxis of the autonomous spectator. Mindful of the original Aristotelian understanding of poiesis as "pro-duction," Agamben raises the consequences of eliding a notion of passion in which artists *bear* truth.

Agamben's argument, I believe, does not support a return to Plato's view of poetry as pure passion in opposition to reason or the establishment of a radical distinction between modes of historical consciousness as "being" and "seeing" (Daniel 1996).[2] The latter view would argue that some people are more committed, for structural rather than directly political reasons, to either "seeing" or "being." In such a model, one might distinguish tomb guardians, oriented toward "being" ("bearing"), from managers and intellectualizing historians, oriented toward "seeing" ("praxis"). An intellectual defender of "seeing" could emphasize the way Jingô, for example, are subjected to positions of "being." Mediums and spirits could be viewed as what Daniel calls "ontic icons," displaying the commitment to being through their respective corporeal transformations and sheer presence. But in fact everyone finds themselves shifting between various perspectives to the past in different contexts.

Their practice enables mediums especially to move in and out of relations conceived more deeply as seeing or being. This lability and their own reflexive understanding of it suggests that rather than opposing "seeing" and "being," it makes more sense to look at historicity within the more moderately phrased Aristotelian framework as a matter of complementary acts of making, doing, and thinking within a context that cultivates specific dispositions and offers (or imposes) particular forms of discipline, tools, and models.

Here, the form of praxis that Aristotle described as the situated, continuous exercise of moral judgment—phronesis—transcends the Platonic oppositions. The concept of phronesis underplays the role of individuated, selfish will or calculating, alienated reason—so common in subsequent theories of praxis—in favor of cultivated dispositions toward doing good. Virtuous practice implies a world in which excellence is established as a recognized means and end, and in which human flourishing is central. I argue that among Sakalava, phronesis is articulated (among other means) by historical poiesis, and conversely, that poiesis is informed by ethical (as well as political) judgment. Without wishing to conflate poiesis and phronesis, to subsume one in the other, or to limit praxis to phronesis, I explore their integration in mediumship.

The Play of Objects and Subjects: From Ironing to Irony

"Tromba Choose Their Mediums the Way Men Choose Women"

People in Mahajanga attribute possession by specific spirits to the spirits' own volition as well as to the astrological character of individual mediums. The suggestion in the heading is not sexual per se but concerns the complexity and serendipity of personal attraction (a combination of action and passion) more generally. I invert the local view to argue that possession by certain spirits, as opposed to others, is a product of the medium's poetic judgment or ethical creativity (passion and action). Possession by specific ancestors is subject to many factors, including the relations, personalities, life experiences, and unconscious motivations of the mediums, the prior distribution of spirits in their families, and the particular relations, characteristics, and interests (as culturally and historically constituted) of the spirits themselves.[3] In the end it is a matter of what makes a fitting complement, deep connection, assertive departure, compelling revelation, wry twist, and so forth.

Possession by particular spirits entails calculation of their own kinship and affinal relations, as well as those of their mediums. Maman' Sera's father had Ndramandisoarivo, and so it was appropriate that his wife should become possessed by Ndramandikavavy. But this forced Maman' Sera's parents to move because her father's brother was possessed by Ndramboay. Ndramandikavavy does not like her brother and cannot live in the same community with him ever since Ndramboay agreed to (or possibly suggested) his sister's sacrifice. (This version of the founding myth of the kingdom is preserved more coherently and consistently by means of the mediums' practice than in narrative.)

Maman' Sera, in turn, could not inherit the senior *tromba* who had possessed her parents once she married Kassim since he was possessed by their son. "*Tromba* would never enter a 'daughter-in-law,'" she said; so they entered Maman' Sera's adult daughter (by a previous husband) instead. They were not put off by the fact that the daughter had received a Catholic education.

Spirits are understood as having a stake in not just a medium but a married couple. Should a couple in which either member is a medium wish to separate, they must inform their spirits. One couple was criticized for breaking up because they were mediums, respectively, of the husband and wife pair, Ndramandisoarivo and Ndramandikavavy. Usually, the arrival of a related spirit in a spouse is a sign that the marriage rests on solid ground. Thus, "happy marriages" from the historical repertoire are re-enacted by mediums or by the presence of both members of the ancestral couple in a female medium.

The fit between mediums and spirits is partly a matter of tacit identifications, unconscious longing, or repression. None of Kassim's spirits drink or smoke; all of his wife's spirits do. Interests often change over the course of a medium's life span, and these may be expressed by the spirits' unexpected arrivals and departures. Maman' Sera once brought the clothes of Ndransinint' to a ceremony and applied kaolin in his patterns to her body in anticipation of his imminent arrival. To her surprise, Ndrankehindraza rose instead. Maman' Sera had had Ndrankehindraza for some time, but he rarely rose in her any more. She admitted she could "no longer take his *zistoires;* I'm too old for dancing."

Yet, "Ndrankehindraza remained a significant presence for Maman' Sera," said Kassim. "Having 'collapsed' as the result of sorcery material placed in a ripe banana, he forbids his mediums to eat them. Many of them beg to have the taboo removed and some are successful. Maman' Sera asked in vain and gets sick each time she visits her family in the countryside where ripe bananas hang in every house. She doesn't eat them, but the mere sight makes her return to Mahajanga with sore eyes." It is no coincidence that the symptom is ocular; Ndrankehindraza's eyes bothered him and whenever he arrives, his medium's right eye is painted with a circle of white clay.

A medium's initial symptoms or sickness often echo those that afflicted the personages during their lifetimes or were associated with their demise, but they may also be deeply motivated on the part of the medium. Each *tromba* causes its medium characteristic pain or discomfort. Ndranaverona choked to death on cashew liquor, and he enters his mediums choking dramatically, staggering, and falling into the crowd until he is revived with kaolin and his chest bound with a strip of cloth. His medium's chest hurts when the trance is over, but the degree of discomfort varies. Comparing notes, one medium found incarnating Ndranaverona much more painful than did another. Mediums of Ndranaverona all refrained from wearing the popular cloth that had a print of cashews on it. Ndranaverona's intimates also know his sexual preference. This has no bearing on his female mediums (though one I knew had suffered from asthma), but his male mediums generally turn out to be homosexual or bisexual. Someone said, "Perhaps mediums are queer (*sarambavy*) because the personages themselves were, but we cannot know for certain."

Ndranaverona was close friends with his brother-in-law (*valahy*) known as Zaman' Ady (Ady's Uncle). Though not royal himself, Zaman' Ady appears as a *tromba,* often in the same mediums as Ndranaverona. A number of people told me his sad story. Shortly after the death of his dear friend, Zaman' Ady climbed a cashew tree and in his grief either jumped or fell off. As a commoner he could not be buried at Betsioko, but Ndranaverona had asked that his friend's mortuary preparations be such that he return as a *tromba*.

On a sudden inspiration, I once asked a spirit who was recounting the story whether they were lovers.

"Who told you?" responded the spirit abruptly, as though I had revealed a confidence. "Someone must have told you! It is true," he then admitted, "but people refer to them as brothers-in-law."

Although many people know about the cashews, the love story is more restricted. This illustrates what I described earlier as the distributive nature of historical memory as well as the way the lives of certain spirits have personal as well as public significance. Put another way, spirits have multiple kinds of significance, and each interpretation has its sphere of social distribution.[4]

Analogous to Ndranaverona is Ankanjovôla, the wife of Mbabilahy. Her appearance is also extremely powerful; she sits on a mat with legs open and outstretched, panting heavily, while another woman (not in trance) supports her from behind. It appeared to me that she was in the act of giving birth.[5] Her taboos concern pregnancy, and she almost certainly "collapsed" in childbirth or as a result of complications. She will not appear in the presence of pregnant women, sometimes gives her mediums false pregnancies, and does not like them to give birth; her mediums are all women who have had no or highly limited reproductive success.

In no case, however, is the association between particular mediums and their spirits mechanically determined or fully predictable. Nor is possession by Ankanjovôla understood as a form of therapy for fertility problems or initiation into a specific "fertility cult."

Spirits are thus distinguished by their various forms of embodiment, each of which realizes and confirms specific historical events, that are not necessarily ever spoken about, while drawing on unconscious motivations of specific mediums and interlocutors for their reproduction. Through possession, mediums rehearse history; they play with the characters who take them on but also draw from them such that the two lives, medium and personage, are deeply entangled. And this extends to those in the mediums' milieux. Ndranavianarivo, a son of Ndramboeny, forbids anyone to stand behind him. Kassim described how he "once saw a girl run behind the spirit to reach her mother. She fell and became unconscious. Unconscious! It's not a joke with him! Also, Ndranavia always walks behind others; even his medium lags behind." Presumably he was stabbed in the back.

The Poiesis of Clothing and Architecture

Services to bathe the relics and restore the shrines are part of a wider set of signifying practices concerning the material artifacts of ancestral personages. Each spirit has particular items of clothing and accoutrements like royal staffs or the toys carried by child spirits. My attempts to understand the ownership of the silver *tsangan'olo* coins (Lambek 2001) illustrate in one small domain the creative flair with which Mme Doso organized the complications of living with a large number of spirits. The dish belonging to her senior spirit, Mbabilahy, contained two old coins. Mbabilahy shared the dish with his wife, Ankanjovôla, who also possessed Mme Doso; one coin belonged to each of them. At the next occasion on which Mbabilahy was called up, I sought to confirm that the dish and coin on the altar table were his. To my consternation, Mme Doso replied that they belonged to another of her spirits, Ndramandaming. But the logic was impeccable. Mbabilahy and Ankanjovôla shared the dish, but as it happened to be a Sunday, which was Ankanjovôla's taboo day, the dish could not be used. So Mbabilahy had to be called up using Ndramandaming's coin. Ndramandaming, in turn, shared his plate with his younger brother, Ndramianta . . . and so on.[6]

The artifacts—clothing, houses, domestic items—and practice of taboos serve to realize and elaborate the personages whose property or demands they are. Clothing and consumption practices provide an ostensible record of times and fashions.[7] They also form a field of play, at once artistic and moral, in which relations between past and present, subjects and objects, are articulated, deployed, and renegotiated. Contributing toward and caring for ancestral material goods entail devotion and servitude but also afford imagination, insight, and pleasure.

Mediums devote themselves to purchasing their spirits' clothing and carefully storing and laundering it. Ndramandaming's costume includes a felt hat and dress shirt. His mediums are always on the lookout to improve his wardrobe. Those who can afford it purchase gold cuff links. Mme Doso was afraid that they would be lost or stolen so she sought a second set, of costume jewelry, for use outside her home. Some mediums wear gold or silver rings or bracelets, belonging to specific spirits, as permanent decoration and protection. Mediums' personal adornment is also affected; as the medium of Zafinimena royalty, Mme Doso is entitled to wear gold ornaments in her hair, but she eschews gold (for her own person, not for Ndramandaming) because she also has Zafinifotsy spirits, for whom gold is taboo.

The more senior ancestors want their laundry neatly folded and not ironed, but for the spirits of the colonial era careful ironing is necessary. As Kassim put it, "They were used to European culture and had already entered '*civilisation*.' Some spirits speak French, while the older spirits can't even count the kind of money presently in use." Mme Doso was particular how the ironing was done, and unlike most people, who used charcoal irons, she had an electric one. She was very fussy about Ndramandaming's handkerchiefs; I once heard her lecturing an assistant on the difference between folding handkerchiefs and table napkins. No doubt she drew here on a short period as a servant in a colonial household, transforming her experience of manual labor into a poiesis of ironing.

Clothing also signifies the relations of participants to the proceedings they attend. Mediums change from Western or Muslim garb into Malagasy wrappers when they begin to supplicate the spirits. It is only when they are dressed in Sakalava clothing that they enter trance, and it is from this clothing that the spirits then change into their own garments. Supplicants, too, must dress correctly if they want to participate in possession ceremonies. Men often simply add a woman's cloth over their trousers but at the shrine they wear just a shirt and wrapper. The further toward the center, the stricter the requirements. For those not used to traditional Sakalava posture or dress, these produce quite intimate forms of subjection; in being forced to attend closely to bodily practice, supplicants are all rendered in something like the condition of mediums.

As for clothing, so it is with furnishing and housing at the shrines. Mme Doso discussed where I would stay if I accompanied her to Betsioko. Mbabilahy's house is empty of any furniture but mats. Younger generations of ancestors maintain houses full of furniture. "Mampiaming, Ndramanisko, and Ndramandaming all have beds, mattresses, buffets—everything! And it is kept clean. But Mbabilahy has none of this; he doesn't like it" she explained. Each house is lived in by the respective custodian (*rañitry*); thus the custodian's style of living is determined by the ancestor for which he is responsible.[8]

On her visits to Betsioko Mme Doso slept in the same house as the custodian of Mbabilahy. She related, "People enter the shrine on a Thursday evening and everyone sleeps at Mbabilahy's. The first night his house is crammed and the overflow sleep on mats outside it. The next day mediums can go to the houses of their respective spirits except those for whose spirits Friday is taboo and who must wait another day. On Friday people rise, bathe, and then pay their respects to Mbabilahy and give him money. They ask his permission to enter (*mangataka varavara*)."

Mme Doso then goes to pay her respects to her other spirits who lie at Betsioko— Ndramianta and Ndramandaming—but she must continue to sleep at Mbabilahy's. She recalled wryly what happened when she slept elsewhere: "A ceremony was being held in the house of Ndramanefy, and the Great Woman ordered everyone to stay the night. The Bemanangy was soon groaning in her sleep. Then I fell asleep. I saw a man glaring at me, just the way my father's mother used to when we went out together when I was a child and didn't stop eating when she did. My grandmother was irascible (*mashiaka*)." She thought about all of this in her dream. She continued, "I saw the man glaring and followed him. He said, 'You were clever to follow even though I didn't ask you to because someone down there is calling for you.' The man held a tall spear from which silver decoration sparkled. He wore just a loin cloth. He was very dark and very strong. He pushed me with the end of his spear and I was very frightened. His eyes were red. I thought he was a brigand (*fahavalo*).

"First the Great Woman was troubled, and then I woke with a start and the custodian of another ancestor tried to calm me. Although it was dark and the middle of the night, the custodian took me immediately to Mbabilahy's. That shows how fierce (*mashiaka*) he is," she laughed.[9]

The Conservation of Authentic Copies: Essence, Existence, and Change

I return here to the issue of the marking of change in the register of objects raised in the last chapter. At first sight, spirits' possessions could be likened to precious heirlooms, and the shrines to museums. But their relationships to the past are very different. This is primarily because the Sakalava world is a Maussian one, that is, a world predicated on the social reproduction of its personages. Personages are continuous from generation to generation as successive biological individuals become their hosts and the trustees of their remains.

Brad Weiss distinguishes clothing as costume from clothing as heirloom (1996: 137): "A coat can be worn as an heirloom in order to represent the values that attend to a certain position, or it can be worn as a costume in order to imitate particular figures who occupy these positions." Such a distinction is not fully appropriate in a performative Maussian universe. *Tromba* clothes are not heirlooms in the sense of having been passed down, yet they are not as alienable or superficial as costume. Clothes are a kind of reproduced heirloom; despite being new, they belong to ancestral spirits and "represent the values that attend to [the] position." As Weiss points out, the play between heirlooms and costumes provides a means for constructing "alternative models of hierarchy and authority, as well as continuity and discontinuity over time" (ibid.: 137). Each arrival of a spirit is indicated by a change of clothing, and each spirit has its characteristic outfit. New mediums purchase appropriate clothing, and the offering of the outfit to the spirit is a significant stage in the legitimation of its presence in the medium. The spirit's clothing is said to resemble the kind of clothes the ancestor liked to wear when he was alive; it is not the deceased's former clothing itself. Likewise, items of clothing are not passed on from mediums to their successors but purchased anew by every medium. The spirit's clothing is channeled to the realm of the dead when a medium dies, sent to the ancestor's guardian at the cemetery (or, if too distant, to any appropriate Ancestor Person) where it can be worn by the guardian or members of his family. Worn out clothing is carefully discarded.

The homes of the former monarchs at the tombs are furnished by the mediums, just as the tombs are refurbished through their endeavors. As storehouses of and monuments to the past, some people explicitly likened shrines and tomb enclosures to museums. But the principles of curation and conservation on which they rest are quite different. For one, shrines and their objects are the products of collective agency. Individual donations are absorbed into larger prestations; ultimately, each person's act of giving becomes objectified in the new walls or implements and people can say with pride, "We built this."[10] For another, whereas one function of a museum is to exhibit to a general public, shrines conceal certain objects.

More critically, in the shrine people do not confront objects rescued from the past and inevitably objectified and decontextualized. Rather, the objects are in their context and constitute a setting in which it is the visitors who are not "at home." Visitors do not gaze at past objects but are confronted with a past that is coterminous with them, that thrusts itself forward along with them, and in whose terms they must view themselves. Instead of continuously restoring old artifacts or attempting to keep disintegrating fragments under glass, Sakalava properly discard and replace them. "Worn" goods—clothing, housing, even the sacred saber—can and should be replaced when they are in bad condition. The drums beaten during the Great Service receive the skins of newly slaughtered cattle every year. Other items are replaced more slowly. The melting of silver to reproduce a saber does not invalidate the fact that it *is* Ndramandikavavy's saber or preclude it from being treated with the same awe and reverence as the former artifact. Authenticity does not depend on identity, antiquity, and originality,

It might be thought that what Sakalava conserve or remember is not the material object per se but, rather, the idea of it—the form. But particular items (even the incarnations of the mediums themselves) do not stand in for a Platonic "ideal;" each manifestation is received as real and potent. The Sakalava world is one of immanent social reality, not transcendent ideals. The act of replacing one materialization with another is itself salient. Indeed, it is the very acts of replacement and practices of curation that conserve the sanctity and identity of ancestral artifacts. Replacements ("copies") are "authentic" insofar as they contain ancestral aura and are the property of ancestral personages. The aura is present if the artifacts have been made, purchased, handled, given, and received correctly.

There is no clear distinction between costume and heirloom precisely because the distinction between original and copy is largely irrelevant. Just as the idea of an "authentic original" can only appear in an "age of mechanical reproduction" (Benjamin 1969b), so the idea of the copy has less salience when performance forms the primary means of reproduction (Taussig 1993). Mediums buy new outfits in old styles (though, to be sure, both the idea of what constitutes the old style and the availability of the means to reproduce it change over time and across space and thus produce some variability among the actual appearances of the same spirit in different mediums). The *tromba* expect fine new clothing, just as living monarchs do. The Sakalava chronotope is not one in which the past remains in the past and is therefore left to decay or can only be represented exactly as it was; rather, the past continues its half-life in the present. In life, the replacement of worn-out clothing is not inappropriate, and a new out-fit is just as much a personal possession as what it replaces.

Hence, Sakalava see no contradiction between conserving the artifacts and dwellings of the ancestors and replacing or renovating them. Yet they cannot do so un-thinkingly, without marking the changes they make. What grants legitimacy to re-placements and renewals is neither the antiquity of the items themselves nor necessarily the antiquity of the models on which the artifacts are based, but the fact that they have been specifically requested or accepted by the ancestors and, frequently, that they are marked by the "bloody parturition" of sacrificed cattle at the threshold. Continuity does not imply stasis or the preservation of "originals." The aura (or sanc-tity) of an object comes not from its antiquity per se but from the fact that it has been correctly offered to, accepted, and perhaps consecrated by an ancestral personage.

This has extensive implications. It means that change is recognized, acknowledged, and accepted; meaningful change itself carries the authority of the past. In this way, it can be said that Sakalava tame or domesticate change, transforming the randomness of sheer change into the meaningfulness of history. As Sakalava understand it (at least, within the politico-ritual frame), change does not happen unconsciously—behind the backs of its subjects—but as the product of self-conscious agents, addressing the con-tingencies of the present with reference to the past, and responding to the address of the past with gentle reminders about the contingencies of the present. It is in this di-alogical space that "historical consciousness" is to be found.

The only things that cannot be replaced are the bones of the ancestors. The bones index what is unchanging in the identity of the personage. The balance of movement to stasis—of existence to essence—is not the same for Sakalava as for much Western historical thought. For Sakalava, existence, as manifest most explicitly in the lively *tromba,* is most salient, though it is guaranteed with reference to the essence of the bones. The diffusion of ancestral essence occurs through blessing, through the admin-istration of kaolin and silver, and of course through the speech and play, that is, through the existence of the lively *tromba.* However, the weight given existence does not mean that Sartre would have approved; existence is precisely *not* predicated on the auton-omy of the individual or the primacy of individual choice. The theme of this book has been that of the collective burden, of subjection to history as the basis of a form of consciousness; consciousness with, through, by means of history. Thus, not the con-sciousness *of* history, as though "history" were a discrete object in a universe of pos-sessive individualism that could be acquired, competed for, chosen, or discarded at will, but *historical* consciousness—consciousness that is itself thoroughly and intrinsically historical.

Covering the Mirror

Like the slippage between original and copy or heirloom and costume, so too the slippage between ancestral spirit and medium, that fleeting space or moment of cre-ation that is neither self-alienated nor self-captivated. The creativity in possession is neither that of the puzzlingly self-conscious state of art (the artist, aesthetics) that has developed in the West (Agamben 1999) nor that state of blind tradition epitomized in Plato's view of poetry as mimesis and passed on ever since as the imaginary "other" to Western reason and philosophy.

When she held a possession ceremony at home, Mme Doso began by carefully covering the mirrors on her wardrobes with long pieces of cloth. It was as if to catch a glimpse of oneself while double would either shatter a fragile illusion or confirm a state from which there could no longer be any return (I received no exegesis). But the very refusal of reflection in the act of covering the mirror is itself a fragment of reflection, a momentary catching up of oneself. In the juxtaposition and play of perspective and voice, possession is deeply, intrinsically ironic. Hence it is never simply mimetic.

If possession—a chief register of history,—is ironic, so is Sakalava historicity. In this sense it may be more "modern" than the literal-minded historical consciousness characteristic of the West.

A Colloquy of Colonials

Contemporary Sakalava, drawing perhaps from school textbooks and national ideology, distinguish three historical epochs: *Fahagasy,* "Malagasy times," is followed by *Fahavazaha,* "European times," which begins not with the first arrival of Europeans but with the imposition of colonial rule; the current period is *l'Indépendance.* Sakalava do not idealize any of them. I was told: "During European times no one Malagasy group dominated the others; Europeans had no 'sister's sons' to give preference to. Today, if a post is announced the Borzan (Merina) always manage to direct it to one of their own. If the Europeans had a preference it was for Comorians, whom they made their *boy* (servants); they liked them because they didn't drink. But we weren't *libre* then. In Malagasy times there were fights among us."

Possession illuminates and voices the overlaps, interpenetration, and simultaneous salience of these periods, realizing the irony of history.

Mme Doso is possessed by the popular and urbane Ndramandaming. He was a provincial "governor" under the French and traveled to France. Mme Doso also incarnates his older brother, Ndramianta, who wears a suit jacket and a red fez with a gold pin to indicate that he is royalty through both parents (*ampanjaka mena;* "red royalty"). Unlike his younger brother, Ndramianta is haughty (*miavong*) and, hence, doesn't enter many mediums. He has a hoarse voice and says little. But Ndramianta is the man who dismissed Ndramisara as "Makoa" (chapter 3). Makoa are people of African ancestry, most originally brought from Mozambique as slaves; Ndramianta thus gives voice to elite snobbery and racism.

Interestingly, the position is enunciated through a personage who was closely linked to the French colonial authorities. The opinions are certainly not those of Ndramianta's mediums. Mme Doso, herself, is dark and probably has some Makoa ancestry. The context is one in which the ideology is presented by one voice activated within a dramatic whole, that is, relative to other voices (cf. Burke 1945)—a whole in which Ndramianta is seen as morose, unsociable, and ultimately isolated. Ndramianta cannot rise together with Ndramisara as it is forbidden for them to look at each other. Many people, including a number of mediums, do not know the reason for the taboo; colonial racism, like other past violence, is remembered with discretion.

In addition to the historical contextualization of racism, it is evident how incarnation imposes situational judgment. Mediums act on the knowledge of when, where, and in front of whom they can or cannot incarnate a specific spirit. Such judgment also shapes how spirits are addressed. I once heard a medium complaining to a spirit from the colonial period (risen in his own daughter) that whereas European governments looked after the poor, only in contemporary Madagascar were they left to their own devices. This was less a piece of colonial nostalgia than the use of the spirit as a foil for political commentary.

Vazaha Kely, a flamboyant and clever woman who favored head scarves and large golden hoop earrings and whose light skin and airs had earned her the name "Little European," was a medium of Ndramianta's grandson, Ndramanisko. She said, "He was raised by Ndramandaming, the provincial 'governor,' and developed expensive French tastes at his table—for Dubonnet, whiskey, and pastis.[11] Ndramanisko was forbidden by his parents to become a soldier or go to France as Ndramandaming and Ndramianta had done, so he studied instead. He is an 'intellectual.'" Although Vazaha Kely, herself, knew little history, she explained, "When offered wine, Ndramanisko speaks expansively in French and recites history (tantara)."

Although some members of the royal descent group were integrated in the colonial bureaucracy, others were not. Ndrankehindraza was threatened with imprisonment for failure to pay the colonial head tax (la tête) and was protected by one of his uncles, who refused to let his kin go to jail. A medium, William, said that a spirit he incarnated was "readily frightened. He lived in the bush and stayed away from the government in order to avoid paying taxes or being harassed. Today the spirit hides every time a jet flies overhead." William drew a strong distinction between those ancestors resting at Betsioko who were civilisés and those who were unsophisticated and relatively unfamiliar with Europeans.

One of the Water-Dwellers provided an antidote to the colonial voices. Speaking through a medium named Lidy, Kôt' Mena said that he did not like people to say that Sakalava "worship idols (manampy sampy)." He continued, "The respectful greeting, koezy tômpin, is like the prayers of Christians or Muslims. We face east, Muslims face north, but it is all the same in principle." Kôt' Mena admonished his young clients and onlookers to ask questions and learn about their traditions. Although Lidy engaged in "traditional" performance, through it he articulated and addressed cultural objectification and the historical perspective of "modernity."

People laughed at the unsophisticated spirits, whereas the spirits themselves were often pleased at their effect. Often this was produced by linguistic play. One spirit asked where I came from and then kept repeating, "Panadá," comically playing the part of someone pretending to be more adept in French than they were. With great timing and verve, the spirit then pronounced, "Ponsoir!" and, evidently very satisfied with himself, repeated it to everyone. Not to be outdone, another spirit declared loudly at his exit, "Bau revoir!"

Junior spirits are often teased by onlookers. When one of the Water-Dwellers pestered me for liquor, a woman suggested I had a rifle (*mbas*) in my bag. Such mutual play generates much pleasure, but at the same time it resonates with historical significance. An educated young Muslim woman from a family that did not engage in possession activities said of the Water-Dwellers, "They don't know anything, they are illiterate. They are very naive when confronted with television, saying things like, 'Aren't you afraid of all those guns? Will they kill me?' Or, 'Look at all those European mediums!' Other *tromba* are very civilized; they can write well, often much better than their mediums. It is *comique*," she concluded, "to observe those who know nothing and those who know a lot."

The humor has to do with the recognition of divergent historical positioning and degrees of knowledge and sophistication on the part of speakers, interlocutors and audiences, as well as how the social hierarchy that they imply shifts over time. Sakalava laugh at themselves, their encompassment by Western "modernity," and the incompleteness and distortions of that transformation. But it is the historical condition, historicity itself, that stands as the ultimate subject of reflection.

ETHICAL PRACTICE AND EMBODIED CONSCIENCE

Although irony in Western art is sometimes taken as a kind of mockery and the evasion of an accountable stance, that is not the implication in the Sakalava context. Possession is ironic because of the playful juxtaposition and ambiguity of voice and the recognition of the insufficiency of any single voice. But if possession is intrinsically ironic it is also deeply informed by moral concerns. Rather than evading accountability, it doubles it.

In *The Prince and the Pauper*, Mark Twain imagined the consequences of monarch and subject exchanging places for a time. Senior spirits provide something of the same experiment—as monarchs incarnated in the bodies of commoners, they meet their subjects face to face and encounter their misery. Conversely, mediums become regal. Through their comportment—their generosity, graciousness, dignity, energy, and compassion—they provide a model for a just kingship, a model for how the living monarch should also behave. Spirits and mediums provide a political conscience.

As the previous chapter showed, the role of spirits is less to balance royal power (as I once thought) or provide a veto when necessary, than to provide a standard against which to evaluate it. Spirits rarely speak out of their own accord; when severely displeased, they strike. Their victims, as we saw, are not usually the powerful themselves. Moreover, the monarch is not fully bound by what they have to say and can always change the messengers. But he cannot silence what is broadly perceived as the truth. Senior spirits enunciate what is right, but they cannot determine conduct.

Mediums mediate the positions of rulers and subjects. As the bearers of demanding *tromba* they epitomize and demand political subjection and are expected to be the most consistent supporters of royals. Yet their burden also enables a regal "bearing": They are expected and expect themselves to exemplify royal benevolence and generosity. Thus, they must give in both directions; supporting royals but also serving as a model for royals as to how to act toward their subjects.

Kassim is someone who takes this doubling very seriously. No one is more careful than he in subscribing to the demands of his exigent spirits, and yet no one carries himself with more dignity. One day as Kassim and I stood at the entrance to the shrine village, we saw a decrepit Ancestor Person. Kassim called him over and gave him fmg500. Seeing this, a woman in an adjacent house was impelled to bring the man a cup of raw rice. Kassim said that he was moved by the man's suffering, but he also referred to his gift as *mosarafa* because he, Kassim, "is the medium of a monarch and is giving to one of his subjects (*olo' nazy*)." Kassim said that he distributed some of his own money each month when he came to pay his respects at the shrine. Walking in town on another day he was accosted by a young woman who was suffering a lengthy illness. He gave her fmg2,000.

It is critical that Kassim attributes his charitable acts to the fact that he is the medium of a royal spirit. He sees himself as a representative or an exemplar. He is acting as ancestors act and as he thinks the living monarch should act. When he ignores his responsibilities he is punished by the spirit. He embodies conscience.

Mme Doso attempted to assist poor women whom she took as companions. She felt "very sad" for the plight of an indigent royal woman, Zaina, who "has young children and an aging parent to support and sometimes goes hungry. There are taboos that limit what kind of work a person of royal status can take up; for instance, she cannot do other people's laundry." So Mme Doso gave Zaina odd jobs such as guarding her house and prepared enough food to share with her. When supplicants presented Mme Doso's spirits with cash (given in "pairs" of bills), she often reserved the smaller one for Zaina. The royal woman was, in effect, the client of the medium who bore some of her powerful ancestors. Mme Doso tried to frame her assistance in a way that would not be demeaning. She told Zaina to think of her as her "grandmother" (*dady*). "After all," she explained to me, "I have Mbabilahy, whose descendant Zaina is."

Mme Doso's point can be generalized to say that spirits, via their mediums, attempt to care for their living relatives. When one well-loved prince fell sick, the spirit mediums not only held a nightlong vigil, they paid for his visits to the clinic and his medicines at the pharmacy. In a sense they were providing ancestral and parental support to one of the closely related descendants of the spirits they hosted, thus in fact exemplifying the way that responsible people should act toward their dependents and living rulers toward their followers.

Although Mme Doso was clear that an *ampanjaka* could not contribute to his or her own curing ceremony, there is some ambiguity as to which direction "respect" should be paid. "In Mahajanga," she explained, "if a monarch is sick or has gone to hospital, people gather at once and start to collect money by the bedside."

"But in Analalava," I was told by others, "if the queen is sick it is forbidden to give her money; that would be to wish her death." That is one reason, some believe, that there are more people who claim royal status in Mahajanga. "Here royals receive money whereas in Analalava they have to give it." In any case, "the living monarch should greet *tromba* respectfully; they are his or her elders (*olo be nazy*)."

Mediums are expected to act on behalf of their spirits to support living people who are close kin of their spirits even when the kin are not royal. I was visiting William, a medium of Ndramanisko, when a messenger arrived to say that Ndramanisko's com-moner half-brother had died. His mediums were expected to attend the funeral or at least to pool and send money. William explained that Ndramanisko himself would not appear—"he would be too sad"—but concurred that his mediums certainly ought to participate.

William, incidentally, is a man who grew up in a community that had a small British meat-canning company. He once asked me whether British monarchs rise as spirits in their subjects—"Queen Elizabeth's father, for example?"

Historical Consciousness as Conscience

Mediums serve as exemplary subjects and exemplary royals. They must live in a man-ner that takes into account the various and sometimes contrary demands of all their spirits. Even small illnesses are diagnosed as caused by the anger of their spirits, and mediums medicate themselves regularly with a plant known as *fangalaheloko,* "removal of anger." Application of the plant washes away inadvertent offense. The same medi-cine comprises the first bath of the initiation ceremony of new spirits, where it is said to remove any anger the spirits might have that it had taken so long to get the cere-mony accomplished. This bath "clears the air."

What spirits generally become angry about is the violation of their personal taboos—taboos that serve as signifiers of their identities and memories of their lives.[12] These taboos draw their import from the lives of the historical personages who are incarnated in the mediums. They are thus of collective rather than merely personal sig-nificance, and some refer to the condition of royalty in general.

Individual *tromba* may be more conservative than living royals and, hence, better representatives of what are generally considered the true values of royalty. Certain mediums thus outdo living royals with respect to their practice. In the past, royalty had to leave the house where someone was dying, abandoning close family members to evade pollution. But in the 1990s a royal woman refused to move when her beloved husband of many years fell sick and died. He was laid out in the house prior to a Catholic burial, and she remained in mourning over the corpse. Her siblings also came to pay their respects. Only the mediums stayed away, and only they were any longer at risk of ancestral punishment for breaking the taboo.[13] Not only did Dadilahy Kas-sim not pay his respects, he—or rather his spirits—required that anyone who wished to visit him after having visited the deceased first bathe and change their clothing. Kassim had petitioned his spirits to be able to act like a good Muslim and participate in the funerals of kin and neighbors, but was refused. In this way the practice of the most charismatic mediums carries on royal tradition more faithfully than does the practice of members of the royal family themselves. Looked at another way, it is only mediums for whom their status in the traditional system remains significant enough that they avoid the neighborly and religious duties of mourning.

Such taboos leave mediums responsible for carrying the burden of tradition, vul-nerable to punishment for evading it, and conscious of being the ones to do so. This was evident in Mbabilahy's treatment of his medium (Mme Doso, as described in the previous chapter) and is recognized by others, who take care to purify themselves before visiting the mediums. Here the mediums are more marked with royal qualities than living royalty themselves. They take on the attributes of royalty as they lived in the past, whereas living royals exist more directly in the present. It is not only that mediums keep past practice contemporary but that their situation constitutes the reminder that makes such adhesion to tradition necessary. We see here a critical feature of committed mediums, namely, their function as *embodied conscience—and not simply consciousness—of history.*[14]

The strongest mediums are thus, in some way, hostage to history, carrying its burdens, drawing the attention of the fierce and dangerous spirits in their direction, and subject to punishments for violations. The mediums combine violence and victim in one (cf. Bloch 1992). And they gain authority from both sides of this relationship—from embodying powerful royal spirits and from suffering under them. Mediums not only work on behalf of the community but serve as historical icons and ethical exemplars. People hold the strongest mediums in awe.

However, I do not want to exaggerate their singularity. Although they observe particular taboos, they are not ascetics. They are well-rounded citizens who enjoy life's pleasures; not prophets divinely inspired or crying in the wilderness, but bearers of worldly figures from the past. Indeed, when Kassim asks permission to act as a good Muslim, his practice does not simply represent the past but articulates multiple layers of culture and history.

Speaking as a Spirit

Ancestral voices were once part of a tightly constituted regime, meant to affirm political decisions and lend authority to ruling descendants. Ancestors were politely informed of decisions that affected them, and their acquiescence was sought. Spirits sometimes gave warning or suggested additional precautions, but they generally agreed with what was proposed. When they spoke independently, it was to assert or affirm their dignity and to see tradition consulted and followed. Over the course of time, society has changed, and the polity and its leaders along with it. As successive generations of royalty "collapse" and are transformed into *tromba,* these spirits in turn invoke the "tradition" (now "modernity") of their own day. Over time, the nature of mediumship has also changed. Among Bemihisatra, possession has become democratized—the king in every man. As the reach of the polity has shrunk and individual spirits become less powerful, so too have mediums become freer of political control.[15]

As mediumship has become more autonomous, spirits speak on a wider range of issues. This speech is never entirely "free;" it is always connected to the authority of the personages, and this requires that spirits stay "in character." But spirits are available to the larger community. Spirits become observers—witnesses and commentators—as well as sanctioners and sanctifiers of change.

This quasi-autonomous stance works because each personage or voice is set off against others. There are many voices and many points of view—often coming from the mouth of a single medium. This ironizes, in Burke's sense (1945), any individual statement and produces conversation, debate, argument, drama, and witticism rather than an authoritative, conclusive, predictable set of pronouncements. Possession is not a relatively closed "local" system or myth machine that routinely transforms "externally" produced events into signifiers for "internal" models. Nor is it a fragile imaginative web ready to collapse in the face of (or be seduced by) an interesting external idea, forceful presence, sudden event, or the pervasive effects of global capital.

Dialogue is characteristic both of the enactment of innovations at the shrines and of the mediums' practice off the main stage. When spirits talk with clients or dispute the practices of their mediums, broader social transformations can be articulated. In character, spirits can also be highly original. They are not simply voices of or from certain periods but voices for certain perspectives in certain contexts. There is a spirit ready to speak for virtually every political persuasion, though spirits are not the direct product of political interest. Possession does not merely enumerate and display ethnic or historical types but appropriates their distinctive voices and perspectives on politics and history, thereby constructing a complex heteroglossia and multileveled historical consciousness. Thus, Sengy, with whose comparison of Sakalava to Israelis I began this chapter, is a voice for European "modernity." From his Muslim medium, this French-speaking, Christian spirit argued that in order to preserve the traditional purity requirements of spirits, Muslim mediums (who at feasts eat from a communal bowl using their right hands) should take up cutlery.

I conclude with accounts of two interventions that demonstrate the inextricable link of poiesis—craft—and phronesis—situated ethical judgment in the practice and performances of strong mediums. The first concerns Lidy and his spirits Kôt' Mena and Sengy; the second, Mme Doso and her spirits Mbabilahy and Ndramandaming. They also illustrate again the dialectical relationship of consciousness and history; spirits change over the course of history but are also able to establish a purchase on and as history.

THE SAILOR'S DISCOURSE ON AIDS

One year at the Great Service I was accosted by Kôt' Mena, a Water-Dwelling Zafini-fotsy spirit. He invited me to visit his medium and gave me his address. Lidy was a man in his thirties who often seemed to spend more time in trance than out of it. All day long Kôt' Mena sat in his house drinking rum and chatting with clients. Lidy lived off the earnings of the consultative practice of Kôt' Mena and his other spirits who seemed to decide what proportion to give the medium.[16] Kôt' Mena had been pushed from his home in Marangivato by invading Merina. Having drowned, he was tabooed sharks and other large fish (who might have consumed his body). He was talkative and youthful. Lidy had also inherited a senior Zafinifotsy spirit, Ndranihaniña (one of the "four men"), from his mother's father. But this happened indirectly; one day the spirit rose in his senior medium, an older woman unrelated to Lidy, called him over, and said he liked him. This was a hint that Ndranihaniña was planning to make Lidy his medium as well. Ndranihaniña subsequently rose in Lidy, infrequently at first but gradually taking over at public events that the elderly medium found tiring. Lidy himself felt more exhausted incarnating Ndranihaniña for a few minutes than Kôt' Mena for many hours and left trance covered in sweat from the intensity.

Lidy's spirits included a sailor, Sengy, originally from Ste. Marie, an island off the east coast of Madagascar that long saw the intermixing of French and Malagasy, producing a sort of Creole culture. Sailor spirits were under the employ of the nineteenth century Sakalava king, Ndramañavakarivo (who fled the Merina to Mayotte), when their boat was wrecked and they drowned between Madagascar and Mayotte. They appeared frequently during the 1970s in Mayotte, where they were called *changizy*, youthful carousers (Lambek 1981), but were uncommon in Mahajanga in the 1990s. Lidy explained, "In the past the sailors were told they weren't welcome at the Great Service because they got drunk and caused too much trouble. Then the majority of the mediums left for the Comores during the violence of 1977."[17] "In addition," he continued, "sailors are fussy about their mediums and only want hosts who are clean and look after their clothes well."

Sengy, also known as Father of Alexandre, served as a popular diviner and conversation partner with whom the relation of past and present could be explored with freedom and good humor. Lidy embodied him brilliantly. When possessed by the Sailor (as I shall henceforth refer to Sengy), he switched from Malagasy to a pungent, idiomatic, properly accented French and put on a blue-and-white striped sailor's shirt and jaunty blue hat with a red pompom. Lidy's bearing and whole demeanor shifted as he bodied forth a different habitus and engagement with the world. In contrast to older Sakalava spirits who chew tobacco and drink rum from the bottle, the Sailor chain-smoked cigarettes (sometimes requesting menthol) and drank rum from a glass.

I quote the following from an edited excerpt from my fieldnotes (27 July 27, 1994).

A young man accompanied me to Lidy's house, ostensibly to show me the way, but also, it turned out, in order to pose his own question. The Sailor lit a cigarette and addressed my friend, Richar' in French. Thereafter their conversation was largely in French; Richar' indicated he was able to follow, but his own contribution was limited as he responded, "*Ah oui,*" or "*c'est vrai*" to the Sailor's various observations.

The Sailor began by remarking, "We can only pray to God for what we want." He added, "While religions are many, there is only the one God, *le Bon Dieu.*" [Both Lidy and Richar' are Muslim.]

Richar' explained that he was having difficulty finding work on one of the small transport boats that plied the coast.

The Sailor inquired whether he had work papers, then picked up a deck of playing cards in order to perform divination. "*Votre nom, monsieur?*" he asked politely.

"Richar'."

"Nothing more?"

"Théophile."

"*Ah, vous avez un joli nom* [You have a nice name]," opined the Sailor. He shuffled the cards, asked Richar' to cut, and then laid them out. After a few more questions he concluded, "You have luck. Wait a little and you'll have work. You think too much about your future, but one of these days it will come."

The Sailor spirit, Sengy, in possession of Dadilahy Lidy.

After a pause the Sailor suggested, this time in Malagasy, "You have a child out of wedlock (*bitiky an tany*)?"

"No. . . . Well, I might, but I don't know," said Richar' somewhat lamely.

"You must use a condom," responded the Sailor. "It's the fashion in Réunion."

"Corn silk worked as a medicine for syphilis," the Sailor went on, "but it won't serve for AIDS (*SIDA*)." Offering us more rum and another cigarette, he began reminiscing about his ostensible experiences in the good old days. "Working in the boats, you travel. I visited Singapore and Johannesburg when I was a sailor. Single men . . . one knows the malady. . . ." But the Sailor's point was the radically changed implications of sexually transmitted diseases and the necessity today for proper protection against AIDS.

The Sailor made these observations in a droll French accent, but seriously, a little world weary. He continued that he was young himself [spirits remain the age at which they died, in his case around thirty-five], but that he saw how young people today tend to forget the customs of their ancestors. "One must respect the ancestors. *C'est eux qui étaient içi avant nous* (it is they who were here before us)."

The Sailor asserted, "I am more intelligent than my medium." He explained also, "Spirits from earlier times could never understand about *SIDA* and condoms."

Richar' brought the conversation back to his main concern, entreating that he might find work on a boat and that things would turn out well for him.

The Sailor replied that this depended on "*la volonté de la personne, la volonté de Dieu* (the will of the person, the will of God). If you strive with force, aim toward your goal, God will help you. You must not be jealous of other people. Do your own work and you can succeed with the force that God has given you."

Warming to his subject and broadening the argument, the Sailor continued, "Malagasy shouldn't think they are better than each other, shouldn't be so competitive." He turned to me. "It's not like that between Toronto and Montreal . . . or Paris and Marseille. That's why this country is in trouble.

"You young Malagasy are lucky; this is a good country (*vous avez [de] la chance; c'est un bon pays*)." He compared it with Kenya where people don't understand each others' languages and have to use English or Swahili. "But here in Madagascar, everyone can understand each other. . . ."

Before we left, the Sailor provided Richar' with more specific assurances about his future and further warnings about his behavior. Richar' having drunk several glasses of the rum that was offered freely during the interview (and that, as was customary, we had provided to the spirit at the outset), the Sailor observed that he should watch his drinking.

Although unexpected, the lecture embedded in the consultation appeared appropriate from the worldly spirit. Indeed, a discussion of sexually transmitted diseases could have found no better spokesperson since, as his conversation made explicit, the Sailor was someone once associated with promiscuous sexual exploits himself. At the same time, it was a highly innovative contribution. The representation of AIDS appears to emerge directly from the disposition of the sailor, the embodied condition giving rise to the discourse or at least having tremendous affinity with it. It is a direct product not of the period at which the spirit lived but of the perspective that the spirit takes to history—of the specific relation to history that he embodies. It is as though the Sailor's life shapes how he has experienced the changes that have taken place since.

In a similar vein the Sailor is able to speak about "Madagascar," to construct "the Malagasy" as an object; he speaks of the nation from the perspective of a partial outsider. In episodes not included in the above excerpt, the Sailor reflected on race: ("God created us all the same except the color of our skins"); on ethnic politics (in his view, the critical problem in Madagascar); on religion ("every religion has its forms of worship, its own sacred day, etc., but God is really One"); and formulated a work ethic. The Sailor is a voice of modernity. Regarding religion, the Muslim spirit medium speaks as a Christian from within the Sakalava idiom of possession. You can hardly get more heteroglossic than this; the ecumenical point is prefigured in the practice.

The Sailor's ability to play with positionality was apparent in a conversation that we held the year before. He explained that Sailor spirits were all pro-French in their politics and had helped ensure the victory of the pro-French forces in neighboring Mayotte.[18] "But," he added, "I'm just a simple sailor and don't mix in politics. . . ."

There is, of course, a wonderfully imaginative weaving of politics and culture in the Sailor's talk. To my ear he seemed to have the tone exactly right, and he was also extremely charming in the role, exuding a confidence and sparkle absent from Lidy's ordinary speech. Indeed, the Sailor went on to apply his insights to the medium, saying that he [the Sailor] had taught Lidy some French. The Sailor announced that Lidy had been an art student in his youth and fetched Lidy's portfolio. As we looked over the rather mediocre drawings, the Sailor said of Lidy, "*Il n'avait pas de talent, peut-être, mais il s'amusait* (He may not have had much talent, but he enjoyed himself)."

It is evident that the "mimesis of alterity" (Taussig 1993) is not a quality of the "mindful body" (Lock and Scheper-Hughes 1987) alone. Although mimesis, in Plato's formulation (Havelock 1963) and subsequent arguments, is opposed to contemplative or discursive reason, possession enables Lidy to take a more distantiated position vis-à-vis society and the self. In Weberian language it is from the very position of enchantment, that the rational discourse of modernity emerges here. The Sailor declares himself more intelligent than his medium and, in commenting on Lidy's art career, engages in remarkable self-objectification. He comments reflectively on contemporary politics and religion; with the cosmopolitanism of a French sailor, he can refer with ease to the anthropologist's home in Canada—and he makes a specific and quite deliberate intervention with regard to his client's sexual practice.

It is quite right that it should be the sailor spirit who propagates AIDS prevention, even though, as the Sailor said, "In our day the problem was only syphilis." The knowing cosmopolitanism of the Sailor contrasts with the fear and chauvinism of several of the older Sakalava spirits. Yet Lidy also embodies some of these spirits and thus at times speaks in different voices. Possessed by a spirit from an earlier epoch, I once heard him stumble playfully over the word for cigarette lighter and fumble as he tried to use one. Each voice emerges from its habitus and is closely connected to it. But too great an emphasis on the smooth unfolding of embodied dispositions would underplay the acuity and artistry of the particular virtuoso.

MEDIUM MEETS MEDIUM: BETWEEN HISTORY AND MEMORY

I return now to Madame Doso and an incident that condenses many of the issues raised in this chapter and epitomizes Sakalava historicity. Remember that among the spirits who possessed her were Mbabilahy, recognized as male founder of the dominant royal faction and alive during the first half of the eighteenth century, and Ndramandaming, a male descendant of Mbabilahy, who lived during the height of the

French colonial period in the twentieth century, served as provincial "governor," and even visited France. One day in July 1996, Mme Doso reminisced with chagrin about the occasion when a satisfied client (or as she put it, *rahalahy,* male-to-male friend) of Ndramandaming's had rewarded her with the gift of a television. She had been delighted but had only been able to keep the set a few days before she gave it to one of her kin. What had happened?

Mbabilahy did not like television. One night when Mme Doso was watching a show, he suddenly rose [in her] and tried to smash the screen. "He was afraid of the violence displayed on the program and the weapons he didn't know. The program I was watching at the time," Mme Doso said, switching momentarily from Malagasy into French, "was about *la guerre en Indochine.* What particularly frightened Mbabilahy were the guns. He lived long ago, in the time of brigands (*fahavalo*), and had no experience of Western (*vazaha*) things. So I had to give the TV away."

This is a complex statement about a complex act. It typifies the historicity of Sakalava spirit performance in that it works forward rather than backward. Rather than consisting simply of people in the present reminiscing about past experience, as Mme Doso herself might do, figures from the past, like Mbabilahy, express an immediate experience of disorientation with respect to the present. In a sense, Mbabilahy could represent all living adults for whom television has been a new and initially disconcerting experience, but he also represents here cumulative apprehension of the entire transformation from the precolonial to the postcolonial era and the disjunctures in experience produced by time and travel and along lines of class and education.

As noted, change is often legitimated for Sakalava by explaining it to the royal ancestors and acquiring their acquiescence and comprehension. So, whereas spirits are sometimes presented as historically naive, at other moments they become the wise assimilators of novelty into the ongoing tradition. Spirits provide reflexive awareness in counterpoint to the ostensible naiveté of mediums and other people who would simply accept what comes. Or rather, the obstinacy of the spirits requires people to take a more reflective stance toward change, not letting things pass unremarked. The spirits, and the past that they come from, demand to be remembered. In her eagerness to enjoy the television, Mme Doso had probably forgotten to explain its nature to her spirits and to ask their permission to keep it in the house.

Thus, ideally, history may be said to be kept alive in a number of senses: alive in that the past is spoken for; alive in the sense that the past continues to provide a series of positions from which to interpret the present (and note that Mbabilahy's perspective here took precedence over the alternate perspective of his descendent, Ndramandaming); and alive in the sense that it is not fixed or stagnant but can acquiesce to change. Ideally, the past is addressed and responds, and acknowledged and acknowledges the present. Voices from the past are both responsive and responsible; the past contextualizes and renders the present meaningful, but it does not fix or limit it. The present does not serve as a memorial *to* the past, kept frozen by its sense of fear or homage (as might be the case in some neurotic illness or ultraconservative ideology) but, rather, changes deliberately, with reference to the past. Indeed, action in the present is sanctified (Rappaport 1999) by the past. The mutual commitment of present to past, and past to present, is crucial.

Mbabilahy's actions speak not only to the present, that is, to the consumption of television, but also to a period that is past relative to the present but future relative to Mbabilahy's life, namely, French colonial aggression. Mbabilahy rejects French colonialism as upheld by Ndramandaming and as visibly exemplified by the war in Vietnam.[19] Modern violence is not opposed to an unambiguously romanticized precolonial past. After all, the spirit responds to the display of colonial violence with his own threat. Mbabilahy, like most senior spirits, is described as *mashiaka,* cruel or violent. And Mme Doso depicts the era in which Mbabilahy lived as a time of brigands, implying a condition of insecurity rather than of order—a state, in effect, of violence of its own. The medium brings to bear a colonial or modern view of the past as anarchic, in some discrepancy to the way she generally defers to Mbabilahy and his pronouncements. Thus, the past is neither homogenized nor idealized. History is multiply situated and multiply voiced.

Insofar as the spirit's reaction to the powerful weaponry is one of fear, the display may be said to ruefully acknowledge the facts of history, the eventual subordination of Sakalava to France after Mbabilahy's time (and, one might add, the success of television). Mbabilahy's response to the danger is one of disquiet—he does not recognize the guns—yet it occurs long after the fact. Hence, the location of this appraisal is both prospective and retrospective. This is curious only with respect to the linearity of conventional history, not if we think in terms of the temporal fluidity and thickness of memory (Casey 1987). In memory the prospective and retrospective speak continuously to each other—anticipation conceived in terms of recollection and vice versa (Loewald 1980, Weiss 1996).

This vignette also demonstrates the artificiality of separating the public from the personal or the subjective from the objective. Although Mbabilahy, like Ndramandaming, is a public figure—a personage—his act of threatening the television takes place in the house of a single medium and is not widely known. (Insofar as it is known, it serves to further legitimate the medium's possession, thicken peoples' understanding of Mbabilahy, and widen reflection on time, history, and change.) It is a novel act, a creative response on the part of the personage, in character (as far as Mme Doso understands that character) but not part of a collective ritual script. That it happened at all and how widely she disseminates the story are aspects of her situated judgment and self-production of her own life as exemplary historical conscience.

Insofar as we see memory as the practical subjectification of the past and history as its discursive and disciplinary objectification, so it becomes difficult to decide whether to refer to Sakalava practices as memory or history. Spirits speak as the conscience of history; subjectively located, this history may also be described as memory. Both terms, history and memory, serve useful rhetorical functions, but here it is their coincidence that I wish to emphasize. In a Maussian universe, like that of the Sakalava polity, the distinction between them is not reified.

More generally, in any cultural context, memory is social when one person stands as witness for an event of collective significance. The mediums, in effect, provide marked acts of witness—acts that link the observation of external events, like the reception of television or military violence to those of their earlier life experiences, and that link events in their own lives to the historical experiences of the spirits whose trustees they are.

Let us briefly consider how the encounter with the television articulates with the circumstances of Mme Doso's own life. Unlike Mbabilahy, Mme Doso had a cosmopolitan outlook. She was a widow, probably in her late sixties at the time, who for many years had been married to a man from the Comoros whose profession was none other than soldier in the French army. With him she moved between military bases throughout Madagascar and in Réunion and could reminisce about sharing Ramadan customs with Turkish and North African wives of her former husband's colleagues. Along the way she must have heard a good deal about French and American intervention in Vietnam. That the fear of guns lies even deeper is suggested by the fact that another of her spirits is a Water-Dweller whose character is epitomized by his death from gunshot wounds (inflicted by Merina soldiers as they were chasing his family down to the Loza River). As all spirits must avoid the things associated with their deaths, Mme Doso withdraws at the climax of the Great Service in anticipation of the rifle fired in celebration.

Mme Doso, herself, does not shun television. She gave up ownership of her set with considerable regret. In 1998, I often found her gazing avidly at the television owned by one of her tenants. In the intervals between her strenuous activities as a medium, housekeeping, and making pickles and cakes to sell for a bit of cash, she watched news, sports events, and French films. She explained that she could watch television so long as the program was not about war. In a way, she speaks to my own ambivalence about seeing the portrayal of suffering on the nightly news. It is as though the spirit functions as a kind of microchip censoring device.[20]

Following my argument about the limitations of embodied mimesis as an explanatory model, I emphasize that the spirit's passionate response to the televised images is counterpointed by the medium's dispassionate retelling of the event. This shift to a more distantiated, wryly objective perspective is critical to what is at issue and typical of possession's ability to produce a recontextualization of its acts. Possession entails a "dialectic of distantiation and appropriation," similar to what Ricoeur describes with respect to the reception of texts (1976: 43–44) or what Loewald refers to as shifts between representational and enactive remembering (1980: 164–65). Although I have been emphasizing the lack of a firm distinction between objective and subjective spheres, here is the germ of a split in which the formation of a perspective we can call "history" is forged from a perspective we can call "memory." This is a process in which events become increasingly abstracted with respect to regnant cultural models and shift into the discursive realm of linguistic representation, becoming available for new kinds of entextualization (cf. Boddy 1989; Besnier 1996; Silverstein and Urban 1996). The inherent communicational triad (Lambek 1980) of possession events—the fact that host, spirit, and interlocutors speak *to* each other and with discrete voices—as well as the necessity for redundancy—that mediums must be told what their spirits said and did when they were in trance—provides the means for increasingly detached and reflexive responses, much as in the successive retelling and inscription of a dream narrative. Moreover, in many instances, it is the spirit who responds dispassionately to the passionate or incoherent utterances of host or clients.

I venture that through the cultivation of this dialogical or polylogical space in which imaginative perceptions are registered and validated, spirit possession provides the means for a kind of collective working through of history, roughly analogous to what psychoanalysis is said to offer individual experience. The medium, qua personage, draws on collective history to develop her personal situation, but likewise her practice serves as a model for collective history-making and assimilation of change. In its emergent reflexivity, her practice is one through which, "Being a person is now understood to be a much more complex business, a dialectic between the 'I' that does things, assigns meanings, makes, honors, and betrays commitments and loyalties, and the 'I' that *knows* some of the things done, meanings assigned, commitments undertaken, honored, and betrayed" (Mitchell 1988: 265).

It is evident that from Mbabilahy's perspective there is no difference between the scene portrayed on television and reality. There is some humor in the fact that he appears to mistake the one for the other. But this, in turn, produces reflection on the ontology of possession itself, its own relation, so to speak, to reality (costume or heirloom?), and ultimately on the relationship between television and spirit possession as alternate, and possibly competing, media for transmitting and juxtaposing images and for inhabiting and provoking reflection on the passage of history.

Conclusion:
Imagined Continuities

Quand tu dis la verité, n'oublie pas aussi le mensonge.

Sakalava proverb, quoted by Charles Poirier (1939: 42)

Whereas contemporary descriptions of "southern" societies frequently describe social breakdown, endemic poverty, loss, violence, and disease, I have portrayed a relatively cohesive cultural system and enthusiastic engagement in a set of practices that continue to produce a coherent, polyvalent, and insightful way of comprehending and engaging social change.[1]

My view of these practices is also more positive than many current analyses of cultural phenomena, seeing them less as reaction or resistance[2] and less as imposed or invented than as cultivated and taken seriously. I have pointed to discontinuities, mystifications, and conflicts, but these take place in or with respect to an ongoing tradition. My focus has been less on the disruptive effects of change than on the meaningful world in which change occurs and to and through which specific changes are appropriated and made meaningful. Similarly, I have tried to complement the weight that contemporary theory places on power and interest, with attention to the ethical. The approach balances structure with practice and subjection with agency. The practice I have in mind is not the instrumental practical reason that Sahlins (1976) argues is specifically Western and capitalist, but moral practice—cultivated judgment informed by local ends and means. Similarly, the agency is not that of the autonomous possessive individual (Macpherson 1962) but more intersubjectively defined (Werbner 2002).

As Feeley-Harnik (1978) has argued, only royals were entitled to Sakalava history. The emergence of other elites is not accounted for in ancestral history, nor are the lives of commoners and servants except insofar as their places in the hierarchy are rationalized. But gradually, this royal history has been popularized. If it is not history *of* the people, it has become history *for* the people—a history that has served their enjoyment and edification, and that has rested partially in their hands. This has occurred with the democratization of mediumship, though mediums pay for the privilege.

The people who appear in this book are not entirely alienated from history but engaged with it—active producers and reproducers of a set of discursive and embodied practices and incisive commentary about the world and public events. They live in and with history. There is a sense of loss, breakdown, encapsulation, and of things not being what they were, but it is advanced and balanced by not only a vivid experience and rehearsal of continuity with the past but an active, imaginative, sometimes frustrated participation in the articulation of change.

RUPTURE

The most recent challenges concern the objectification and commoditization of Sakalava practice. Those with direct access to the money accumulated at the shrine may come to see themselves less as ancestral subjects than as owners of ancestral property, and this in turn begins to alienate the majority of devotees.[3] The sanctity of the relics is undermined, existence threatens to give way to essence, and the maze becomes transformed into display cases of a museum.

In the summer of 2001, I returned to Mahajanga for the Great Service. This time, I vowed, I would observe the relic bathing, and I made arrangements behind the scenes for permission to attend. My machinations proved to be entirely superfluous. Following the procession of monarchs and dignitaries unimpeded through the blood red gate and in the sacred southern door to observe the bathing were a regiment of Merina pilgrims from Antananarivo, dressed in identical T-shirts advertising the *Fanompoa Be* of 2001, and several European tourists. I found myself outside the temple door alongside several devoted Sakalava practitioners who were prepared to await the bathing in the shade along the exterior southern wall. Faced with their embarrassed demeanor at the sight of the doorkeepers admitting masses of strangers, and stunned myself by what was happening, I could not bring myself to enter after all. Later, after we had joined hands and protected the newly bathed and rewrapped relics as they circumambulated the outer perimeter of the temple, the observer side of my anthropological persona reasserted itself and I joined the small crowd awaiting the chance to follow the monarchs into the *zomba vinda* to see the relics in their resting place. Three by three, without apparent discrimination, those who wished, were let into the constricted space of the miniature house to prostrate themselves before the ancestral remains.

Since my previous visit, policy had changed abruptly: The sacred reliquaries were disclosed. According to some, Ampanjaka Désy had merely accepted a fait accompli. Since descriptions of the bathing were widely publicized (as in the newspaper article quoted in chapter 8), there was no secret in any case. According to others, the shrine manager was well paid by the organizer of the Merina pilgrims for the privilege of entry.[4] The relaxation of the rules reflected Bemihisatra security over their possessions, their openness relative to Bemazava reserve, and a further step in their long-term policy of conciliation or assimilation with Merina within a single, unified nation.[5] Moreover, it was only in Mahajanga that the relic bath had been conducted in such secrecy; in Menabe, among southern Sakalava, relics are bathed in the sea and the procession passes in full view of large crowds.

Most older Sakalava respected "tradition" and remained outside as the "four men" were bathed. Their silence, pallid faces, and downcast eyes before the foreigners eagerly pushing their way in appeared to condense the sum of Merina/Sakalava relations since 1824. The scene was a sad, even tragic one. And yet, all was not lost. Those who stayed outside maintained their dignity and their principles; the relics remained hidden to them. Each still had their own form of service to perform and the sense of purpose that entailed. They were certain that those who had committed errors (*kôsa*) by entering where they should not have would be punished. The mediums retained their own resources, their means of maintaining purity, alternate sites of devotion, and sources of sanctification. Despite national attention to the shrine and despite its evident importance, much ancestral activity in Mahajanga, as I have shown throughout the book, takes place at arm's length from it. Mediums continue to have broad scope to exercise their principled creativity.

POLYPHONY AND CHANGE

Ancestral practices enable both a polyphonic historical consciousness and an energetic means to influence how change takes place. Creative production and ethical judgment are closely intertwined. Kassim, on whose understanding of the changing organization of religious practice I have much relied throughout this book, once told me that young people were abandoning the ways of their parents and that he expected the system to be dead within the next decade or two. The very next day Ndramboeniarivo rose in what, as always, was an energetic performance. The monarch gave a short speech in which the gist of Kassim's remarks was repeated. He said that in the past there had been eight clans of Ancestor People who worked hard. Recently their children had abandoned ancestral practice and no longer knew how to serve. "However," Ndramboeny now confidently concluded, "the younger generation would learn how to serve and their reason would return (*jery hipody*)."

The statements of medium and spirit are not equivalent. Kassim's can be considered primarily locutionary—a descriptive account of the state of affairs as he sees them. But Ndramboeny's statement is predominantly perlocutionary and illocutionary—both persuading its audience and bringing into being the state of affairs it announces. However members of the young generation subsequently behave, the standard against which their behavior can be measured is set by statements like Ndramboeny's. The ancestral order remains morally significant, and it will remain so for as long as ancestral spirits continue to speak and people accept them.

I cannot adjudicate which of these conjoined predictions is correct; however, although practices are changing, the fact of change is nothing new. Kassim and his generation of mediums migrated to Mahajanga from the surrounding Sakalava countryside. Today, many newer mediums are Tsimihety, and the city produces its own cohort. One Sakalava medium, a generation younger than Kassim, alternates his time between Mahajanga and a lucrative practice in Mayotte and has recently expanded operations to Paris. From a historical cohort that had more trouble than Kassim in acquiring decent wage employment, he seeks alternate forms of entrepreneurial opportunity and brings different ideas to the shrine. Younger generations of mediums undoubtedly make new compromises with living royalty, shrine managers, and government officials. Many would say that the overseers of the Doany Ndramisara have made even greater compromises in their attempts to establish their exclusive control of the relics on a more sound economic and political basis irrespective of tradition and popular opinion.

The mediums maintain their source of authority in the embodiment of the royal ancestors, whereas the shrine managers ultimately depend on the relics for their power and prestige. Both retain a public that looks to the relics for health, prosperity, guidance, blessing, and intervention and that is willing to contribute generously to the shrine as a result. Already in Kassim's generation many of the mediums and their contributions to ancestral service have been heavily subsidized by Franco-Malagasy expatriates living in Réunion or metropolitan France. Unless the national economy revives, this trend will continue. But it is compelling that emigrants and their offspring continue to orient themselves from overseas to the shrine of Ndramisara, drawing from and adding to its power.

Conversely, there is also a "modern" self-consciousness within Mahajanga and among even the most committed mediums; this comes to light in phrases that compare the shrines to museums or in the wishes that I write a book that can be read by their children. Meditating on the fact that he had not taught traditional practices to his children, one medium worried that they would be lost. "Sakalava have long been literate, but recording ancestral practices in writing," he explained, "has been forbidden (fady)." There is thus consciousness of a potentially significant objectification that would both constitute and help restore a break within tradition.

In 2001, concerned about possible lapses of discretion, I asked whether I could reveal what I had learned about the ancestors. One person said to publish things as I saw them; another said that because what I knew was already histoire (tantara) I was free to publish. After some reflection, Kassim replied that it was not up to him to say and that I should ask permission from the ancestors themselves. I posed the question to Mbabilahy at the main shrine, and he told me to proceed. A few days later, Ndramboeniarivo was called up at Kassim's shrine. Maman' Sera more or less repeated my query to the spirit, and the case was taken up by a visiting official from Bezavo who explained that my book would be a good source for future generations. In response, Ndramboeny admonished me that the next time I had a request like this I should call upon several ancestors together [rather than assuming that one of them could speak for what should be a collective decision]. Nevertheless, he told me to go ahead, "since so many taboos have already been broken."

Tradition: Transmissibility and Irony

With respect to "tradition," and its demise, Agamben writes with great acuity:

> In a traditional system, culture exists only in the act of its transmission, that is, in the living act of its tradition. There is no discontinuity between past and present, between old and new, because every object transmits at every moment, without residue, the system of beliefs and notions that has found expression in it. To be more precise, in a system of this type it is not possible to speak of a culture independently of its transmission, because there is no accumulated treasure of ideas and precepts that constitute the separate object of transmission and whose reality is in itself a value. In a mythical-traditional system, an absolute identity exists between the act of transmission and the thing transmitted, in the sense that there is no other ethical, religious, or aesthetic value outside the act itself of transmission.
>
> An inadequation, a gap between the act of transmission and the thing to be transmitted, and a valuing of the latter independently of the former appear only when tradition loses its vital force, and constitute the foundation of a characteristic phenomenon of non-traditional societies: the accumulation of culture. For, contrary to what one might think at first sight, the breaking of tradition does not at all mean the loss or devaluation of the past: it is, rather, likely that *only now the past can reveal itself with a weight and an influence it never had before.* Loss of tradition means that the past has lost its transmissibility, and so long as no new way has been found to enter into a relation with it, it can only be the object of accumulation from now on. In this situation, then, man keeps his cultural heritage in its totality, and in fact the value of this heritage multiplies vertiginously. However, he loses the possibility of drawing from this heritage the criterion of his actions and his welfare and thus the only concrete place in which he is able, by asking about his origins and his destiny, to found the present as the relationship between past and future. For it is the transmissibility of culture that, by endowing culture with an immediately perceptible meaning and value, allows man to move freely toward the future without being hindered by the burden of the past. (Agamben 1999: 107–8, my emphasis).

I have quoted this passage at length because of its remarkable precision in summarizing a body of Western ideas. Agamben's picture of the Other ("a mythical-traditional system") grasps, more elegantly than I have been able, something about the nature of Sakalava historicity and its possible breakdown while, at the same time, remaining within the kind of dualistic discourse from which I have been trying to escape.[6] I think it is clear that Agamben idealizes (in both senses) his picture. Agamben's notion of "transmissibility" captures exactly a point that I have been trying to make throughout the book with respect to Sakalava historicity; as long as mediums continue to be subject to ancestral possession, transmissibility is retained. Yet I have argued that transmissibility was itself represented and experienced as a great weight, and that there have been partial discontinuities and objectifications within ancestral tradition. The point is that these things have been constituted and apprehended through Sakalava means, within their chronotopes, polyvocally, through three kinds of divisions of labor (and knowledge and interest) and through the multiple and potentially discordant registers of narrative, architecture, and embodied performance. Most critically, I have attended to an ironic stance within, even at the heart of, tradition, a possibility that Agamben's dualistic picture seems to overlook.

Irony is an effect of multiple voicing. If the medium says one thing and the spirit another, or if one ancestor says one thing and another ancestor something different, it becomes evident that truth is not self-evident, that interpretive responsibility is returned to the addressee, that the "said" of any given utterance is always accompanied by the "unsaid," and that consciousness must include a recognition of its limits. Irony evades simple reductions and easy choices. If spirit possession is sometimes compared to theater, the crux of the matter may lie less in "performance" per se than in the putting into play of multiple voices. As Kenneth Burke observed, it is precisely from the totality of a multivoiced situation where each character offers a perspective and comments on the others that irony is generated. "The dialectic of this participation produces (in the observer who considers the whole from the standpoint of the participation of all the terms rather than from the standpoint of any one participant) a 'resultant certainty' of a different quality, necessarily ironic, since it requires that all the sub-certainties be considered as neither true nor false, but *contributory*." (1945: 513).

Irony draws from the fact that if Sakalava tradition is not without its objectifying and deobjectifying tendencies, neither is it internally undifferentiated, rigidly bound, static, or pristine. Sakalava practices have undergone a process of continuous change in a history forged in encounters with both Muslim and Western worlds, moving from Arab- to European-dominated phases of mercantile activity, through the Anglo-Franco competition that assisted the rise of the Merina and their defeat of the Sakalava, to colonialism, national independence, and the ambiguities of postcolonialism. Each phase further complicated the social differentiation characteristic of the region, producing overlapping class and status groups distinguishable on the basis of knowledge and interest and with respect to their relationships to the ancestral practices described in the book. Some of these historically located perspectives retain their salience by means of the ancestral spirits.

Agamben writes of tradition enabling movement forward "without being hindered by the burden of the past," and yet this book has referred continuously to such a burden. What Agamben does not notice, because he writes in abstractions is, first, that such a burden may in certain contexts be intrinsic to the practice of tradition and, second, that it may be distributed among the members of a society, and distributed unevenly. But, in closer approximation to Agamben, it has also been at least implicit in my account that the past is much more than just a burden for Sakalava—that the burden is self-transcending. It is not the pile of accumulated wreckage mournfully observed by Benjamin's angel (or the average museum goer). Sakalava historicity is a world in and through which its subjects live, fully accompanied by their "angels." It is simultaneously religious, artistic, political, therapeutic, intellectual, and ethical. Within the ancestral sphere, acts of the living are guided and judged not according to abstract, universal principles but by means of specific voices of past monarchs. Past and present voices address and answer each other in supplication, persuasion, affirmation, and argument. Judgment is supple and discussion lively insofar as individual ancestors personify distinct positions and perspectives on the world, articulating shifts in historical consciousness. The present is understood with respect to multiple juxtaposed precedents.

SANCTITY AND MONARCHY

Agamben's model can be modified by suggesting that precedents are organized within a temporally graduated hierarchy of sanctity such that smooth transmissibility at some levels enables change—and irony and self-consciousness—at others. The sanctity of core truths both offsets and is protected from the irony. This hierarchy enables a partial autonomy for political action and conversely protects the sacred from corruption by the worldly (Rappaport 1999).[7] What could cause a radical break is disruption of this hierarchy itself.

The sacred is a property possibly less restricted to discourse than Rappaport proposed. Among Sakalava, *hasing* inheres in bones, artifacts, persons, and places no less than in words and is established by acts of setting apart, enclosure, concealment, giving, curation, and honoring. As "ultimate sacred objects" ancestral bones may provide durability and meaningfulness equivalent to that which Rappaport locates in "ultimate sacred postulates." Acts of sanctification thus entail not only speech but movements of some kind outward from the sacred core. Aspersion of water from a white dish containing a silver coin and white clay is the primary means; coinage, kaolin, whiteness, and hardness being materializations of ancestral bone and force (Lambek 2001). Conversely, people approach the material ancestral remains to "pay respect." As discrete objects of limited number, the question has been in whose hands the bones and relics rest, producing a whole system of custodianship and "constitutional balance of powers" (Ndramisara's distributive model) no less than arguments about inheritance, placement, theft, and responsibility.

Sakalava do not, if they ever did, comprise a fully "divine kingship," yet it is instructive to compare the position of royalty with the series of variations (symbolic transformations) identified by de Heusch in his rehearsal of the Frazerian account of African sacred kingship (1997; cf. Feeley-Harnik 1985). de Heusch speaks of the reigning king as transformed into a "fetish-body," that is, "a living person whose mystical capacity is closely tied to the integrity of his physical being" (1997: 214). Among Sakalava at the present time it is rather the former monarchs who have become fetishes—in the form of the relics, somewhere between "body" and "object." de Heusch says of Bemba regicide, "In the end it is the power of giving life, inherent in the very body of the king, which is recovered at the point of death" (ibid.: 216). For Sakalava, the power is recovered not through regicide, not through what happens, so to speak, at the moments preceding or producing death, as much as through what happens to the royal body immediately following death and through the rehearsal of that death in the medium's performance.[8] It is the transformation of collapsed royal bodies that ensures the lives of their subjects. Living monarchs are sacred only in anticipation.

In any case, as in all of de Heusch's examples, "it has been necessary to avoid the *natural* death of the king and to impose on him a *cultural* end, as if the society sought to reappropriate control over the cosmic forces incarnated in him" (ibid.: 216). Among Sakalava, this is accomplished by denying the very fact of death, euphemizing it, and then separating decaying flesh from perduring bones. It is also accomplished by removing all connection between living monarchs and the dead. As among Nyoro, "the king's body is also required to be in a state of ritual purity, and can never come into contact with a corpse. Why should this be so if the Nyoro king were not . . . a reservoir of life?" (ibid.: 217).

de Heusch also notes how at installation the king often has to commit a "real or symbolic transgression of some fundamental rule of social life" (ibid.: 230), such as incest, which "cuts the sovereign off from the ordinary framework of kinship . . . and places him in an ambiguous zone where he is looked upon either as a sacred monster or as a dangerous sorcerer" (ibid.: 217). The Sakalava monarch, on the other hand, to draw from Kassim's description of Amina's inauguration, was "elevated (*nampitsanga*) to rule by the people of Mitsinjo. Two months later, they 'drove' her (*naness*) to Bezavodoany to announce (*mañambara*) their choice to the ancestors there: 'This is your descendant who will lead your polity now.' The ancestors agreed. They wouldn't have agreed if she were false; she wouldn't have lived to such a ripe age." Present-day Sakalava project transgression back onto early monarchs who married incestuously, murdered their wives or offspring, and could murder again. Monstrous cruelty is represented not as a regular feature of installation, that is, as part of the order of ritual, but as pertaining to the character, life circumstances, and actions of individual monarchs, that is, as part of the order of history. Deceased, but not living monarchs are referred to routinely as violent (*mashiaka*).

de Heusch remarks on the common practice of a substitute victim for the ritual regicide. In Sakalava understanding, the kingdom is founded by an original act of real yet substituted regicide (a mother, for her son). Thereafter, full substitutes (first Ancestor People, then cattle) have been used at the "bloody red gates." As in other societies discussed by de Heusch, these substitutions take place "paradoxically, after the natural death of the sovereign (when they are an element of the funeral rituals)" (ibid.: 219). The substitutions not only emphasize royal violence and privilege but counter or even deny the royal death itself, separating the transient from the enduring elements and purifying the emergent ancestor of death. Substitute victims, lying on the ground, raise royalty above it.[9]

Finally, de Heusch emphasizes that the symbolic structure of kingship ought not to be equated with the ideological superstructure of the state, which may be superimposed upon and draw from it. Just as sacred chiefs or priests are characteristic of egalitarian West African societies like Tallensi or Samo (culturally related to the Mossi state), so within Madagascar, ethnographers of Sahafatra (Wooley 1998), Tanala (Hanson 2000), Betsimisaraka (Cole 2001), and Karembola (Middleton 2001) refer to *ampanjaka* in what are otherwise relatively egalitarian contexts. It is almost as if "royalty" refers to a functional element necessary to the very idea of society in Madagascar, unless that society is constituted in direct opposition to it (Wilson

1992; Astuti 1995). Many people in Mahajanga informed me that each ethnic group had its *ampanjaka*. Indeed, some argued that the *ampanjaka* across all of these groups are descendants of a single macro-clan. de Heusch asserts that violence is only associated with the sacred in the state form (1997: 225) and that may have been true in Madagascar as well.[10]

How separate the sacred and political should be from one another is certainly of contemporary practical and intellectual debate, but it is undoubtedly a longstanding issue. It is interesting that the shrine is referred to as Ndramisara's. Ndramisara is the holy diviner who established the very possibility of kingship and set its terms, and who understood the need for and oversaw the royal sacrifice that both provided the "entry" or "implantation" of Zafinimena royalty into the region of Boina and limited succession to a single line. But even though he established the shrine and it is in his name, he does not have political seniority. As a senior medium put it, "Ndramisara advised Ndramandiso at every step of the way, what days to travel, when to stop, and where to settle, and in return Ndramandiso gave him the shrine. However, Ndramandiso is the proprietor (*tompintany*) and takes precedence." When the two are present together, Ndramandiso walks ahead and sits to the north of Ndramisara.

Ndramisara's domain is that of the meta-monarchy and meta-history, that is, the sacred. It provides the frame within which history happens and from which it can be viewed or authorized. This is not the objectivist, detached space of Western science, colonial gaze, or deconstructionist criticism, but a space nevertheless at arm's length from the immediate forces and direct impact of historical events. It is what enables historical *perspective*. Thus, Sakalava do not simply either react to events or transmit tradition; their consciousness is partially insulated from new historical forces but not determined by old ones. Historical consciousness can issue a measured response from its own ground, can appreciate or absorb change, and is not simply absorbed or buffeted by it. Sakalava can continue to assert with a degree of assurance and moral authority how change should take place.

The sanctity of kingship is complicated by the distinctions among royal ancestors. The generational relations observed among the ancestors, not simply by means of a formal genealogical hierarchy or sterile list but through their moral connections, gives history depth and tradition resonance. Senior generations of ancestors sanctify, authorize, bless, and protect their descendants. Junior generations honor and serve on behalf of their ancestors. In this gradation there is no sharp break between life and death (cf. Kopytoff 1971) or between the political and the sacred. Living humans and more junior generations of ancestors work together to produce the ceremonies honoring more senior ancestors. Senior ancestors and living servants cooperate to assure the proper treatment of those in the process of becoming ancestors. Relative spatial, temporal, and moral distances are always recognized and maintained. The marking of successive enclosures and the observance of their respective obligations and taboos are practices that constitute and reproduce the system. The core of Sakalava (and, I venture, more generally, of Malagasy) religion is not sacred kingship per se but sanctification of the present by the past. Insofar as the past does so, Sakalava continue to live in a sanctified universe.

POWER AND POTENCY

I argued that Sakalava historical consciousness and memory are dispersed over the landscape and distributed among different kinds of people. So, too, is the force this form of historicity or social imaginary holds over people. Reigning monarchs were once perhaps linchpins of the system and relatively autonomous leaders but now merely play their part in the reproduction of a more diffuse power; they are torn between the demands of the ancestors and those of the state and between the system of reciprocity and the need for cash. The primary adherents and exponents of ancestral historicity are the mediums. They are subjected to it, but they are also engaged in a kind of self-production of their own power. That is, the power within each of them is intrinsically linked to external forces that authorize it and that they sustain. This can be seen in the way spirits sanction each other's appearances and in the way mediums support the shrines that, in turn, legitimate them. Mediums are not the conservative remnant of a former system. On the one hand, their numbers have been growing; and on the other, they are positively engaged with modernity and issues of contemporary well-being.

Kassim once corrected me when I referred to *tromba* as alive. "How can someone dead return to life?" he asked. It struck me that it is really the mediums who are full of life and activity. They take charge, make and implement plans, and work on behalf of clients; they are always on the go. The "democratization" of spirit possession has meant not only the spread of the "taxation" base but the creation of a corps of collectors. Mediums promote the flow of knowledge, interest, and "service." They generate wealth and direct it to the various shrines; but even more, they generate enthusiastic subjects. As Kassim observed on another occasion, "The younger generations of *tromba* serve (*manompo*) together with the public. If people didn't see *tromba* serving they wouldn't feel like doing it either. It is also the lot (*anjara*) of mediums to contribute on behalf of the ancestors of the *tromba* who possess them; if they left off they would fall sick." Mediums thus have links and obligations to multiple shrines and serve as a model in this respect for the nonpossessed. Tomb guardians, in contrast, stay still; their role is less an active, generative one, than one of separating and absorbing what is used up, finished, and dead.[11] They remove the pollution of death and set the limits to life, providing the immovable ground against which mediums move and generate movement.

When mediums, rather than tomb guardians, are placed at the center of the portrait of Sakalava society, the focus on death (Feeley-Harnik 1991a) is revised. However, the social activity of mediums needs to be understood against the background of deeper conceptualizations of potency. In an interesting discussion drawn from Merina material, Graeber (1996) distinguishes the power to act directly on others (a potential) from the power to move others to action by display. These powers are symbolized, respectively, as concealed and visible. This distinction fits nicely the contrast between the potency of the relics and senior *tromba,* on the one side, and the efforts of the living to influence them, on the other. Ancestors act directly, by striking out; this is what earns them the epithet of violence. The living exert their powers of persuasion on the ancestors through public displays of giving money and clothing, song, dance, sexual humor, and drink, thereby putting the recipients and addressees into good humor. Here, "display" is closely linked to "play." It is as though one must show oneself celebrating life in order to receive life.

The contrast occurs, recursively, at multiple levels. The relics are a form of display relative to the bones in the tombs but are, in turn, hidden potency relative to the performances of spirits. Senior spirits are more reserved than junior ones, and all spirits are less accessible than humans. Discretion and concealment are values of those with greater potency or higher status, of elders relative to youth, of royalty relative to their servants. We find ourselves again in the chronotope of the maze. When one turns a corner, things that were hidden become revealed, and vice versa. In the maze, things are never revealed all at once, in the round. From one perspective, certain people are likely to feel unhappy about revelations in my book. At another turn, however, the book is mere display—the rattling of a half-empty container—and discretion remains assured. Indeed, a book is a paradox; the more accessible it is, the less authoritative. This is perhaps why the books that many shrines are said to contain are kept hidden.

Rather than fixed categories, the logic produces a structure of relations such that

hasing of the ancestors	:	vitality of the living
	::	
inside	:	outside
concealment	:	display
stasis	:	movement
past	:	present
bones	:	relics
relics	:	spirits
spirits	:	mediums
mediums	:	nonpossessed people
elders	:	youth
older generations	:	younger generations
higher status	:	lower status
Ancestor People	:	commoners
legislation	:	execution
singularity	:	plurality
exclusivity	:	complementarity
punishment	:	pleasure
direct action	:	persuasion
potentiality	:	actualization
potency	:	power

This relationship also approximates Aristotle's distinction between dynamism and energy (Agamben 1999).

The place of gender is somewhat complex:

female spirits	:	male spirits
but,	::	
men	:	women

Graeber distinguishes the potency of money as "the potential for future specificity" from the persuasive power of specific objects of display "whose value is rooted in past actions" (1996: 20). But what Graeber calls potency can be referred to as sanctity (*hasing*). Sakalava sanctity, as pointed out, increases as one works back in generational time; worldly knowledge increases as one works forward. Historical consciousness is the product of this texture.

RATIONALIZATION

Many of the points I make in this book about how things work, what the "rules" are, and so forth, were not gleaned easily. There was no omniscient teacher, and most people attended to Sakalava injunctions of discretion. Practices on the ground, as well as explanations for them, were not entirely consistent or tidy. Sakalava tradition has been under continual change, of course, and at different rates in different places and among different segments of society. But more than that, it has made virtues of ambiguity, polyvalence, irony, and opacity, the very sorts of things the bureaucratic state tries to weed out. Indeed, perhaps there are no "weeds" here, just multiple locations, interests, interpretations, and forms of practice.

This is "enchantment" in the Weberian sense—thought and practice that has been subject to less systematization than modern political constitutions, operating blueprints, religious rule books, or bio-politics. There *is* an underlying model, there *are* many rules, and certainly it is *not* true to say that for Sakalava "everything—or anything—goes." But significantly, injunctions are both distributed according to the multiple divisions of labor and often phrased negatively, by means of taboos. It is clear what specific things one cannot do and, left open, what one *can* do (Lambek 1992). Bilateral descent is similarly open-ended.[12] The irony generated by multiple voices and multiple positions undercuts the determinism and reductionism of objectifying structures and rationalizing procedures. But significantly, irony also undercuts the determinacy and reductionism inherent in ideas of autonomous agency, free will, and rational choice.

The relative weakness of rationalization enhances the communicative value of interactions. These move slowly, leading the parties gradually to understand whether they are on the same ground. The social component of action (small talk, courtesies, hospitality, and so forth) is highly valued. Social juniors acquire dispositions and discernment by means of their encounters; the slowness and indirection of the maze enhance the value of both the journey and what is to be found within. The ends of human flourishing and sociality are thus to be found within the means.

IMAGINED CONTINUITY AND THE COMMUNITY OF TRADITION

James Ferguson (1999) has drawn attention to the power yet inadequacy of the metanarrative of modernity with respect to postcolonial African experience, at least in the urban Copperbelt of Zambia. At independence, Zambians strove for "modernity," and anthropologists were ostensibly complicit in seeing modernity both as a realizable goal and as the only possible human direction globally. Ferguson argues that Africans have been hurt by this meta-narrative insofar as its goals have proved unreachable.

Much of this has been true of Mahajanga, though with a good deal less force. Urban Malagasy had high hopes for "modernity" in the sense of raised standards of living, improved health care and education, and decent, steady employment. Some individuals have been successful, but collectively the community has faced decline. Some would argue that things were better in 1960 when the French left than forty years later. People have difficulty finding work, they are frustrated by inadequate and run-down health and education facilities, and they resent corruption. However, in Mahajanga, "modernity" has never been the only meta-narrative or chronotope available. I have portrayed an alternate one. Its relationship to the dominant meta-narrative and hegemonic social forms is complex and not directly oppositional. Viewed from some distance, it appears socially and politically marginal, but the force of its poetic imagination is such that it is able to subsume "modernity" within the deeper "ancestral" world. In the Sakalava social imaginary, modernity is a part of the big picture, not its sum, frame, or endpoint.

Drawing on Anderson (1991), Taylor distinguishes two forms of social imaginary. In the first, social space is imagined as hierarchical; access is mediated so that "one belonged to this society via belonging to some component of it" (1998: 39). In contrast, in the modern nation-state, citizenship is direct and all individuals qua citizens are uniform and equal. Taylor argues that whereas the form of social imaginary in which hierarchical order is central draws on a notion of a "higher time" in which the past founds and institutes the context of present actions (much as Ndramisara founded Sakalava political order), the modern form of belonging draws on the imagination of simultaneity, a homogeneous time in which events are primary. As Taylor realizes, these are ideal typical and no modern secular form of time can exist in a pure state. It is also implicit in his argument that social imaginaries of either kind will not be identical to one another.

What I have tried to show in this book is the form of "imagined continuity" characteristic of Sakalava mediated belonging and the kind of mediated belonging characteristic of the ancestral chronotope.[13] Both the nation-state and the global forces and institutions that impinge on it are powerful and pervasive, but the social imaginary presented here provides a unique and privileged stance from which many Sakalava view them. In providing a dense, meaningful context for action and debate, the Sakalava ancestral world exemplifies the concept of a live, ongoing historical tradition as developed by thinkers like Gadamer (1985) and MacIntyre (1990).

I have demonstrated the richness, liveliness, and strength of a tradition, albeit one whose primary register is not textual. When the "local" articulates with the "global," it is not a matter of mere reaction, however inspired but, rather, the considered response of a tradition that draws upon its own resources. "Tradition" here is not something to be directly contrasted or opposed to "modernity" but, rather, a shared body of practices and meta-narratives through which knowledge is handed down and rendered available for reflection and debate within a moral community. Tradition is "the community of understanding that the participants in a dialogue share through language" (Bernasconi 1999) and, I would add, practice. Tradition, by definition, has depth; it is multilayered and multistranded. Those who work within a tradition may argue with one another, and inevitably will argue if the tradition is dynamic; but at the same time, a viable tradition will provide means to adjudicate, authorize, establish, and affirm value, maintaining the very conditions for further moral and political debate.

Sakalava ancestral tradition is fully historical, not only situated at a particular historical moment, but richly sedimented, conscious, and able to draw on its own history to shape the present; indeed, unable to do otherwise. It is in this sense that "bearing" history is an unequivocally positive phenomenon. A society that does not bear its history cannot respond with the same consideration, resonance, or autonomy to the contingencies of the present.

Having begun this book with a well-known quotation from Marx, I end it with a less cited one from Freud. Departing from the archaeological metaphor of concealed layers of ruins, he proposes an alternate model of mind:

> Suppose that Rome is not a human habitation but a psychical entity with a similarly long and copious past—an entity, that is to say, in which nothing that has once come into existence will have passed away and all the earlier phases of development continue to exist alongside the latest one. This would mean that in Rome the palaces of the Caesars and the Septizonium of Septimius Severus would still be rising to their old height on the Palatine and the castle of S. Angelo would still be carrying on its battlements the beautiful statues which graced it until the siege by the Goths, and so on. But more than this. In the place occupied by the Palazzo Caffarelli would once more stand—without the Palazzo having to be removed—the Temple of Jupiter Capitolinus; and this not only in its latest shape, as the Romans of the Empire saw it, but also in its earliest one, when it still showed Etruscan forms. (1985: 257–58).

Although Freud called this picture a fantasy, "unimaginable and even absurd" (ibid.: 258), the architectural imagery captures the texture of Sakalava historicity. Its materialization, however, can be only intermittent. For a brief moment each year the four ancestors emerge and are borne triumphantly around the temple. Though it is located far outside the shrine precincts, dare I say that this book has attempted to share in that burden and participate in that celebration? All too quickly, however, the past slips back into its place of concealment and the gate is closed.

Notes

Chapter 1 Notes

1. G. Feeley-Harnik is writing of Sakalava in Analalava. The theme of ancestral burden recurs elsewhere in Madagascar; see, e.g., Astuti (2000).

2. "Zaka. Qu'on peut porter, supporter, gouverner, réglementer, dont on vient à bout, qu'on peut, dont on est capable." (Abinal et Malzac 1987: 866). I am indebted to the interesting discussion by Wooley (1998: 151).

3. I owe the concept of polyvalence and its centrality to Sakalava thought to conversations with Emmanuel Tehindrazanarivelo.

4. *Fanompoa* includes but exceeds notions of corvée and contains strong religious connotations; it has been the main form of subjection to the Sakalava monarchy (Feeley-Harnik 1982, 1991a).

5. I am indebted to Karen Middleton for pointing out the association.

6. The massacre of Comorians that took place in 1977 is (as far as I know) omitted from the public ancestral "record," presumably because it did not touch on matters intrinsic to the Sakalava polity or involve acts of royalty, though Ampanjaka Moandjy held a ceremony to cleanse the town of blood. The Vichy regime and the English invasion of 1942 brought hardship but are likewise absent. Any of these events could be brought up by spirits who lived through them. Much that directly concerns royalty is also omitted.

7. The combination of the necessary and desirable evokes Turner's Durkheimian analysis of ritual (1967). See also the remarkable discussion of pain and imagination in Scarry (1985).

8. White (1973) applies these genre categories drawn from Frye (1957) to modes of historical writing.

9. To caricature somewhat, structuralists like Sahlins (1985) ask how the "inside" shapes its reception of the "outside," whereas materialists like Wolf (1982) see history happening as the "outside" reshapes the "inside." Historical analysis ought to draw dialectically upon the insights of both schools (e.g., Comaroff and Comaroff 1992, 1997). I have not set myself such an ambitious project here (but see Lambek 2001).

10. My use of "consciousness" shares with the Comaroffs an attention to poetic form and a recognition of "the implicit language of symbolic activity" (1992: 157). Like them, I

am interested in developing an "ethnography of the historical imagination" (1992: 31). Indeed, "imagination" might be a more appropriate word than "consciousness" since I do not wish to imply a measure of Sakalava recognition of what Western leftist intellectuals take to be the reality of their situation. I have also been inspired by Parmentier (1987); by Boddy (1989), De Boeck (1994, 1998), James (1988), Peel (1984), Stoller (1995), and Werbner (1991), among other Africanists; and specifically on Madagascar, by Baré (1980), Bloch (1986, 1989a, 1989b), and Feeley-Harnik (1991a). Bayart (1993) critiques both modernization and underdevelopment theories for their lack of attention to African historicity.

11. I learned the distinction in Aram Yengoyan's lectures on Australian aboriginal myth.

12. "A Klee painting named 'Angelus Novus' shows an angel looking as though he is about to move away from something he is fixedly contemplating. His eyes are staring, his mouth is open, his wings are spread. This is how one pictures the angel of history. His face is turned toward the past. Where we perceive a chain of events, he sees one single catastrophe which keeps piling wreckage upon wreckage and hurls it in front of his feet. The angel would like to stay, awaken the dead, and make whole what has been smashed. But a storm is blowing from Paradise, it has got caught in his wings with such violence that the angel can no longer close them. This storm irresistibly propels him into the future to which his back is turned, while the pile of debris before him grows skyward. This storm is what we call progress" (Benjamin 1969a: 257–58). Benjamin himself held another view—one rather more similar to Sakalava, in which techniques of disclosure enable engagement with the past and redemption of suffering (Honneth 1993). I am indebted to Roger Simon for the reference.

13. For modalities that may exemplify greater alienation, see Cole (2001) and Sharp (2002). On the juxtaposition of incommensurable discourses, see Lambek (1993).

14. Sakalava historicity thus transcends the division between the two historical forms of power articulated by Foucault (e.g., 1978).

15. The difference between "historicity" and "history" may be grasped in part by Gadamer's conceptualization of the fusion of horizons (1985). For Gadamer, the past "appears as an inexhaustible source of possibilities of meaning rather than as a passive object of investigation" (Linge 1976: xix).

16. The *kula* exchange of the Trobriand archipelago provides a classic example of totality. Malinowski (1961) attempts to see it from all sides, in relation to economic gain, political competition, magic, religion, art, and adventure. He emphasizes the romance of engaging in *kula;* it is what people are passionate about, what moves them, and gets them moving.

17. Mauss's personage exemplifies not a distinctive intellectual structure, nor subjectivity but a distinctive social structure and cultural ideology.

18. Drawing on Vico, Rappaport refers to this as "verity" as opposed to "veracity" or "certainty" (1999: 293–97). Philosophers are skeptical of this "elevated" form. Hacking (1999: 229) cites Austin's "fine remark" that "*In vino,* possibly, '*veritas*', but in sober philosophical symposium, '*verum*'."

19. Under "savage" I lump together various conceptions of the primitive, undomesticated, prerationalized, etc. If, over time, anthropologists found themselves in the position of defending and illuminating the traditional, the savage, the enchanted, and the irrational and developing ever more clever ways to do so, contemporary thinkers seek ways of avoiding the entire set of dichotomies. Thus Bloch (1998) turns to cognitive science and Kapferer (1997) to phenomenology—different directions to escape the same intellec-

tual thicket. Regarding the contrast between the Platonic and Aristotelian legacies for anthropology, see Lambek (2000a).

20. An account of the verbal register, historical narrative, is regretfully postponed for reasons of length. Eventually, I hope to address the interrelationship of gender, myth, and sacrifice.

21. I have omitted a number of other topics owing to space constraints. Kinship and the exigencies of urban life are addressed for related Sakalava polities in Baré (1980, 1986), Feeley-Harnik (1991a), and Sharp (1993, 2002). Of course, the circumstances and practices in and around Mahajanga during the 1990s are not identical to those in these neighboring regions during the various periods described in these works. On the production of new *tromba* and the fetishization of sacred power, see Lambek (2001).

Chapter 2 Notes

1. But never Sakalava—Sakalava consider it beneath their dignity to pull a *pousse-pousse* and immoral to be carried by the brute labor of another human being.

2. On the founding and history of the old town, see Vérin (1986).

3. Somapêche fishing boats have Japanese captains and engineers, and Malagasy crews working in poor conditions for low wages. The main product is shrimp, which are exported to Japan.

4. It is difficult to acquire reliable population figures. I heard estimates ranging from 200,000 to 500,000 and think that 250,000 inhabitants in 1998 is approximately correct.

5. I learned this term from McElhinny (1998).

6. Sarah Gould is carrying out fieldwork in Mitsinjo.

7. In 1839, British merchant Leigh described all of his financial transactions in Mahajanga in American dollars (Allibert 1999).

8. Malagasy deictics cannot be uttered without temporal inflection: e.g., *eto* "right here and now;" *teto* "right here and then." Conversely, as temporal markers they indicate spatial distance as well.

9. On narratives of place, see Basso (1996).

10. See Grandidier and Grandidier, Vol. 1: 228–33, Vol. 2: 245–46 on the Jesuits, and Vol. 6: 127–28 on the Dutch, as cited by Feeley-Harnik (1991a: 73–74); cf. Vérin (1986).

11. The relics reported were teeth, nails, and a lock of hair from the "grandfather" [ancestor] of Andriantsoly, whom Leigh had met in Mayotte.

12. On planting, see Feeley-Harnik (1991a).

13. The maze thus contrasts with the nineteenth-century European ideal of unobstructed gaze and panoramic perspective, as gained, e.g., by looking down from above (Mitchell 1988).

14. In any religion the sacred has little specific semantic content, but as Rappaport (1999: 256) puts it, is informationless. The sacred core makes the rest meaningful but simultaneously gains its value and is set apart precisely relative to the enclosures surrounding it, the obstacles on the path toward it, and the noisy pilgrims outside the gates. It is these external forms and acts that are material and meaningful—that can be talked about.

    As with the religious quest, so too with the ethnographic.

15. Many shrine residents grew rice at some distance from the shrine and held wage employment in town.

16. Changes are gradually introduced. By 2001, the house at which the reigning monarch held court during the Great Service, had a cement floor covered with carpeting.

17. Leigh was publicly scolded by a woman in 1839 for sitting with the soles of his boots pointing toward the relics (Allibert 1999: 151), as I was in 1993.

18. The fact that the posts are sharpened distinguishes the enclosure from ordinary pens. Along the western "front" the fence had some 130 posts south of the big gate and 40 posts north of it. The western door was made of wood—taller than a man and divided into two sets of four panels. (In 2001, the palisade was rebuilt with cement posts in imitation of the sharpened wooden ones.)

19. Muslims do so as well on entering a mosque. However, Sakalava exit with the left foot, whereas Muslims always lead with the right.

20. I was advised to entreat the ancestors to help find my stolen camera. In my case, this proved futile.

21. One consultant explained the mark as a sign of acceptance of the rule of the royal dynasty, in effect of subjection.

22. A number of descriptions can be discovered in the literature (e.g., Estrade 1985: 170), as can photos of the relics from Menabe (e.g., Lavondès 1967). Writing of Sakalava royalty generally, Grandidier says that the relics include, "one of the vertebrae of the neck, a nail, and a lock of hair [which are] placed in the hollow of a molar tooth of a crocodile" (1891: 312). "The tooth," he says, "must be taken from a living animal; they choose one of the largest size, and bind it firmly with strong cords; then they insert between its jaws, at the desired place, a burning potato, and after a quarter of an hour, the coveted tooth can be easily extracted. The animal is then set free" (ibid.: note).

23. *Manantany* (land-holder) and *fahatelo* (third [after *ampanjaka* and *manantany*]) are titles and offices that are also found in political contexts outside the shrine. By 1998, there was a *manantany* present but no intercessor, and the *fahatelo* continued to play the key role.

24. The intersection of sanctity and hierarchy became clear to me from Rakotomalala et al. (2001).

25. The image is that of "Chinese boxes" or "Russian dolls," or, for that matter, of Malagasy fiber containers. When the shrine was located at Mahabibo, it appears to have had two successive fenced enclosures (Rusillon 1922–1923: 180); in 1926, Linton described an inner stockade of reeds "neatly lashed together" (Mosca 1994: 45 no. 90). See also the layout of Bezavodoany (Appendix to chapter 9).

26. I asked whether there was possession by *zar* spirits in Saudi Arabia, and he replied that both his brother and his nephew had them.

27. Many Sakalava, especially Muslims, do not follow this common Malagasy practice of returning "home" for burial.

28. Two other mediums of Ndramisara started to imitate Somotro in wearing red and braiding their hair; but they did not always dress in his clothes and their hair was much shorter.

29. Ralibera was one of three mediums of Ndramisara at the shrine. One of the others was from Tsaratanana; the other a local Sakalava. Ralibera had first appeared two years earlier and his presence may have been a factor in Somotro's departure. He was not evident in 2001.

30. Many Malagasy also found him, if not exotic, dangerous and intriguing.

Chapter 3 Notes

1. On the division of labor see Feeley-Harnik (1978, 1982, 1984, 1991a) and chapters 4 through 6, below.

2. In fact, Ndramaro's fear and savagery is based not only on when but on *how* he lived; as a rustic, hunting in the forest with only his dogs for company. His "wildness" is thus overdetermined.

3. "Heteroglossia" is Bakhtin's term (1981), well applied to spirit possession by Besnier (1996). On the "coincidence" of temporalities see also the profound analysis by Becker (1979).

4. See Baré (1980), Ramamonjisoa (1984), Sharp (1993), and chapter 7 below.

5. Expanding from the central highlands since the early nineteenth century, Merina were rivals of the Sakalava but more similar to them than most people now imagine.

6. This is a paraphrase from Hayden White's essay (1987: 172) on Ricoeur's *Time and Narrative* (1984).

7. "Imagined community" is a term borrowed from Anderson (1991).

8. "Poiesis" overcomes some of the insufficiencies that others have seen in the text model, in large part because they have mistaken the idea of fixed texts for textual production.

9. I also recognize that the Sakalava constitution of the past is itself historically situated and undoubtedly different now from what it was in the various times of and from which Sakalava speak (cf. Feeley-Harnik 1984, 1991a; Sharp 1995).

10. Lévi-Strauss's "science of the concrete" (1966) provides a notable attempt to mediate Plato's dichotomy and is thereby an instance of the very process that Lévi-Strauss takes to be central to human thought.

11. Aristotle's mimesis is a general term for the representational aim of aesthetics, not opposed to reason.

12. White is concerned with narrative understood in the conventional sense as writing; it seems a small step to include various kinds of performances within his concept of discourse.

13. White draws on the typologies of broadly Aristotelian critics such as Burke (1945), Frye (1957), and Pepper (1942).

14. It is one of the virtues of White's structuralism that it distinguishes different styles or emphases *within* a given cultural tradition. Western modes that approach Sakalava possession include the presentation of history at tourist sites like Edinburgh Castle and Shakespearean representations of antiquity composed in Elizabethan idiom and transformed to a nineteenth-century setting for a twentieth-century audience (Wendy James, pers. comm. and Maggie Sackville-Hunt, pers. comm.). There is every reason to expect variation among Sakalava approaches to history as well.

15. The place is Androntsy and was visited by Catat in 1889–90 (1895: 259–60).

16. Auerbach distinguishes the nineteenth-century French novel as a kind of apogee.

17. Bakhtin's concept of "chronotope" must be supplemented by his insistence on "heteroglossia;" any given social group will produce multiple chronotopes that issue productive challenges to one another.

18. Clearly the chronotope itself shifts and no longer bears the same relationship to the dominant social institutions that it held in earlier periods.

19. This supports Connerton's argument concerning the significance of embodied and performative memory (1989).

20. This is in contrast, e.g., to the Greek adventure story that is a chain of random contingencies in which, "the nature of a given place does not figure as a component in the event" (Bakhtin 1981: 100) and where spaces are interchangeable just as the individual escapades could be rearranged in a different temporal order. These distinctions are prefigured in Aristotle's comparison of tragedy and epic poetry (1947: 1449b).

21. All of this is quite different from Mayotte (Lambek 1981), where the arrival of particular spirits is more haphazard and spirits routinely move between mediums during a single performance. The ability to ensure the presence of particular spirits lies in the fact that Mahajanga is a large city with a relatively high number of mediums to draw upon.

22. Moreover, the material described in this paragraph serves as a reminder of the artificiality of any firm boundary between poiesis and practice (chapter 11).

23. Posthumous names summarize some striking aspect of the deceased's character or agency, but often the meanings of the names and stories that lie behind them are forgotten. Nicknames tend to single out a kinship or age status that epitomizes the deceased—mother's brother, small child, etc.

24. The quotation concerns the characters found in Apuleius. As Geertz (1973) observed, temporalization and personalization are closely linked. Bakhtin refines the point.

25. Aristotle distinguished six parts to every tragedy: plot, character, diction, thought, spectacle, and melody. He acknowledged that pity and fear could be aroused by the spectacle but felt the superior way was by means of the plot.

26. The juxtaposition of death with life and the striking nexus of vitality, boisterous exuberance, and power are found in other Malagasy societies, notably in contexts of funerals, reburials, and circumcisions. Their relationships have exercised a number of interpreters (Huntington 1973; Bloch 1982, 1986; Astuti 1995; Graeber 1995; Middleton 1995a, 1995b).

27. Drawing on the root metaphors of Pepper (1942), White proposes four alternate modes of historical argument: formist, organic, mechanistic, and contextualist (White 1973: 13).

28. New experiences are represented by the spirits of successive generations.

29. The portrait of Ndramandaming as the precedence demanding governor suggests Taussig's arguments about the colonial mirror and that "what's being mimicked is mimickry itself" (1993: 241; cf. Kramer 1993; Stoller 1995).

30. White distinguishes four modes of emplotment as set out by Frye (1957): romance, comedy, tragedy, and satire.

31. It is worth quoting White's preceding remarks on the comic mode, as epitomized in the historiographical style of Ranke (White 1973: 28):

> Those "forms" which Ranke discerned in the historical field were thought to exist in the kind of harmonious condition which conventionally appears at the end of a Comedy. The reader is left to contemplate the coherence of the historical field, considered as a *completed* structure of "Ideas" (i.e., institutions and values), and with the kind of feeling engendered in the audience of a drama that has achieved a definitive Comic resolution of all the *apparently* tragic conflicts within it. The tone of voice is accomodationist, the mood is optimistic, and the ideological implications are Conservative.

32. The allusion is to Freud. Similarly it fits a Ricoeurian interpretation, having "as its 'ultimate referent' nothing other than temporality itself." (White 1987: 52).

33. Aristotle, himself, distinguished history from poetry with characteristic succinctness: "One describes the thing that has been and the other the kind of thing that might be" (1947: 1451b).

34. For a profound reading of possession as allegory, see Boddy (1989).

35. This is not to deny the presence of skepticism or that the knowledge of spirits in key mediums is tested (Feeley-Harnik 1978: 412; chapter 9, below).

Chapter 4 Notes

1. "Ancestral order" refers to the total social phenomenon that is at once political and religious. I am indebted to Tehindrazanarivelo (1997) for my usage.

2. The shrine (*doany*) at Mahajanga has a different status from the clear opposition between royal residences of the living (*doany*) and royal burial compounds (*mahabo*) characteristic of Analalava.

3. I use masculine pronouns since the current ruler is a man. However, he succeeded his mother. Feeley-Harnik (1991a) has argued that women began to reign only once Sakalava power was compromised by Merina incursion. Early monarchs were men, but the pattern was broken by Queen Ravahiny (Ndramamelong) who reigned for many years when Sakalava power was at its height.

4. I have carried out little archival research nor reviewed most primary sources; what follows relies heavily on Vérin (1986) and Lombard (1988), who provide fuller accounts than space here allows.

5. See Campbell (1981, 1989, 1996).

6. Lombard (1988) argues that when Sakalava conquered Boina they overran a number of chiefdoms. They preserved this structure so that the great Sakalava *ampanjaka* ruled over fiefs governed by lesser *ampanjaka* who recognized him as elder.

7. Southall developed his model with respect to the East African Alur (1956). Sakalava are frequently referred to as a state society and their rulers as kings and queens, or princes and princesses. An alternate intellectual tradition would distinguish chiefdoms as a political type from states. Sakalava polities might be compared to chiefdoms of the Polynesian or African sort, but having lost power and autonomy over nearly two centuries they represent the devolution of a more centralized system. What has changed, however, is less the form of the segmentary state than the mode of production supporting the elite.

8. Traditional territorial units do not correspond precisely to those of the Malagasy state, though Sakalava have sometimes fought to keep administrative divisions in line with their own.

9. I do not mean to reify the definition of shrine. The word *doany* is applied quite broadly. Feeley-Harnik translates it as "royal compound."

10. A cemetery that no longer has guardians is known as *kalambay*.

11. Ndramañavaka's son, Ndranihotsy, and most of his descendants lie at Ankarañanjy, whereas Amina's predecessor, Manjakareniarivo (Moana Salimo), requested burial at a new site at Ansakoamanera.

12. This distinguishes them from *togny*, sacred trees or vertically planted rocks, at which people address prayers (*mijoro*) but do not enact service (*manompo*).

13. For example, the village of Bemilolo contains the shrine of Ndramandikavavy, which is maintained by her descendants, some of whom reside there. An annual Service sees the

arrival of more descendants and mediums to witness the purification (polishing) of her artifacts.

14. E. Tehindrazanarivelo (personal communication).

15. Tehindrazanarivelo reinforces the structural aspect by referring to them as the "four corners" (personal communication).

16. I am indebted to the late Mady Simba for information on Maromadiniky. Another source emphasized two branches alternating authority over the tombs or, rather, the rights to cattle brought to the tombs by devout subjects.

17. Although people in Mahajanga refer to Ndramboeniarivo (as do those in Nosy Be [Baré 1980]), the name represents a shift from Ndramboninarivo (Lord Who Is Above Thousands or Who Dominates Thousands) as reported in Guillain (1845).

18. The more elaborate account collected by Guillain (1845) suggests an extended process. Between Ndramahatindry's death somewhere between 1750 and 1763 and Ravahiny's ascension to the throne around 1785, there may have been six intervening rulers, all close kin. Maka, the grandson of a daughter of Ndramahatindry who abdicated in favor of Ravahiny, was offended at not succeeding Ravahiny (ca. 1811) and started a rebellion among his followers, known as Bemazava, on the west bank of the Betsiboka. Ravahiny was herself a granddaughter of both Ndramahatindry (through her mother) and the latter's brother, Ndrananilitsy/Mbabilahy (through her father). Guillain points out that according to Sakalava theory, Maka had a stronger right to rule than Ravahiny's grandson, Tsimalome, who did succeed, because he was descended from royalty through his mother and Tsimalome was not. Essentially, Bemazava could show a pure line of descent through Ndramahatindry's daughter and granddaughter. Bemazava concerns were kept alive because Tsimalome, followed in 1822 by his brother, who became known as Andriantsoly, suffered Merina attacks. The accounts that I heard in Mahajanga in the 1990s linked the Bemazava to Ndramahatindry and the Bemihisatra (Ravahiny's descendants) to Ndrananilitsy, conflating much of the succession prior to Ravahiny.

19. These dates are very tentative as sources do not agree. Guillain says she ruled for forty-two years, possibly including the years she was married to two of her predecessors.

20. The oxen were raised by citizens but available for her use (*tsimirango*). Royal cattle were identified by ear markings and possibly by the color and pattern of their coats. The slave trade was active during her reign. Like her predecessors, Ravahiny is depicted by Europeans as holding absolute power.

21. Ndramfefiarivo's identification is problematic. Feeley-Harnik identifies her with Agnitsaka, granddaughter of Ravahiny and ancestress of the Analalava line, and says she died in 1822. My consultants listed this person as Ndrantsirehanarivo (as does Rusillon 1922–1923). Feeley-Harnik's genealogy leaves Ravahiny no descendants through her daughters and omits the entire Ambatoboeny/Betsioko line (1991a: 85). On the other hand, some Bemihisatra consultants referred to Ndramfefiarivo as Rasalimo, actually a wife of Radama from Menabe.

22. Rusillon (1922–1923) mentions Ramboatofa (Andriamahavitarivo), a niece of Ratarabolamena, who returned from Antananarivo in 1861 and died in 1913 at Betsioko at the age of 90. She is indicated as a childless daughter of Andriamifefiarivo on the written genealogy that Philibert Tsiaraso kindly showed me.

23. Bemazava say that they refused to permit her half-Merina descendants to cross the Betsiboka to bury their dead at Doany Mahabo.

24. Bemazava trace their ancestry to his older brother, Ndramahatindry, alone. I have not heard mention nor seen the *tromba* of Ndramahatindry's daughters, to whom (following Guillain but not present knowledge) the split could also be traced.

25. Citing the same source (Bénévent 1897: 53), she adds that the relics were reinstalled in Majunga by General Metzinger and "guarded by Sakalava under the surveillance of local [French] authorities" (1991a: 123). She dates this to October 1895 (citing Ferrand 1902, vol. 3, p. 66).

26. See note 22 here.

27. Dates were provided by Fahatelo Dzaovita of the Doany Avaratra in 1993 and confirmed by other sources.

28. Most Christians are of central plateau (including Merina), Tsimihety, or slave origin.

29. The name White Lord has nothing to do with Europeans, who were described derisively as Red Ears, (*Menasofing*) or Gray Eyes (*Garamaso*). Clan names were originally Zafinibolamena and Zafinibolafotsy.

30. Of course, versions recited in the Zafinifotsy polity of Antankaraña are quite different (Walsh 1998).

31. Foreigners are placed at the origin of all groups, especially royalty. As one man said, "What do you expect when Madagascar is an island; only the plants are autochthonous."

32. In some genealogies he is identified as a childless brother of a ruler of Menabe (Rusillon 1922–1923; Lombard 1988). In another tradition (Poirier 1939) Ndramisara is the posthumous name of Ndramboay, who in turn is considered the brother of Ndramandikavavy. This adds another twist to her sacrifice.

33. This is not to reduce the relics to this function nor to use the function as an explanation for the shrine's continued presence.

Chapter 5 Notes

1. I follow Sakalava here in identifying Ndramisara as a historical individual responsible for establishing Sakalava order. However, this sort of "great man" account should be offset by historical and comparative ethnographic evidence suggesting a structural basis for the model.

2. Tsimihety means, literally, "who do not cut their hair." It is said to refer to those who refused to follow the normal mourning practices on the demise of a Sakalava ruler.

3. This ethnicization is often attributed to French colonial misapprehension of indigenous categories and the reification of ethnic groups as part of their own strategy of rule. There is much truth to this; broad "kinds" of Malagasy people became the "tribes" or "ethnic" units of which the nation state was composed, but there were surely other forces at work besides the colonial vision. The ethnicization of Sakalava also undermines my previous definition of it as a particular form of political and historical subjectivity. I thank Solofo Randrianja for clarifying the point.

4. Sharp (1993) documents how non-Sakalava in Ambanja became Sakalava through *tromba* possession.

5. Several writers on Madagascar prefer to speak of "ancestries" rather than clans; this is probably more precise, if a little obscure. For Southall, "If descent is a rule whereby 'groups exclude and recruit' then most Malagasy peoples do not have descent, since they do not have bounded kin groups " (1986: 275). Lombard, in contrast, speaks of

284 SM The Weight of the Past

southern Sakalava clans (*firazañana*) as characterized by the following five features: ancestral histories, ancestral proscriptions, cattle ear marks, tomb location, and location of the ceremonial post at the base of which ancestral sacrifices, circumcisions, etc., take place (1988: 80). *Firazañana* is sometimes used for clan, and *karazaña* reserved for more encompassing units—"*tribus*" as some translate it.

6. Short, three-four generation, socially significant lines of descent ("lineages") are referred to as *taranaka*.

7. These sentences summarize a long account by Amana Ibrahim, who emphasized ear markings on cattle as signs of the common origins of groups who today may go under different names in different regions.

8. In his excellent discussion of social organization among Sakalava of Nosy Be, Baré (1986) emphasizes stated rules of patrilineality and patrivirilocality that are often disregarded in practice. To the south, Dina (2001) and Astuti (1995) provide accounts of the *soron'anake*, a ceremony necessary to move a child's identity from the mother's to the father's ancestry, irrespective of the nature of the union between the parents or, for that matter, the genitor. Lavondès (1967) provides a comprehensive exposition of kinship in a southern village.

9. Originally the royal line was ideally endogamous (Guillain 1845). The children of royal women were undisputedly royal, whereas the children of royal men with commoner women were not, since paternity could never be proven. Lombard (1988: 80) argues that royalty were wife takers but not wife givers. The legitimate offspring of married royal men were always considered royal. To evade paternal claims by commoners, royal women did not officially marry men of other clans. Children born to them are royal and the genitors can make no claims on them. The story about the boy who died as a result of following royal custom indicates the limits of exclusive claims; the Antifañahy clan had to be treated as an exception to the rule. Although inequality in rights to offspring is a mark of high status, it is hardly characteristic of caste systems, which assign offspring to the lower ranked of the parents. Marriage rules for royalty have undoubtedly undergone several transformations since the formation of the Sakalava monarchies (Lavondès 1967), becoming much more open.

10. As someone in the port city of Antsiranana once responded memorably to my question about his *karazaña*, "We are cosmopolitans." Vérin writes, "The most striking characteristic of the civilizations of the Madagascan échelles [ports-of-call], then, is the heterogeneous nature of their population. At the same time, it may be contrasted with the cultural homogeneity brought about by the seafaring, commercial and religious life of the people" (1986: 380).

11. Another consultant gave a different interpretation, saying that *kosilim* meant an adult convert to Islam, e.g., as a result of marriage. He added that Ndrankehindraza was a real Muslim, Silamo, but the Sakalava refused him Muslim funerary rites when he died.

12. See Tehindrazanarivelo (1997) for a Malagasy intellectual who activates multiple dimensions of identity yet sees the constraints imposed by them as a significant difference between himself and me.

13. The broad reference of the term "*ampanjaka*" is characteristic of Mahajanga. In Analalava the term was restricted to the ruler herself.

14. Compare Walsh's discussion of the Antankaraña monarch, Ampanjaka Isa, who adjudicates disputes and addresses every detail of royal ritual down to the calculation of firewood (1998).

15. Land ownership is a complicated issue, depending in part on whether it was registered and, if so, by whom. Certain rulers have private estates; Ampanjaka Désy inherited from a former ruler able to accumulate while serving as an official in the colonial regime.

16. This is not strictly true, or perhaps only true for the public sphere; other groups do recognize ancestors and may supplicate them (*mijôro*). As Kassim put it, "Ordinary people cannot communicate with their descendants through trance. Commoners might see their ancestors in dreams, and the dead person might request a drink or tobacco if, say, she/he liked them when alive. The descendant would bring it to the tomb."

17. Her own power would be in dangerous conflict with the power emanating from the Doany Ndramisara. This is because both of them own a *vy lava tsy roy*, literally a "long iron that is not two," i.e., the "single" ancestral staff that limits legitimate succession to one person.

18. Not all restrictions index modernity; substances forbidden at the Doany Ndramisara include chicken, pigs, peanuts, and kerosene.

19. In September 1995, rice cost fmg1,500 per kilogram, fish fmg2,000/kg, and a loaf of bread fmg500.

20. In April, honey is brought to the shrine from the constituencies (Ambatolampy and Ambalakida) whose obligation it is to provide it. This uncooked honey, stored in clay pots, is divided into three portions: one for Ndramisara, one for living rulers, and one for Ancestor People. The portions of the living rulers (Ampanjakas Désy, Amina, Amada, and Soazara) are handed to the Great Women (*bemanangy*), who are their respective representatives in Mahajanga. They, in turn, divide it among those associated with the rulers. They send a portion to resident *ampanjaka* and offer another portion to the mediums of royal ancestors whose branch they represent. The Ancestor People's portions are divided among representatives of eight clans and the workers at the main shrine. Ndramisara's portion is used to make the mead. *Katrafay* is a hardwood once reserved for royal constructions (Lavondès 1967: 132) and ground to stun sweetwater fish. There may be additional ingredients.

21. Mediums of senior spirits told me that the cow that they supply is for the first event of the season when the plants that have filled the enclosure during the wet months are cut or trampled by cattle, two months prior to the honey cooking. They said that the cow of the honey cooking was provided by the shrine.

22. These clans were Zafindramahavita, Antankoala, Tsiarana, Jingô, Sambarivo, Andraramaeva, Vôrong Mahery, and an eighth, possibly Antavarabe. Not everyone agrees that each of these kinds of people came north with Ndramisara. Clan names and functions also vary somewhat across the north (cf. Baré 1980, 1986; Feeley-Harnik 1991a; Jaovelo-Dzao 1996: 95–6).

23. My data on slaves and abolition are insubstantial. However, "upward promotion" is neither unprecedented nor unauthorized. Dina's analysis of the Fihereña Andrevola of southern Madagascar suggests that the granting of specific ritual tasks may have been a means of ascension for favored *karazaña* (2001: 22).

24. Tsiarana means "Different." Tsiarana are identified with Sakalava Mañoroñomby, a clan who used to make burnt offerings of cattle sculpted from fat, and who, like Tsiarana, have a special relationship to the sacred drums. Tsiarana appears to be the name more commonly used in Mahajanga.

25. See Feeley-Harnik (1991a) and Jaovelo-Dzao (1996) on royal funerals.

26. Compare Baré (1986: 358), who also details functions of Jingô in Nosy Be that I did not find in Mahajanga.

27. See Walsh (2002) for a fascinating discussion of the *lohateny* relationship.

28. I heard variable accounts of the division of labor at the preparation of royal corpses, but Jingô clearly play a central role. One man said they prepare the upper half of the body while Sambarivo bear responsibility for the lower half.

29. In comparing Jingô to the sacrifice of cattle who fall or are pushed over, a consultant explained that the Jingô volunteer would just fall over dead. The *viarara* was used to cut his throat once he fell.

30. And Jingô were sacrificed each time the shrine moved location.

31. A Jingô consultant was clear that *marovavy* were Jingô (cf. Baré 1986). The "child" spirits common in Mayotte are a transformation of these figures (Lambek 1981). Childhood refers to dependent status more than chronological age. Male counterparts of *marovavy* are known as *fihitry*.

32. See Feeley-Harnik (1991a: 343–44) for a similar account.

33. These people lived at the north end of the shrine village whereas Ampanjaka Désy's "people" lived toward the south—a microcosmic representation of the geographical and hierarchical relations between the domains.

34. Masikoro and Vezo of the southwest coast suggest what the underpinnings of Sakalava society might have been like before the imposition of political hierarchy. Dina (2001), Lavondès (1967), and Marikandia (2001) illustrate descent groups that maintain ancestral authority in the form of the *hazomanga,* a wooden pole at which ancestors are invoked and which becomes their objectified sign. With the imposition of the Sakalava polity the autonomy of lineage ancestors was usurped and the ancestors of each group replaced by those of the royal clan; the *hazomanga* were in turn replaced by royal relics. But Sakalava gave the various descent groups roles to play in the veneration and care of the relics rather than excluding them.

Chapter 6 Notes

1. Tsiarana are singled out to guard their ancestress Ndramandikavavy.

2. These houses may be compared with the royal houses that rested atop Merina tombs and with the house understood as a coalescence of ancestral substance among Zafimaniry. I am indebted to Rita Astuti for the point.

3. This depends on collective interest and the availability of susceptible and competent mediums.

4. The literature on spirit possession has gone beyond worrying about how to define or explain it; here I simply accept its presence and see how it is structured and practiced in a given cultural and historical setting (cf. Lambek 1981, etc.; Boddy 1989, 1994).

5. On curing and initiation, see Jaovelo-Dzao (1996), and for neighboring Mayotte, see Lambek (1981). Aspects of the relationships of mediums to their spirits are described also in Sharp (1993) and Lambek (1988, 1993, 2000a).

6. Their nature is captured by association with the moon—a pale, reflected light that negates the negation of the day (Lambek 2001).

7. In contrast to royalty and mediums, Ancestor People are immersed in death and cannot maintain the pretense of avoiding it.

8. The same spirits manifest with lesser intensity in female mediums. When possessed by senior female ancestors, women mediums also display tremendous concentrated energy, notably in the case of the ancestress who appears to be in childbirth.

9. This is a contrast of roles, not persons. Guardians can be mediums as well.

10. A comparison to the *kula,* the great game of exchange that occupied the Trobriand islanders and their neighbors in which valuable shell ornaments circulated in a ring some 200 miles in diameter, may not be inappropriate. Malinowski writes (1961: 83):

> The Kula is thus an extremely big and complex institution, both in its geographical extent, and in the manifoldness of its component pursuits. It welds together a considerable number of tribes, and it embraces a vast complex of activities, interconnected, and playing into one another, so as to form one organic whole.
>
> Yet it must be remembered that what appears to us an extensive, complicated, and yet well ordered institution is the outcome of ever so many doings and pursuits, carried on by [people who] . . . have no knowledge of the *total outline* of any of their social structure.

11. Coherence need not imply social cohesion or stasis. On the contrary, my argument is, in large part, how the ancestral system enables historical consciousness and agency and, hence, the ability to address novelty and accept change responsibly.

12. Conversely, one of the ways to seize power and, hence, to "alter history" has been to kidnap and displace the relics; the subject of competing claims and the index of real power, the relics are always vulnerable to appropriation and theft (cf. Feeley-Harnik 1991a). On Merina architecture, see Kus and Raharijaona (2000).

13. For example, Giles (1987, 1999) reports the presence of Malagasy spirits in Zanzibar where the original significance has become so far displaced that it no longer makes sense to speak of Sakalava history.

14. In fact, the trip is not quite so difficult.

15. Recall my encounter with the shy female retainer (*marovavy*) spirit reported in chapter 5.

Chapter 7 Notes

1. Money collected for the *fanompoa* is known as *paria* or *paré.*

2. His full name is Mbabilahimanjaka, or Ndrananilitsiarivo.

3. This is the Bemihisatra version of the genealogy.

4. Great Woman Safia explained that other spirits from Nosy Be rarely appear because they are very old and need fancy armchairs—*fauteuils* or even *chaises boeing* (presumably recliners). They speak so softly that they can barely be heard.

5. Ndrankehindraza is an older "brother" (first cousin) to Ampanjaka Amina. He passed away when Mme Doso was a girl.

6. It is taboo (*fady*) to bring just one item to a *tromba* event; things must be in even numbers. When there was only one beer, a fmg1,000 note was sent along as its "companion."

7. Unlike what transpires in Mayotte (Lambek 1981), spirits generally do not "hop" from one host to the next but stay in a given host for the duration of their participation at a given event.

8. In denser neighborhoods the intake at a fund-raiser for a popular spirit was as much as fmg150,000.

9. Mbabilahy's older sibling, Ndramahatindriarivo, is identified with the Bemazava faction.

10. Ndransinint', known as Adan'Añilo or Zinôro when alive, is a descendant of Ndramandisoarivo's son Ndramamonjy and, therefore, a member of the junior lineage that rules over the burial grounds of the senior line.

11. Ndrankehindraza's energy and popularity is used to collect money for both Bezavo (where he is buried) and Nosy Be (where his father is buried). In 1998, I went to a fund-raiser for Nosy Be similar to those at Mme Doso's. There were several musicians, including a male singer and female drummer, but only one spirit, Ndrankehindraza. Women dropped in and handed money ranging from about fmg500 to fmg2,500 to Bemanangy Safia, who laid it carefully on a saucer, put wet kaolin on the palms of the donors, and distributed cigarettes and a little beer.

12. People wishing to give on behalf of a spirit more distantly related to those who possessed Mme Doso found an appropriate medium to pass it on to the Great Woman.

13. Part of this money is used for renovations at Bezavodoany (chapter 9).

14. Interestingly, many of the participants were migrants from the south and east of Madagascar.

15. Maromalahy died as a child by falling in mud and drowning. When I asked Mme Doso later about the simultaneous presence of two incarnations of the same spirit, she said good-humoredly, "Be quiet! (*Mangia!*)" Such encounters were not uncommon on this day, and in 2001 I heard that Ndrankehindraza even had special dispensation for multiple simultaneous appearances.

16. She also had to negotiate her relationship with Ampanjaka Désy rather gingerly. To a degree, her position was dependent on his approval; yet his actions were not always ones of which Mbabilahy approved (chapter 10).

17. She is assuming that a husband would be earning money. Her previous husbands both were reasonable providers. She has never been rich but never needy either. It would be difficult to serve as a senior medium without sufficient means.

18. When royalty give, it is called *takitaky* rather than *manompo*.

19. In 1995, the eleven full-time guardians (five men, six women) received only fmg15,000 each after the Great Service. A servant who complained to Désy was told that he had allocated fmg300,000, so the managers must have removed some. The servants assumed the senior officials each received around fmg500,000.

20. In fact, Ampanjaka Désy had many hectares of cotton, grew beans and rice, and was reputed to own a bar.

21. The display of honor, caring, and the circulation of women's wealth for public good is reminiscent of Weiner's account of Trobriand social reproduction (1976).

22. *Mosarafa* is presumably derived from the Arabic *saraf,* to spend.

23. The mystification entailed in this ostensibly two-way exchange has been analyzed by Bloch (1989a).

24. Conversely, money is also of long-standing import (Lambek 2001).

25. Chickens are taboo to royal ancestors and their mediums. Goats are taboo to some descent groups, and no other animal has the saliency of cattle.

26. A 'tether cord' accompanies all kinds of royal ancestral prestations. The stated reason is that gifts must be given in pairs or even numbers; the smaller gift, like North American gift wrapping, reinforces the intentionality, voluntary quality, formality, and commitment of the act (cf. Keane 1997). Cattle, themselves, cost anywhere from fmg250,000 in the countryside to fmg800,000 for a relatively small animal in the city.

27. In other words, Sahlins's continuum of reciprocity (1965) can be applied to redistribution.

Chapter 8 Notes

1. See Rappaport (1999: 47) on the frequent appellation of ritual by words meaning "work."

2. Compare Csordas (1997) on prophesy among North American charismatic Catholics.

3. She has sacrificial sabers at other shrines as well and her own shrine at Bemilolo, near Betsioko.

4. The drums are known as "great ennoblers" (*mañandria be*) or "long bamboo" (*valiha lava*).

5. Mme Doso brought fmg13,000, another woman 32,000, and so forth.

6. Amina's Great Woman was not available as she was embodying Ndramañonjo.

7. I had trouble following, but the next day a friend remembered the figures as fmg6,500,000 from Great Woman Twoiba, which included fmg3 million from France as well as money from both Mahajanga and Mayotte. Fmg115,000 was announced by the Great Woman of Nosy Be; fmg 500,000 plus a cow from Analalava; fmg100,000 plus a cow from Ndransinint'; and fmg150,000 from Antananarivo. Other towns announced included Ambato, Amboanio, Boanamary, Mampikony, Port Bergé, and Toamasina.

8. Sakalava recounted memories of much larger crowds and of people designing new jewelry for the fete.

9. I heard later that the provincial governor offered fmg1 million.

10. It is not the original weapon (chapter 11). In the past, the *viarara* would have been used in human sacrifice.

    Despite the local explanation, *viarara* are found in other Sakalava polities as well. The royal *jiny*, or ancestral relics of the southern Fihereña Andrevola kings (who claim to have adopted the custom from Menabe), comprise "a saber (*vy arara* or *viarara*, also called *vy lava*) . . . which served the king when sacrificing cattle during the royal relic bath . . . two long drums (*hazolahy*) used exclusively by royal families" as well as a sea conch, incense burner, water pitcher and "former sovereigns' old guns and assegais" (Dina 2001: 17). Almost the same list comprises the sacred artifacts in Mahajanga.

11. I am indebted to Andrew Walsh for the clipping. The translation from French is mine. Rusillon (1912) provides an earlier description; in 1910, the procession of the bathed relics took place on the Friday.

12. Furthermore, the liquor that humans drink at the *Fanompoa* is always received as generosity *from* spirits; one does not purchase it for oneself. This overlooks the fact that the spirits received the liquor as gifts from humans in the first place and is exactly the transposition that Bloch (1989a) describes with respect to *hasina*.

13. The word "playground" is Tehindrazanarivelo's (1997). On the carnivalesque, see Stallybrass and White (1986).

14. Here it was *tromba* who displaced themselves for the living rather than vice versa, but these were not the most senior ancestors.

15. Relic bathing in Mahajanga used to involve a dunking in the middle of the Betsiboka estuary; it still uses Betsiboka water brought to the shrine.

16. The happy crowd in the plaza is also a noisy crowd, and I suspect that at some level, like the blasts on the conch and the beats of the drum, this is the noise of transition and liminality, at once covering, marking, and fueling what transpires inside during the bath.

17. Interestingly, although Mbabilahy is proprietor (*tômpin*) of the keys, Mme Doso says she herself has never seen them. Mbabilahy carries them on crossed arms under a cloth. In his depiction of the Great Service of 1910, Rusillon (1912) depicts both key and drum originating at Doany Mahabo and passing through Betsioko and Marovoay.

18. One consultant, educated in constitutional politics, suggested that the two factions maintained "legislative" and "executive" functions, respectively.

Chapter 9 Notes

1. He was an older first cousin of Ampanjaka Amina who died in the mid-twentieth century.

2. The tension I describe here can be translated directly into Rappaport's hierarchical levels of sanctity and sanctification. The threat to the earlier generations is equivalent to what Rappaport sees as the pathologies of religion in which utterances of lesser sanctity and greater specificity come to take the place of those with greater sanctity (1999).

3. Kassim likened Bevony to Jingô and his grandfather to the *fahatelo*. However, there were no living Zafinifotsy monarchs to whom the servitors were responsible. On the occasion of the annual *Fanompoa*, both the goods—and the mediums—were plunged into the turbulent waters of the Loza.

4. In 1994, this was around fmg80,000 per month.

5. This man lived and died around Analalava and never saw either Mahajanga or Bezavo.

6. *Tromba* did not rise in children. This mix of prospection and retrospection is typical of the way accounts of mediums' lives are formulated.

7. In fact, Lakolako is not very junior either. The seniority of Kassim's spirits limits his activities as a healer since he cannot adopt the humble position necessary to placate recalcitrant spirits. Lakolako most frequently acts as a diviner and helps clients locate appropriate curers, who sometimes collaborate with Kassim on a case.

8. Dubourdieu (1989: 465) offers a brief, instrumental explanation for the origins of the sky spirits among Makoa in Ankáboka around 1958. She sees them as a cult of diviners substituting for the dynastic cult.

9. A Jingô person who was also a medium would not take on the task of removing death pollution. Kassim has petitioned his spirits to be allowed to pay condolence visits to no avail. Mediums with less stringent taboos bathe in water to which a little honey and kaolin are added after a condolence visit.

10. Kassim described the snake as yellow and black; it is large and does not bite, but not a boa.

11. "Our firstborn." The cardinal direction, the song, and, Kassim supposed, the signs, would differ for other spirits undergoing certification.

12. Fines at Bezavo are not of money or live cattle, Kassim explained, but of blood sacrifice: "Blood must run on the ground."

13. On another occasion she furiously criticized her own guardian for staying at her son's house (i.e., Kassim's) rather than hers (that of any of her mediums) in Mahajanga. The guardian shifted residence forthwith.

14. Kassim raises another kind of domestic duck (*gana*). Since the whole line of spirits are tabooed chicken, this is the main form of poultry eaten in the household. Occasionally, he fattens a turkey.

15. Conversely, he would not enter trance where a medium with whom he had quarreled was likely to embody Ndramandikavavy in order to avoid having her place her hand upon his head.

16. Hearing Kassim's wife refer to this woman as her mother-in-law, for a long time I assumed they really were kin. Maolida Kely is now deceased

17. The fact that Ndramandikavavy and Ndramañonjo quarreled has no bearing on their joint presence in a single medium. The quarrel took place long ago and was resolved. This is not always the case; some disputes (especially those that involved living ancestors) are commemorated by the mutual avoidance of spirits. Among Maolida Kely's

other spirits was Gida, a concubine of Ndramañonjo, said to be buried outside the enclosures at Bezavo.

18. In some instances, the *manantany* is conceptualized as representative of the people; at others, as the person who delivers and administers the ruler's decisions to the people. Kassim called his assistant *manantany* because it was he who told the latter what to do. In this way, Kassim was comparing his own position to that of a ruler.

19. The Bezavo service need not take place every year nor does it amass anywhere near as much money as the Doany Ndramisara. Moreover, the *fanompoa* does not support those who live at the shrine. The guardians keep cattle and grow their own food. They earn cash selling dried fish, eel, and crocodile fat and keep the money brought by pilgrims paying their respects at the shrine.

20. Unlike Doany Ndramisara, there is no fixed month for celebrating at Bezavodoany. It is done when the work is completed but before it gets too wet.

21. In 1997, they brought twenty-seven sacks.

22. The captain of the barge also happened to be possessed by Ndramboeny.

23. He explained that unlike Merina, Sakalava do not roll new cloths over older wrappings but take the latter and throw them in a pond or the sea. The rewrapping is carried out by male Ancestor People during the night while the mediums hold an all-nighter outside the burial precincts. At dawn Amina and her Great Women are ushered in to see that the work has been completed. By this point the tombs have been closed up; I think the second *sobahiya* must lie visible atop the outside of each tomb.

24. The man had succeeded to the position of a childless kinsman. The remaining guardians at Bezavo are Morarivo according to the guardian I spoke with; with one exception, they are all male and succeed to their posts patrilineally. The exception is the guardian of Ndranantanarivo, the daughter of Ndramboeny who was married to the Muslim leader of the Antalaotra trading community—subsequently, all descended from Ndramboeny and Ndramandiso as well.

25. The description may be compared to that of Hébert and Vérin (1970) and Vérin (1986: 286–7), who visited the site in 1968. They described an outer wall (*valamena*), 40 meters to the side made of stakes 2 meters high; a middle wall of stones bonded with mortar some 20 meters square, 2.5 meters high, and 1 meter thick; and an inner enclosure 6 meters square containing the coffin of Ndramandisoarivo on the north side and Ndramandikavavy on the south. Various other tombs and coffins were inside the stone enclosure along with some Islamic pottery, a Chinese dish, and other artifacts. The names of the deceased coincide with Kassim's, with the addition of Ndramitanjakarivo and a *manantany*. Vérin adds "It would also appear" that the site contains "the funeral monument of a woman who was buried alive; a *moas* seems to have called for this sacrifice for the prosperity of the kingdom" (1986: 287). This is an alternative to Ndramandikavavy's story.

26. Not everyone agrees on the identity of these two figures.

27. Ndramañaring, who "holds the key," has no house, but must be visited in Mitsinjo in order to gain permission to proceed to Bezavo.

Chapter 10 Notes

1. Shrine appointments were usually made at the Great Service in the midst of large numbers of people and selected by the public from a short list of candidates. The doorkeeper had originally been selected by a show of hands. Voting for a *manantany*

was accomplished by dropping a stone into one of two piles. Both candidates were from the Vôrong Mahery clan and understood to be Ampanjaka Désy's men.

2. Mediums saw their authority deriving in part from having personal roots far from the shrine or ruler in whose affairs they participated. One medium said, "Being from so far away, when the *tromba* speak knowledgeably they are seen to be true (*maring*)."

3. On the other hand, the same medium attributed Désy's evident success with government officials to his "good heart" (*tsara fo*).

4. The Sakalava view approximates the model of the relationship among ritual, sanctity, and truth developed by Rappaport (1999), which I evoke here, substituting an "ultimate sacred substance" for Rappaport's "ultimate sacred postulate."

5. My sources for the following sketch include copies of legal files that I was shown by various parties plus oral accounts from Amana Ibrahim, Rafidimanana Botrolahy, Ampanjaka Moandjy, Ampanjaka Tsialaña, and others.

6. Although the relics can be seen as the sort of sacred property that, in its very essence, dissolves the boundaries between persons and things (Mauss 1966: 42–3) and that Weiner (1992) dubbed inalienable, they have been appropriated by conquerors, stolen by rivals, hidden, and fought over for decades. There is, in sum, disagreement as to *whose* inalienable possessions they are. Moreover, as a result of thefts and fire there is some doubt as to whether all of the reliquaries remain filled. Among the forty-seven items declared stolen from the northern shrine were two silver reliquaries, two large and two small drums, two sabers, one large and five small gold coins, seven *sobahiya* cloths, and various other coins, chains, cloths, and weapons.

7. One consultant suggested that in prayer no one would ever omit the names of the ancestors on the competing line.

8. Christian mediums can attend church but not take Communion or cross themselves with holy water. The mode of royal burial may be struggled over. Muslim rulers receiving Sakalava burial do appear as *tromba*.

9. I speak here only of Bemihisatra and Bemazava around Mahajanga. Farther north, Bemihisatra are the more conservative of the two factions (Feeley-Harnik 1991a, Sharp 1999a).

10. Bemihisatra select the name the day after burial (at the culmination of several weeks of mortuary process).

11. If an incarnation appears before a name is announced, the identity remains unauthorized.

12. Thus, Bemazava *ampanjaka* told me that in collecting for tomb renovations they have to lead by example, a role that among Bemihisatra is taken by mediums.

13. Details of the case have been altered. Comparison may be made to the ambivalent and at times highly disturbed way in which corpses are greeted during the *famadihana* (reburial) ceremony among Merina (Bloch 1982; Graeber 1995).

14. No one believed the *fahatelo's* explanation that the earlier date had fallen too close to June 26, the national holiday (Sharp 2001).

15. The president was Ampanjaka Désy; conseils techniques, various notables of Mahajanga; and the treasurers were the *fahatelo* and *manantany*. The medium, whose mother had once worked in the shrine as a Jingô, had received a good education and became wealthy trading and serving as a *tromba* healer in Mayotte.

16. Construction workers observed certain ancestral injunctions like wearing waist wraps instead of trousers, but they were not selected on the basis of clans that had the right to do such work and they received a wage rather than providing service. Managers rationalized the hiring of labor by saying that the job would be finished more quickly.

17. Mbabilahy's position at the Doany Ndramisara is analogous to that of Ndramañonjo at Bezavodoany (chapter 9). As Mme Doso explained, "Ndramandiso rises and gives permission, but Mbabilahy concludes (*manapaka*)." She compared it to me sending my daughter, Nadia, to the kitchen to tell "the cook" what to prepare for dinner. If I had requested meat and the cook asked how to prepare it, Nadia would make the decision. Ndramandiso thanks officials for holding a Service, but Mbabilahy elaborates the instructions. At Betsioko, Mbabilahy takes the role of gracious proprietor and his son, Ndramarofaly, takes charge.

18. This referred back to an earlier debate since the shrine had been composed of a metal roof and raffia plank walls since at least the reconstruction in 1984.

19. The late Ampanjaka Tsialaña, a member of the royal descent group more rooted in tradition than his distant cousin Désy and who served as the royal representative at the shrine, was an active participant.

20. Désy succeeded his mother as reigning monarch. According to the principles of succession, the post should have gone to Tsialaña or one of his immediate kin, but at the time Désy was preferred because, as *chef de canton,* he had "a knowledge of paper" that Tsialaña lacked.

21. Plates and coins are placed under the foundations of temples housing material remains of royal ancestors. Three of the coins were supplied by the mediums of the "four men," the Water-Dwellers, and the Betsioko spirits, respectively. The fourth came either from the shrine's supply or from the Water-Dwellers. On the significance of the coins, see Lambek (2001).

22. These were *mazava loha* (black with a white head), *vato ampetraka* (white spots), *makio* (white with red head and rump, literally "shark"), and *mazava loha volavita* (white head, legs, chest, and hump, literally, "bright head, full moon"). Cattle so patterned are all royal; it is not a matter of specific patterns pertaining to specific ancestors.

23. Kassim offered a good reason for this. Herd sizes are generally much smaller than in the past, and people depend on each animal for subsistence.

24. She refers to the characteristic complacency of Ndramandiso, rather than the particular medium.

25. It is unclear to me whether more traditional materials actually require more work. Roof mats need annual replacing, but hardwood fence posts are very durable. Fashion was certainly an additional factor; it was argued that shrines had to look respectable for external visitors.

26. Royal service might also have been a basis for claiming rights to land or integrating communities of migrants. See the suggestive, if sketchy, argument by Dubourdieu (1989), as well as Sharp's analysis (1993) of internal migrants establishing Sakalava identity by means of possession.

27. This might be compared to constitutionally legitimated change within democratic states.

28. For a relevant illustration from Mayotte see Lambek (2000a).

29. Of course, each generation may have idealized the past and had similar worries concerning their present.

Chapter 11 Notes

1. This position bears comparison with the analysis of the "I" and the "me" developed by G. H. Mead (1962).

2. Daniel's discussion of the mythic element in Sri Lankan history is perhaps a contemporary version of Platonic dualism (1996: chapter 2).

3. See the rich analyses of the personal aspects of possession developed by Crapanzano (1977) and Obeyesekere (1981). Boddy (1988, 1989) provides a particularly subtle analysis of the relationship of "self" to "person" among female spirit mediums in Northern Sudan where, however, the "cultural overdetermination of selfhood" generates rather different psychodynamics from those found in Sakalava possession (cf. Kenyon 1999). On mimetic aspects of possession in Africa, see Colson (1969), Kramer (1993), Stoller (1995), Behrend and Luig (1999), and Masquelier (2001). My perspective is not concerned with why people are possessed in general as much as that they are possessed by specific spirits rather than others (Lambek 1988, 1993, 2000a).

4. The cashews are especially salient in Ambanja because of conflict over cash-cropping (Sharp 1995).

5. No one actually confirmed, this but they did not offer alternate interpretations, either.

6. Mme Doso showed me a larger fifty-franc *tsangan'olo,* dated 1977, that Mbabilahy had given to Toñtông, another of her spirits, who was going to have it attached to a band to wear across his forehead in the style of former Sakalava warriors and to mark the wound where he was felled by a rifle shot.

7. Sakalava photos from a century ago portray both male and female subjects with elaborate hair dressing. Women in old photographs wear rich silks and heavy silver jewelry; warriors have round discs tied across their foreheads (resorting to slices of manioc if nothing better was available); slaves are bound in iron neck rings; and some figures wear cloths rolled across the chest, bandolier style, and turned around the waist. Overall, it is apparent that *tromba* clothing pales by comparison to original styles, though a good effort.

8. Because there are no distinct caretakers for individual ancestors at the Doany Ndramisara, spirit houses there are not permanently inhabited.

9. Mme Doso said she did not see Mbabilahy in her dreams; presumably the man was one of his warriors.

10. The continuous, communal dedication to building is more reminiscent of Gothic cathedrals than modern museums.

11. Spirits who like expensive imported drinks often rely on clients returning from overseas to bring them "duty-free."

12. On the significance of taboos generally, see Lambek (1992) as well as the essays in Middleton (1999).

13. The house and the royals were subsequently purified by a Jingô.

14. Cf. Lambek (2000a, n.d.); the French word *conscience* perhaps serves better.

15. On the diversification of urban mediumship, see Sharp (1993) and Sotoudeh (1997).

16. Spirits have various financial arrangements with their hosts.

17. During several days of violence, hundreds of Comorian residents of Mahajanga were murdered or brutalized and many survivors fled. The perpetrators were not Sakalava. Many Comorians have since returned. Lidy is part Muslim but not of Comorian background.

18. This referred to struggles in Mayotte to ensure special status with France rather than incorporation in the republic of the Comores (cf. Lambek 1995, 2000a).

19. Vietnam might serve as a symbolic displacement of the violent quelling by the French of the Malagasy revolt of 1947 (Cole 2001), but this was not explicit nor, I think, terribly significant here.

20. Emmanuel Tehindrazanarivelo (pers. comm.). By 2001, Mme Doso had convinced Mbabilahy to let her own a television.

Chapter 12 Notes

1. As of June 2002, there have been six months of a national political standoff in Madagascar and much tension, yet each step has been followed toward the next Great Service (Sarah Gould, pers. comm.). Sharp (2002) provides an extensive account of the experience of social change on the part of Sakalava youth, whereas Cole (n.d.) paints a disquieting picture of young people in Toamasina.
2. If the objects of resistance were suddenly removed, and the practices of "resistance" with them, the picture provided by many analysts would leave nothing behind to inform how people would continue to live their lives.
3. Even stronger tensions pertained in the highlands (Rakotomalala et al. 2001).
4. The organizer was a wealthy medium working in the capital. His group brought over fmg4 million to the shrine publicly and possibly an additional amount behind the scenes. On the interest in Sakalava ancestral spirits in Imerina see Rakotomalala et al. (2001).
5. There is some evidence that when Bemazava controlled the relics they were the less conservative of the two factions (Estrade 1985: 59).
6. I mean a discourse that sees "modernity" and "tradition" (or West and Rest) as equivalent yet mutually exclusive and all-encompassing conditions.
7. As living monarchs took on political functions in a world with Europeans, so perhaps they had to become somewhat desanctified, thereby embedding sanctity ever more firmly in the realm of the ancestors.
8. There are echoes of regicide in Sakalava accounts of Zafindramahavita, "descendants of the royal finishers off'," the clan appointed the power of killing a seriously ailing monarch, though I have not heard of a specific instance.
9. See Lambek (2001). The image of doubling the king and killing the substitute is particularly marked among Antankaraña. The king is doubled by the royal mast that is destined to fall at least once during his reign. The collapsed mast illustrates the "collapse of a king," and is deposited in a sacred lake. Until the planting of the new mast the reigning king is symbolically dead, re-emerging triumphantly at midday as soon as the mast is erect. The displacement of death by renewed life and potency extends to the mass circumcision of boys that evening (Lambek and Walsh 1999, Walsh 2000).
10. In addition to all of these similarities is the presence, in continental Africa, notably among Shona (Garbett 1969; Lan 1985), of ancestral royal spirits. Central African "positional succession" without trance also produces a historicity not entirely unlike the one described here. However, I remain agnostic about questions of diffusion. The model of the personage stems from Mauss's discussion of North America (1985). And the similarity of Inca royal oracles (Cobo 1991; Gose 1996) to Sakalava possession should sound a note of caution!
11. The exception proves the rule. When Folo, the ritual purifier of the Antankaraña, moved there from Sakalava country he did so precisely at the instigation of a spirit.
12. I do not wish to discount the fact of hierarchy—social status is often painfully marked—but it is not insurmountable. Marriages, love affairs, commoners made *tromba* out of royal affection, etc., all recur.

13. As indicated previously, inhabitants of Mahajanga draw simultaneously on various so-
    cial imaginaries, all of which have their occlusions and exclusions. It could be argued
    that in opening the relics to general view Ampanjaka Désy was trying to accommodate
    a national imaginary and homogenous mode of belonging. On the history of the relics
    see also Marie-Pierre Ballarin, *Les Reliques Royales à Madagascar* (Paris: Karthala, 2000),
    a source that only became available to me as this book went to press.

Bibliography

Abinal, F. G. and V. Malzac
1987 [1888] *Dictionnaire Malgache Français.* Fianarantsoa.

Agamben, Giorgio
1999 [1994] *The Man Without Content.* Translated by Georgia Albert. Stanford: Stanford University Press.

Allibert, Claude
1999 "Le Journal de J.S. Leigh (1836–1840) à Bord du 'Kite.'" *Etudes Océan Indien* 27–28: 61–170.

Anderson, Benedict
1990 [1972] "The Idea of Power in Javanese Culture." In *Language and Power: Exploring Political Cultures in Indonesia,* pp. 17–77. Ithaca: Cornell University Press.
1991 [1983] *Imagined Communities: Reflections on the Origins and Spread of Nationalism,* 2nd ed. London: Verso.

Aristotle
1947 *Poetics.* Edited by Richard McKeon. New York: Modern Library.
1976 *The Ethics of Aristotle: The Nichomachean Ethics.* Edited by Hugh Tredennick. Translated by J.A.K. Thomson. Harmondsworth, UK: Penguin.

Astuti, Rita
1995 *People of the Sea: Identity and Descent among the Vezo of Madagascar.* Cambridge: Cambridge University Press.

Aucoin, Pauline McKenzie
2000 "Blinding the Snake: Women's Myths as Insubordinate Discourse in Western Fiji." In *Articulated Meanings: Studies in Gender and the Politics of Culture.* Edited by P. M. Aucoin. *Anthropologica* XLII: 11–27.

Auerbach, Erich
1968 *Mimesis: The Representation of Reality in Western Literature.* Translated by W. R. Trask. Princeton: Princeton University Press.

Bakhtin, M. M.
1981 *The Dialogic Imagination: Four Essays.* Edited by M. Holquist. Translated by C. Emerson and M. Holquist. Austin: University of Texas Press.

Baré, Jean-François
1977 *Pouvoir des vivants, langage des morts. Idéo-logiques Sakalava.* Paris: Maspero.
1980 *Sable Rouge: Une monarchie du nord-ouest malgache dans l'histoire.* Paris: L'Harmattan.
1986 "L'organisation sociale Sakalava du Nord: Une récapitulation." In *Madagascar: Society and History.* Edited by C. Kottak et al., pp. 353–92. Durham, NC.: Carolina Academic Press.

Basso, Keith
1996 *Wisdom Sits in Places.* Albuquerque: University of New Mexico Press.

Bayart, Jean-François
1993 *The State in Africa: The Politics of the Belly.* Translated by M. Harper, C. and E. Harrison. London: Longman.

Becker, A. L.
1979 "Text-Building, Epistemology, and Aesthetics in Javanese Shadow Theater." In *The Imagination of Reality: Essays in Southeast Asian Coherence Systems.* Edited by A.L. Becker and A. A. Yengoyan, pp. 211–43. Norwood, NJ: Ablex.

Behrend, Heike and Ute Luig, eds
1999 *Spirit Possession: Modernity and Power in Africa.* Oxford: James Currey.

Bénévent, Charles
1897 "Etude su le Boueni." *Notes, Reconnaissances, et Explorations* 1: 355–379, 2: 49–77 (Cited by G. Feeley-Harnik 1991a).

Benjamin, Walter
1969a [1940] "Theses on the Philosophy of History." In *Illuminations.* Translated by Harry Zohn. Edited by H. Arendt, pp. 253–64. New York: Schocken.
1969b [1936] "The Work of Art in the Age of Mechanical Reproduction." In *Illuminations.* Translated by Harry Zohn. Edited by H. Arendt, pp. 217–51. New York: Schocken.

Bernasconi, Robert
1999 Gadamer, Hans-Georg. *The Cambridge Dictionary of Philosophy,* Second Edition. Edited by R. Audi. Cambridge: Cambridge University Press.

Besnier, Niko
1996 "Heteroglossic Discourses on Nukulaelae Spirits." In *Spirits in Culture, History, and Mind.* Edited by J. Mageo and A. Howard, pp. 75–97. New York: Routledge.

Blanchy, Sophie
1995 *Karana et Banians: les communautés commerçantes d'origine indienne à Madagascar.* Paris: L'Harmattan.

Bloch, Maurice
1971 *Placing the Dead: Tombs, Ancestral Villages, and Kinship Organization in Madagascar.* London: Seminar Press.

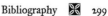

1982 "Death, Women, and Power." In *Death & the Regeneration of Life.* Edited by M. Bloch and J. Parry, pp. 211–30. Cambridge; Cambridge University Press.

1986 *From Blessing to Violence: History and Ideology in the Circumcision Ritual of the Merina of Madagascar.* Cambridge: Cambridge University Press.

1989a [1977] "The Disconnection between Power and Rank as a Process." In *Ritual, History and Power: Selected Papers in Anthropology,* pp. 46–88. London: Athlone.

1989b [1987] "The Ritual of the Royal Bath in Madagascar: The Dissolution of Death, Birth and Fertility into Authority." In *Ritual, History and Power: Selected Papers in Anthropology,* pp. 187–211. London: Athlone.

1992 *Prey Into Hunter: The Politics of Religious Experience.* Cambridge: Cambridge University Press.

1995 "The Resurrection of the House amongst the Zafimaniry of Madagascar." In *About the House: Lévi-Strauss and Beyond.* Edited by J. Carsten and S. Hugh-Jones, pp. 69–83. Cambridge: Cambridge University Press.

1996 "Internal and External Memory: Different Ways of Being in History." In *Tense Past: Cultural Essays in Trauma and Memory.* Edited by P. Antze and M. Lambek, pp. 215–33. New York: Routledge.

1997 Foreword to *The Time of the Gypsies* by Michael Stewart, pp. xiii–xv. Boulder: Westview.

1998 *How We Think They Think: Anthropological Approaches to Cognition, Memory, Literacy.* Boulder: Westview.

1999 "'Eating' Young Men among the Zafimaniry." In *Ancestors, Power and History in Madagascar.* Edited by K. Middleton, pp. 175–90. Leiden: Brill.

Boddy, Janice

1988 "Spirits and Selves in Northern Sudan: The Cultural Therapeutics of Possession and Trance." *American Ethnologist* 15: 4–27.

1989 *Wombs and Alien Spirits: Women, Men, and the Zar Cult in Northern Sudan.* Madison: Wisconsin.

1994 "Spirit Possession Revisited: Beyond Instrumentality." *Annual Review of Anthropology* 23: 407–34.

Bourdieu, Pierre

1977 *Outline of a Theory of Practice.* Translated by Richard Nice. Cambridge: Cambridge University Press.

Burke, Kenneth

1945a *A Grammar of Motives.* New York: Prentice Hall.

1945b "Four Master Tropes." Appendix to *A Grammar of Motives.* New York: Prentice-Hall.

Campbell, Gwyn

1981 "Madagascar and the Slave Trade, 1810–1895." *Journal of African History* 22: 203–227.

1989 "Madagascar and Mozambique in the Slave Trade of the Western Indian Ocean, 1800–1861." In *The Economics of the Indian Ocean Slave Trade in the Nineteenth Century.* Edited by W.G. Clarence-Smith, 166–93. London: Frank Cass.

1996 "The Origins and Demography of Slaves in Nineteenth Century Madagascar." In *Fanandevozana ou esclavage.* Edited by François Rajaison, pp. 5–38. Antananarivo: Musée d'Art et d'Archéologie de l'Université d'Antananarivo.

Casey, Edward

1987 *Remembering: A Phenomenological Study.* Bloomington: Indiana University Press.

Catat, Louis
1895 *Voyage à Madagascar (1889–1890)*. Paris: Hachette.

Chazan-Gillig, Suzanne
1991 *La Société sakalave: Le Menabe dans la construction nationale malgache*. Paris: Karthala-ORSTOM.

Cobo, Bernabe
1991 *Inca Religion and Customs*. Austin: University of Texas Press.

Cole, Jennifer
2001 *Forget Colonialism? Sacrifice and the Art of Memory in Madagascar*. Berkeley: University of California Press.
n.d. "Sex and Money: Coming of Age in Global Tamatave." Paper presented at the American Anthropological Association, Washington D.C., November 2001.

Colson, Elizabeth
1969 "Spirit Possession among the Tonga of Zambia." In *Spirit Mediumship and Society in Africa*. Edited by J. Beattie and J. Middleton, pp. 69–103. London: Routledge and Kegan Paul.

Comaroff, John L. and Jean Comaroff
1992 *Ethnography and the Historical Imagination*. Boulder CO: Westview.
1997 *Of Revelation and Revolution: The Dialectics of Modernity on a South African Frontier*, Vol. 2. Chicago: University of Chicago Press.

Connerton, Paul
1989 *How Societies Remember*. Cambridge: Cambridge University Press.

Crapanzano, Vincent
1977 Introduction to *Case Studies of Spirit Possession*. Edited by V. Crapanzano and V. Garrison, pp. 1–40. New York: Wiley.

Csordas, Thomas
1997 *Language, Charisma, and Creativity*. Berkeley: University of California Press.

Daniel, Valentine
1996 *Charred Lullabies: Chapters in an Anthropography of Violence*. Princeton: Princeton University Press.

De Boeck, Filip
1994 "Of Trees and Kings: Politics and Metaphor among the aLuund of Southwestern Zaire." *American Ethnologist* 21: 451–73.
1998 "Beyond the Grave: History, Memory and Death in Postcolonial Congo/Zaire." In *Memory and the Postcolony*. Edited by R. Werbner, pp. 21–57. Zed: London.

de Heusch, Luc
1997 "The Symbolic Mechanisms of Sacred Kingship: Rediscovering Frazer." *Journal of the Royal Anthropological Institute* n.s.3: 213–32.

Delorme, Yves et Raymond Grellier
1948 *Monographie du District d'Ambato-Boéni.* Imp. Paoli-Fakra. (SOAS Hardyman Madagascar Collection).

Dina, Jeanne
2001 "The Hazomanga among the Masikoro of Southwest Madagascar: Identity and History." In *Emerging Histories in Madagascar.* Edited by Jeffrey C. Kaufmann. *Ethnohistory* 48: 13–30.

Decary, Raymond
1960 *L'Ile Nosy Bé de Madagascar: Histoire d'une Colonisation.* Paris: Editions Maritimes et d'Outre-Mer.

Dubourdieu, Lucile
1989 "Territoire et identité dans les cultes de possession de la basse Betsiboka." *Cahiers des Sciences Humaines* 25: 461–67.

Du Maine
1810 "Idée de la côte Occidentale de Madagascar.[en 1792]." *Annales des Voyages,* Vol. XI, pp. 20–52. Paris.

Dumont, Louis
1970 *Homo Hierarchicus: The Caste System and Its Implications.* Chicago: University of Chicago Press.

Durkheim, Emile
1915 *The Elementary Forms of the Religious Life.* Translated by Joseph Ward Swain. New York: Free Press.

Edkvist, Ingela
1997 *The Performance of Tradition: An Ethnography of Hira Gasy Popular Theatre in Madagascar.* Uppsala: Department of Cultural Anthropology, Uppsala University.

Estrade, Jean-Marie
1985 *Le tromba: Une culte de possession à Madagascar.* Paris: L'Harmattan.

Evans-Pritchard, E. E.
1940 *The Nuer.* Oxford: Clarendon.
1962 [1948] "The Divine Kingship of the Shilluk of the Nilotic Sudan." In *Social Anthropology and Other Essays,* pp. 192–212. New York: Free Press.

Feeley-Harnik, Gillian
1978 "Divine Kingship and the Meaning of History among the Sakalava of Madagascar." *Man* (n.s.) 13: 402–17.
1982 "The King's Men in Madagascar: Slavery, Citizenship, and Sakalava Monarchy." *Africa* 52: 31–50.
1984 "The Political Economy of Death: Communication and Change in Malagasy Colonial History." *American Ethnologist* 11: 1–19.
1985 "Issues in Divine Kingship." *Annual Review of Anthropology* 14: 273–313.

1988 "Sakalava Dancing Battles: Representations of Conflict in Sakalava Royal Service." *Anthropos* 83: 65–85.

1991a *A Green Estate: Restoring Independence in Madagascar.* Washington D.C.: Smithsonian Institution Press.

1991b "Finding Memories in Madagascar." In *Images of Memory.* Edited by S. Küchler and W. Melion, pp. 121–40. Washington D.C.: Smithsonian Institution Press.

2000 "Ancestry, Birth, and the Custody of Children: Comparative Perspectives from Madagascar." In *Rethinking "la femme malgache": New Views on Gender in Madagascar.* Edited by S. Fee, et al. *Taloha* 13: 135–72. (Antananarivo: Revue du Musée d'Art et d'Archéologie)

Ferguson, James
1999 *Expectations of Modernity: Myths and Meanings of Urban Life on the Zambian Copperbelt.* Berkeley: University of California Press.

Ferrand, Gabriel
1902 *Les Musulmans à Madagascar et aux îles Comores.* Paris: E. Leroux.

Fortes, Meyer
1969 *Kinship and the Social Order.* Chicago: Aldine.

Foucault, Michel
1978 *The History of Sexuality.* Vol. 1. New York: Vintage.

1983 "The Subject and Power" and "On the Genealogy of Ethics." In *Michel Foucault: Beyond Structuralism and Hermeneutics.* Edited by H. Dreyfus and P. Rabinow, pp. 208–252. Chicago: University of Chicago Press.

Freud, Sigmund
1985 [1930] *Civilization and It Discontents.* The Penguin Freud Library, Vol. 12, pp. 251–340. London: Penguin.

Frye, Northrop
1957 *Anatomy of Criticism.* Princeton: Princeton University Press.

1991 *The Double Vision: Language and Meaning in Religion.* Toronto: University of Toronto Press.

Gadamer, Hans-Georg
1985 [1960] *Truth and Method.* New York: Crossroad.

Garbett, Kingsley
1969 "Spirit Mediums as Mediators in Korekore Society." In *Spirit Mediumship and Society in Africa.* Edited by J. Beattie and J. Middleton, pp. 104–27. London: Routledge and Kegan Paul.

Geertz, Clifford
1973 "Person, Time, and Conduct in Bali." In *The Interpretation of Cultures,* pp. 360–411. New York: Basic Books.

Gell, Alfred
1995 "Closure and Multiplication: An Essay on Polynesian Cosmology and Ritual." In *Cosmos and Society in Oceania.* Edited by D. de Coppet and A. Iteanu, pp. 21–56. Oxford: Berg.

Giles, Linda

1987 "Possession Cults on the Swahili Coast: A Re-examination of Theories of Marginality."
   Africa 57: 234–58.

1999 "Spirit Possesion & the Symbolic Construction of Swahili Society." In *Spirit Possession: Modernity and Power in Africa*. Edited by H. Behrend and U. Luig, pp. 142–64. Oxford: James Currey.

Gose, Peter

1996 "Oracles, Divine Kingship and Political Representation." *Ethnohistory* 43: 1–32.

Graeber, David

1995 "Dancing With Corpses Reconsidered: An Interpretation of Famadihana (in Arivoni-
   mamo, Madagascar)." *American Ethnologist* 22: 258–78.

1996 "Beads and Money: Notes Toward a Theory of Wealth and Power." *American Ethnologist* 23:
   4–24.

Grandidier, Alfred

1891 "Funeral Customs among the Malagasy." *Antananarivo Annual* IV: 304–318.

Grandidier, Alfred and G. Grandidier

1908–28 *Ethnographie de Madagascar* Vol. 4. Paris: Imprimerie Nationale.

Grant, Damian

1970 *Realism*. London: Methuen.

Guillain, M.

1845 *Documents sur l'histoire, la géographie et le commerce de la partie occidentale de Madagascar*. Paris:
   Imprimerie Royale.

Hacking, Ian

1996 "Memory Sciences, Memory Politics." In *Tense Past: Cultural Essays in Trauma and Memory*.
   Edited by P. Antze and M. Lambek, pp. 67–87. New York: Routledge.

1999 *The Social Construction of What?* Cambridge MA: Harvard University Press.

Hanson, Paul

2000 "Women in Action, Councils in Change: The Productivity of Women's Speech Styles in
   Madagascar's Ranomafana National Park." In *Rethinking "la femme malgache": New Views on
   Gender in Madagascar*. Edited by S. Fee, et al. *Taloha* 13: 263–94. (Antananarivo: Revue du
   Musée d'Art et d'Archéologie)

Havelock, Eric

1963 *Preface to Plato*. Oxford: Basil Blackwell.

Hébert, J-C and P. Vérin

1970 "Le Doany de Bezavo." *Bulletin de Madagascar* 287: 373–76.

Hobsbawm, Eric and Terence Ranger, eds

1983 *The Invention of Tradition*, Cambridge: Cambridge University Press.

Holquist, Michael

1981 Introduction to *The Dialogic Imagination*, by M. M. Bakhtin, pp. xv–xxxiv. Austin: University of Texas Press.

Honneth, Axel
1993 "A Communicative Disclosure of the Past: On the Relations between Anthropology and Philosophy of History in Walter Benjamin." *New Formations* 20: 83–94.

Huntington, Richard
1973 "Death and the Social Order: Bara Funeral Customs." *African Studies* 32: 65–84.

Jackson, Michael
1995 *At Home in the World.* Durham: Duke University Press.

James, Wendy
1988 *The Listening Ebony: Moral Knowledge, Religion and Power among the Uduk of Sudan.* Oxford: Clarendon Press.

Jaovelo-Dzao, Robert
1996 *Mythes, rites et transes à Madagascar; Angano, Joro, et Tromba Sakalava.* Antananarivo: Editions Ambozontany, Paris: Karthala.

Kapferer, Bruce
1997 *The Feast of the Sorcerer.* Chicago: University of Chicago Press.

Keane, Webb
1997 *Signs of Recognition: Powers and Hazards of Representation in an Indonesian Society.* Chicago: University of Chicago Press.

Kent, Raymond
1970 *Early Kingdoms in Madagascar, 1500–1700.* New York: Holt, Rinehart, and Winston.

Kenyon, Susan
1999 "The Case of the Butcher's Wife: Illness, Possession, and Power in Central Sudan." In *Spirit Possession: Modernity and Power in Africa.* Edited by H. Behrend and U. Luig, pp. 89–108. Oxford: James Currey.

Kopytoff, Igor
1971 "Ancestors and Elders in Africa." *Africa* 43: 129–142.

Kramer, Fritz
1993 *The Red Fez: Art and Spirit Possession in Africa.* Translated by Malcolm Green. New York: Verso.

Kuhn, Thomas
1970 *The Structure of Scientific Revolutions,* 2nd ed. Chicago: University of Chicago Press.

Kus, Susan
1997 "Archaeologist as Anthropologist." *Journal of Archaeological Method and Theory* 4 : 199–213.

Kus, Susan and Victor Raharijaona
1990 "Domestic Space and the Tenacity of Tradition among Some Betsileo of Madagascar." In *Domestic Architecture and the Use of Space.* Edited by S. Kent, pp. 21–33. Cambridge: Cambridge University Press.

2000 "House to Palace, Village to State: Scaling Up Architecture and Ideology." *American An-thropologist* 102: 98–113.

Lambek, Michael

1980 "Spirits and Spouses: Possession as a System of Communication among the Malagasy Speakers of Mayotte." *American Ethnologist* 7: 318–31.

1981 *Human Spirits: A Cultural Account of Trance in Mayotte.* New York: Cambridge University Press.

1988 "Spirit Possession/Spirit Succession: Aspects of Social Continuity in Mayotte." *American Ethnologist* 15: 710–31.

1992 "Taboo as Cultural Practice among Malagasy Speakers." *Man* 27: 19–42.

1993 *Knowledge and Practice in Mayotte: Local Discourses of Islam, Sorcery, and Spirit Possession.* Toronto: University of Toronto Press.

1995 "Choking on the Qur'an and Other Consuming Parables from the Western Indian Ocean Front." In *The Pursuit of Certainty.* Edited by W. James, pp. 252–75. ASA Monographs. London: Routledge.

1996 "The Past Imperfect: Remembering as Moral Practice." In *Tense Past: Cultural Essays in Trauma and Memory.* Edited by P. Antze and M. Lambek, pp. 235–254. New York: Routledge.

1997 "Pinching the Crocodile's Tongue: Affinity and The Anxieties of Influence in Fieldwork." In *Fieldwork Revisited: Changing Contexts of Ethnographic Practice in the Era of Globalization,* Edited by J. Robbins and S. Bamford. *Anthropology and Humanism* 22: 31–53.

1998a "Body and Mind in Mind, Body and Mind in Body: Some Anthropological Interventions in a Long Conversation." In *Bodies and Persons: Comparative Perspectives from Africa and Melanesia.* Edited by M. Lambek and A. Strathern, pp. 103–23. Cambridge: Cambridge University Press.

1998b "The Sakalava Poiesis of History: Realizing the Past through Spirit Possession in Madagascar." *American Ethnologist* 25: 106–27.

2000a "Nuriaty, the Saint, and the Sultan: Virtuous Subject and Subjective Virtuoso of the Post-Modern Colony." *Anthropology Today* 16: 7–12.

2000b "The Anthropology of Religion and the Quarrel between Poetry and Philosophy." *Current Anthropology* 41: 309–20.

2001 "The Value of Coins in a Sakalava Polity: Money, Death, and Historicity in Mahajanga, Madagascar." *Comparative Studies in Society and History* 43: 735–62.

in press "Memory in a Maussian Universe." In *Regimes of Memory.* Edited by Katharine Hodgkin and Sussanah Radstone. London: Routledge.

n.d. "Rheumatic Irony: Questions of Agency and Self-Deception in a Relational Universe." In *Illness and Irony.* Edited by M. Lambek and P. Antze.

Lambek, Michael and Jacqueline Solway

2001 "Just Anger: Scenarios of Indignation in Botswana and Madagascar." *Ethnos* 66: 1–23.

Lambek, Michael and Andrew Walsh

1999 "The Imagined Community of the Antankaraña: Identity, History, and Ritual in Northern Madagascar." In *Ancestors, Power and History in Madagascar.* Edited by K. Middleton, pp. 145–74. Leiden: Brill.

Lan, David

1985 *Guns and Rain: Guerillas and Spirit Mediums in Zimbabwe.* Berkeley: University of California Press.

Lavondès, Henri
1967 *Bekoropoka: Quelques aspects de la vie familiale et sociale d'un village malgache.* Paris: Mouton. Cahiers de l''Homme NS VI.

Leach, E. R.
1954 *Political Systems of Highland Burma.* Boston: Beacon.

Leiris, Michel
1980 [1958] *La possession et ses aspects théâtraux chez les Ethiopiens de Gondar.* Paris: Le Sycamore.

Lévi-Strauss, Claude
1966 [1962] *The Savage Mind.* Chicago: University of Chicago Press.

Lienhardt, Godfrey
1961 *Divinity and Experience.* Oxford: Clarendon Press.

Linge, David
1976 Introduction to *Philosophical Hermeneutics* by Hans-Georg Gadamer, pp. xi–lviii. Berkeley: University of California Press.

Loewald, Hans
1980 *Papers on Psychoanalysis.* New Haven: Yale University Press.

Lombard, Jacques
1988 [1973] *Le Royaume Sakalava du Menabe: Essai d'analyse d'un système politique à Madagascar, 17è–20è.* Paris: Éditions de l'ORSTOM.

MacIntyre, Alasdair
1990 *Three Rival Versions of Moral Enquiry.* London: Duckworth.

Macpherson, C. B.
1962 *The Political Theory of Possessive Individualism.* London: Oxford University Press.

Malinowski, Bronislaw
1961 [1922] *Argonauts of the Western Pacific.* New York: Dutton.

Maniry, Steve
1999 "'Fanompoambe 99' A Majunga 'Andriamisara Efadahy' Prennent Leur Bain Annuel." *L'-Express de Madagascar.* Mardi, 3 Août, p. 10.

Marikandia, Mansaré
2001 "The Vezo of the Fiherena Coast, Southwest Madagascar: Yesterday and Today." In *Emerging Histories in Madagascar.* Edited by Jeffrey C. Kaufmann. *Ethnohistory* 48: 157–70.

Marx, Karl
1963 [1852] *The Eighteenth Brumaire of Louis Bonaparte.* New York: International Publishers.
1977 [1887] *Capital,* Vol. 1. Moscow: Foreign Languages Publishing House.

Masquelier, Adeline
2001 *Prayer Has Spoiled Everything: Possession, Power, and Identity in an Islamic Town of Niger.* Durham: Duke.

Mauss, Marcel
1966 [1925] *The Gift: Forms and Functions of Exchange in Primitive Societies.* Translated by I. Cunnison. London: Cohen and West.
1985 [1938] "A Category of the Human Mind: The Notion of Person, the Notion of Self." In *The Concept of the Person.* Edited by M. Carrithers et al., pp. 1–25. Cambridge: Cambridge University Press.

Mauzé, Marie
1994 "The Concept of the Person and Reincarnation among the Kwakiutl Indians." In *Amerindian Rebirth.* Edited by A. Mills and R. Slobodin, pp. 177–91. Toronto: University of Toronto Press.

McElhinny, Bonnie
1998 "Genealogies of Gender Theory: Practice Theory and Feminism in Sociocultural and Linguistic Anthropology." *Social Analysis* 42: 164–89.

Mead, George Herbert
1962 [1934] *Mind, Self and Society.* Chicago: University of Chicago Press.

Middleton, Karen
1995 "Tomb-Work, Body-Work: Gender, Ancestry, and Reproduction in Karembola Ritual." Paper presented at the Satterthwaite colloquium of African Religion and Ritual. Satterthwaite, Cumbria.
1999 Introduction to *Ancestors, Power and History in Madagascar.* Leiden: Brill.
2001 "Power and Meaning on the Periphery of a Malagasy Kingdom." In *Emerging Histories in Madagascar.* Edited by J. Kaufmann. *Ethnohistory* 48: 171–204.

Mitchell, Stephen
1988 *Relational Concepts in Psychoanalysis: An Integration.* Cambridge MA: Harvard University Press.

Molet, Louis
1956 *Le Bain Royal à Madagascar.* Tananarive: Imprimerie Luthérienne.

Mosca, Liliana
1994 *Ralph Linton Nel Madagascar (1925–1927): Una Fonte Per La Conoscenza Della Grande Isola Dell' Oceano Indiano.* Napoli: Luciano.

Nehamas, Alexander
1998 *The Art of Living: Socratic Reflections from Plato to Foucault.* Berkeley: University of California Press.

Nerina-Botokeky, E.
1983 "Le fitampoha en royaume du Menabe: bain des reliques royales." In *Les souverains de Madagascar.* Edited by F. Raison-Jourde, pp. 211–19. Paris: Karthala.

Obeyesekere, Gananath
1981 *Medusa's Hair: An Essay on Personal Symbols and Religious Experience.* Chicago: University of Chicago Press.

Parmentier, Richard
1987 *The Sacred Remains: Myth, History, and Polity in Belau.* Chicago: University of Chicago Press.

Parry, J. and M. Bloch
1989 Introduction to *Money and the Morality of Exchange*, pp. 1–32. Cambridge: Cambridge University Press.

Peel, John
1984 "Making History: The Past in the Ijesha Present." *Man* 19: 111–32.

Pepper, Stephen
1942 *World Hypotheses.* Berkeley: University of California Press.

Pickersgill, W. Clayton
1893 "North Sakalava-Land." *The Antananarivo Annual* 5, Part I: 29–43.

Poirier, Charles
1939 [1927] "Les royaumes Sakalava Bemihisatra de la côte nord-ouest de Madagascar." *Mémoires de l'Académie Malgache* 28: 41–101.

Prud'homme, Lt. Colonel
1900 "Considérations sur les Sakalaves." *Notes, Reconnaissances et Explorations.* Imprimerie Officielle, Tananarive 6: 1–43.

Rakotomalala, Malanjaona, Sophie Blanchy, and Françoise Raison-Jourde
2001 *Madagascar: Les Ancêtres au Quotidien.* Paris: L'Harmattan.

Ramamonjisoa, Suzy
1984 "Symbolique des rapports entre les femmes et les hommes dans les cultes de possession de type tromba à Madagascar." *Bulletin de l'Académie Malgache* 63: 99–110.

Rappaport, Roy
1999 *Ritual and Religion in the Making of Humanity.* Cambridge: Cambridge University Press.

Ricoeur, Paul
1976 *Interpretation Theory.* Fort Worth: Texas Christian University Press.
1984 *Time and Narrative.* Translated by I. K. McLaughlin and D. Pellauer. Chicago: University of Chicago Press.

Rusillon, Henri
1912 *Un Culte Dynastique avec Evocation des Morts chez les Sakalaves de Madagascar: le "Tromba."* Paris: Alphonse Picard.
1922–1923 "Notes explicatives à propos de la généalogie Maroseranana Zafimbolamena." *Bulletin de l'Académie Malgache* 1922–1923: 169–80.

Sahlins, Marshall
1965 "On the Sociology of Primitive Exchange." In *The Relevance of Models for Social Anthropology.* Edited by M. Banton, pp. 139–235. London: Tavistock.
1976 *Culture and Practical Reason.* Chicago: University of Chicago Press.

1985 *Islands of History.* Chicago: University of Chicago Press.

1995 *How "Natives" Think, About Captain Cook, For Example.* Chicago: University of Chicago Press.

1999 "Two or Three Things That I Know About Culture." *Journal of the Royal Anthropological Institute* 5: 399–421.

Scarry, Elaine

1985 *The Body in Pain: The Making and Unmaking of the World.* Oxford: Oxford University Press.

Scheper-Hughes, Nancy and Margaret Lock

1987 "The Mindful Body: A Prolegomenon to Future Work in Medical Anthropology." *Medical Anthropology Quarterly* 1: 6–41.

Sharp, Lesley

1993 *The Possessed and the Dispossessed: Spirits, Identity, and Power in a Madagascar Migrant Town.* Berkeley: University of California Press.

1995 "Playboy Princely Spirits of Madagascar: Possession as Youthful Commentary and Social Critique." *Anthropological Quarterly* 68: 75–88.

1999a "Royal Difficulties: A Question of Succession in an Urbanized Kingdom." In *Ancestors, Power and History in Madagascar.* Edited by K. Middleton, pp. 103–41. Leiden: Brill.

1999b "The Power of Possession in Northwest Madagascar: Contesting Colonial and National Hegemonies." In *Spirit Possession: Modernity and Power in Africa.* Edited by H. Behrend and U. Luig, pp. 3–19. Oxford: James Currey.

2001 "Youth, Land, and Liberty in Coastal Madagascar: A Children's Independence." In *Emerging Histories in Madagascar,.* Edited by J. Kaufmann. *Ethnohistory* 48: 205–36.

2002 *The Sacrificed Generation: Youth, History, and the Colonized Mind in Madagascar.* Berkeley: University of California Press.

Sibree, James

1885 [1878] "The Sakalava: Their Origin, Conquests, and Subjugation." *Antananarivo Annual* 1: 456–68.

Silverstein, Michael and Greg Urban, eds.

1996 *Natural Histories of Discourse.* Chicago: University of Chicago Press.

Solway, Jacqueline

2002 "Navigating the 'Neutral' State: 'Minority' Rights in Botswana." *Journal of Southern African Studies* 28(4): 707–25.

in press "Reaching the Limits of Universal Citizenship: 'Minority' Struggles in Botswana." In *Ethnicity and Democratization in Africa.* Edited by B. Berman, D. Eyoh, and W. Kymlicka. Oxford: James Currey, and Bloomington: Indiana University Press.

Sotoudeh, Shirin

1997 *Lebende Stimmen - Bewegte Koerper: Interaktion und Transformation in der tromba-Besessenheit in Westmadagaskar.* Ph.D. dissertation. University of Bern.

Southall, Aidan

1956 *Alur Society.* Cambridge: Heffer.

1986 "Common Themes in Malagasy Culture." In *Madagascar: Society and History.* Edited by C.Kottak, et al. Durham, NC: Carolina Academic Press.

1999 "The Segmentary State and the Ritual Phase in Political Economy." In *Beyond Chiefdoms: Pathways to Complexity in Africa*. Edited by S. K. McIntosh, pp. 31–38. Cambridge: Cambridge University Press.

Spyer, Patricia, ed.
1998 *Border Fetishisms: Material Objects in Unstable Places*. New York: Routledge.

Stallybrass, Peter and Allon White
1986 *The Politics and Poetics of Transgression*. Ithaca: Cornell University Press.

Steiner, George
1997 *Errata: An Examined Life*. London: Weidenfeld & Nicolson.

Stocking, George
1966 "Franz Boas and the Culture Concept in Historical Perspective." *American Anthropologist* 68: 867–82.

Stoller, Paul
1989 *Fusion of the Worlds: An Ethnography of Possession among the Songhay of Niger*. Chicago: University of Chicago Press.
1995 *Embodying Colonial Memories: Spirit Possession, Power, and the Hauka in West Africa*. New York: Routledge.

Taussig, Michael
1980 *The Devil and Commodity Fetishism in South America*. Durham: University of North Carolina Press.
1993 *Mimesis and Alterity: A Particular History of the Senses*. New York: Routledge.

Taylor, Charles
1998 "Modes of Secularism." In *Secularism and Its Critics*. Edited by Rajeev Bhargava, pp. 31–53. Delhi: Oxford University Press.

Tehindrazanarivelo, Emmanuel
1997 "Fieldwork: The Dance of Power." In *Fieldwork Revisited: Changing Contexts of Ethnographic Practice in the Era of Globalization*. Edited by J. Robbins and S. Bamford. *Anthropology and Humanism* 22: 54–60.

Turner, Victor
1957 *Schism and Continuity in an African Society*. Manchester: Manchester University Press.
1967 *The Forest of Symbols: Aspects of Ndembu Ritual*. Ithaca: Cornell University Press.
1969 *The Ritual Process: Structure and Anti-Structure*. Chicago: Aldine.

van Gennep, Arnold
1960 [1908] *The Rites of Passage*. Chicago: University of Chicago Press.

Vérin, Pierre
1986 *The History of Civilization in Northern Madagascar*. Rotterdam: A. A. Balkema.

Walsh, Andrew
1998 *Constructing 'Antankaraña': History, Ritual, and Identity in Northern Madagascar*. Ph.D. dissertation, University of Toronto.

2001a "When Origins Matter: The Politics of Commemoration in Northern Madagascar." *Emerging Histories in Madagascar.* Edited by J. Kaufmann. *Ethnohistory* 48: 237–56.

2001b "What Makes (the) Antankaraña, Antankaraña? Reckoning Group Identity in Northern Madagascar." *Ethnos* 66: 27–48.

2002 "Responsibility, Taboos and 'the Freedom to do Otherwise' in Northern Madagascar." *Journal of the Royal Anthropological Institute* 8 (3).

Weiner, Annette

1976 *Women of Value, Men of Renown: New Perspectives in Trobriand Exchange.* Austin: University of Texas Press.

1992 *Inalienable Possessions: The Paradox of Keeping-While-Giving.* Berkeley: California.

Weiss, Brad

1996 "Dressing at Death: Clothing, Time, and Memory in Buhaya, Tanzania." In *Clothing and Difference: Embodied Identities in Colonial and Post-Colonial Africa.* Edited by H. Hendrickson, pp. 133–54. Durham, NC: Duke University Press.

Werbner, Richard, ed.

2002 *Postcolonial Subjectivities in Africa.* London: Zed Books.

White, Hayden

1973 *Metahistory: The Historical Imagination in Nineteenth-Century Europe.* Baltimore: Johns Hopkins University Press.

1987 *The Content of the Form; Narrative Discourse and Historical Representation.* Baltimore: Johns Hopkins University Press.

White, Leslie

1969 *The Science of Culture.* New York: Farrar, Straus & Giroux.

Wilson, Peter

1992 *Freedom by a Hair's Breadth.* Ann Arbor: University of Michigan Press.

Wolf, Eric

1982 *Europe and the People without History.* Berkeley: University of California Press.

Wooley, Oliver

1998 *Earth-Shakers of Sahafatra: Authority, Fertility, and the Cult of Nature in Southeast Madagascar.* Ph.D. dissertation, London School of Economics.

Index